Probability, Signals, Noise

Probability, Signals, Noise

by

Jacques Dupraz

Technical Directorate of CIT-Alcatel
Professor at ENSEA, ENSTA and ESE

Translated by

A. Howie

McGraw-Hill Book Company

New York St. Louis San Francisco Auckland
Bogota Hamburg Johannesburg London Madrid
Mexico Montreal New Delhi Panama Paris
Sao Paulo Singapore Sydney Tokyo Toronto

English translation © 1986 North Oxford Academic Publishers Ltd

Original French language edition
(Probabilités, Signaux, Bruits)
published by Editions EYROLLES and C.N.E.T.
© Editions EYROLLES and CNET-ENST 1983

Revised and updated 1986

All rights reserved. No part of this book may be reproduced or transmitted in any form or by any means electronic or mechanical, including photocopying, recording or by any information storage and retrieval system, without permission in writing from the publisher.

English edition first published 1986
by North Oxford Academic Publishers Ltd,
a subsidiary of Kogan Page Ltd, 120 Pentonville Road,
London N1 9JN

Published in the U.S.A. by
McGraw-Hill Inc.

ISBN 0-07-018330-9

Printed and bound in Great Britain

Contents

Foreword ix

General introduction xi

Chapter 1. Random variables 1
  1. Probability space 1
    1.1. Definitions 1
    1.2. General properties 7
    1.3. Conditioning on an event 9
    1.4. Notion of independence 13
    Exercises 19
  2. Random variables 23
    2.1. Definitions and general properties 23
    2.2. Probability law of a real random variable 26
    2.3. Distribution function of a real random variable 35
    2.4. Moments of a random variable 38
    2.5. Characteristic function and generating function 45
    2.6. Conditioning of a real random variable by a sub-Borel field 47
    Exercises 55
  References 63

Chapter 2. Random vectors 65
  1. Random vectors of finite dimension 65
    1.1. Definitions and general properties 65
    1.2. Probability law of a real random vector 67
    1.3. Covariance and correlation 74
    1.4. Characteristic function of a real random vector 78
    1.5. Notion of independence 81
    Exercises 87
  2. Infinite sequences of random variables 96
    2.1. Modes of convergence 96

2.2. Comparison of the modes of convergence 98
2.3. The weak law of large numbers 100
2.4. The strong law of large numbers 102
2.5. The central limit theorem 103
Exercises 105
References 107

Chapter 3. Deterministic signals 108
1. Generalities 108
   1.1. Definitions 108
   1.2. Fundamental rules of calculation in the space $\mathscr{S}'$ 109
2. Complex spectrum 111
   2.1. General definitions 111
   2.2. Band-limited deterministic signals 112
   2.3. Time translation 114
3. Types of deterministic signal 114
   3.1. Integrable deterministic signals 115
   3.2. Finite energy deterministic signals 117
   3.3. Deterministic signal of finite mean power 120
   3.4. Periodic deterministic signals 123
   Exercises 126
4. Linear filtering 127
   4.1. Impulse response 127
   4.2. Transfer function 128
   4.3. Response to an arbitrary deterministic signal 128
   Exercises 135
5. Sampled deterministic signals 139
   5.1. Sampling of a band-limited deterministic signal 139
   5.2. Numerical sequences 145
   5.3. Numerical filtering 148
   Exercises 155
References 163

Chapter 4. Random signals 164
1. Definitions and general properties 164
   1.1. General definitions 164
   1.2. Probability law of a real random signal 170
   1.3. Pairs of random signals 173
   1.4. Equivalence of two random signals 176
   1.5. Events generated by a random signal 178
   1.6. Moments associated with a pair of random signals 180
   Exercises 183
2. Stationarity 192
   2.1. Strong stationarity 192
   2.2. Weak stationarity 193

3. Random signals of finite mean power 197
   3.1. Generalities 198
   3.2. Continuity in mean square 198
   3.3. Differentiability in mean square 199
   3.4. Integrability in mean square 204
   Exercises 207
   References 208

Chapter 5. Weakly stationary random signals 210
1. Random signals in continuous time 210
   1.1. Properties of the correlation functions 210
   1.2. Spectral properties 212
   1.3. Weakly equivalent representation in (x, y) of a random signal 221
   1.4. Differentiability in mean square 224
   1.5. Integrability in mean square 230
   1.6. Ergodicity 232
   1.7. Analogue linear filtering 234
   Exercises 241
2. Random signals in discrete time 254
   2.1. Properties of the correlation functions 254
   2.2. Spectral properties 255
   2.3. Numerical linear filtering 258
   2.4. Sampling of a random signal 261
   Exercises 266
   References 269

Chapter 6. Elements of measure theory 270
1. Measurable spaces 270
   1.1. Simple measurable spaces 270
   1.2. Measurable product space 274
   Exercises 277
2. Measurable functions 280
   2.1. Definition 280
   2.2. Properties 281
   Exercises 285
3. Positive measure 287
   3.1. Simple measure 287
   3.2. Product measure 294
   Exercises 297
4. Integration 298
   4.1. Integration over an arbitrary space 298
   4.2. Integration over a product space 314
   Exercises 318
   References 321

Appendix 1. Table of Fourier series 323

Appendix 2. Table of Fourier transforms 330

Appendix 3. Table of Laplace transforms 334

Appendix 4. Table of numerical values of the function $\Phi(x)$ 337

Index 339

Foreword

This work is intended for engineers and physicists who are interested in probability and in signal theory. Its aim is to provide the basis necessary to deal successfully with the most specific and most complex problems. It has resulted from a course given for more than ten years at the National Higher School for Aeronautics and Space (ENSEA), at the National Higher School for Advanced Techniques (ENSTA) and at the Higher School of Electricity (ESE).

The author offers his sincere thanks to D. Lombard, Chief Engineer at the National Centre for Telecommunication Studies and Professor at the National Higher School of Telecommunications, for the constant friendly support which he has received during the writing of the book. He thanks very sincerely his colleague J. Y. Delabbaye, Professor at ENSTA, to whom he is indebted for numerous fruitful discussions. He wishes to thank also Mr Bic and Mr Goldlewski, engineers at CNET and at ENST respectively, who have read the manuscript and whose suggestions have contributed to its improvement.

General Introduction

It is a truism that we live in an age of information. This information is extremely varied, its importance ranges from the insignificant to the dramatic, and it forms an ever-denser fabric of the relations between men. Starting from the immediate post-war period, this spectacular development has been stimulated by needs which are never satisfied and by the fertility of the human imagination. It has been supported by rapid and continuous technological progress which has exploited systematically the fundamental discoveries of quantum mechanics, providing engineers with the tools which they need.

It is convenient to include within the domain of telecommunications everything which allows the exchange of information at a distance. Thus radar, sonar, ... are telecommunications techniques, as are telephony, television, and telemetry. This area is at present undergoing rapid expansion parallel to that of data processing. Who has not heard of telematics, 'emancipated daughter' of telecommunications and of data processing? This work is devoted to these subjects but is deliberately placed at a higher level than specific techniques, so as to be concerned only with what they have in common.

First of all, what is a piece of information? Certainly it is something not known in advance, contingent on something and uncertain. There is a link between the notion of information and probability. It was the American Shannon who first exploited this connection when he founded in 1948, in a publication which has remained a classic, information theory.

Information, at least on the technical level, is an abstract mathematical notion. More and more frequently, it is a sequence of binary numbers provided with a structure which renders it intelligible.

To transmit information between correspondents, the information must be given a concrete physical support, suitable for reproducing it. It is the signal, on which the information is 'printed', which plays this role of carrier. A signal is clearly random in so far as it contains a part unknown to the receiver for whom it is intended; a signal is analogue when it consists of the information itself, and it is numerical when it can be dissociated from the information which it carries.

During their propagation, the signals traverse physical media which are extremely varied, always imperfect, and often hostile: the atmosphere, a metallic cable, an optical fibre, ... On arrival at its destination, the link between the signal and the information has become fuzzier because of the alterations, we should say more specifically the distortions, undergone by the signal. In addition, a supplementary signal, noise, which can be important to a greater or lesser degree, perturbs the useful signal and complicates things by 'shuffling the cards'. Noise also is a random signal but it is a parasite against which we must defend ourselves.

What must be done so that the recipient will be able, in spite of everything, to recover in the best possible condition the greater part, if not the whole, of the information intended for him? This is the problem which is presented to the engineer. To solve it he must learn to recognize the phenomena with which he is dealing, those which he must support, and those which he must oppose. This knowledge is provided to him by means of models: a model of the source which generates the information and the associated signal, a model of the medium within which the signal is propagated, and a model of the receiver which recovers the information. These are probabilistic models; they are quite unable to conceptualize the whole of the real situation but are nevertheless necessary for an understanding of the underlying mechanisms and for the development of techniques. Two complementary activities must be distinguished: the construction of models and the use of them. We shall not concern ourselves here with finding how to construct models sufficiently representative to be profitably exploited. Our purpose is simpler: to introduce systematically and progressively the notions and the tools enabling us to understand and to exploit models. This is the reason why we are concerned to a large extent, in what follows, with probabilities, signals, and noise.

In five chapters, the work proceeds from probability spaces to random signals. A supplementary chapter, of mathematical results, completes the whole. It is important in so far as it is a source of reference for the rest of the work, but nevertheless it can be omitted since it contains only reviews of known results. Why then include this chapter? Modern telecommunications systems will become more and more complex, as much in their construction as in their equipment. To master them, the engineer will have to have a certain familiarity with the domain of probability and with measure theory. We have tried to collect in this last chapter material which seems to us essential for this familiarity. We should like to emphasize that we are by no means introducing mathematics for its own sake but because it is useful to the physicist and the engineer. This is particularly true with probability calculus, where it is very easy, if one misuses intuition, to commit major errors.

The first two chapters are devoted to the notions of random variables and vectors. The third chapter is rather distinct: here we are concerned not with probabilities but with distributions (as they are defined by Laurent Schwarz, *Théorie des Distributions*, Hermann, Paris, 1966). Its subject is deterministic signals. Random signals are studied in the two following chapters.

The work ends with four appendixes containing in turn a table of Fourier series, a table of Fourier transforms, a table of Laplace transforms, and a table of numerical values of the error function. The author hopes in this way to ease the task of his readers by enabling them to avoid searching for results which are widely dispersed.

We should add that all the chapters are illustrated by numerous examples and contain sets of exercises which are not academic; an indication of their solutions is given with them. They should enable the reader to test his understanding before tackling more complicated problems.

Chapter 1
Random variables

The notion of probability space is fundamental. It is developed in the first part of this chapter and occurs continually in the sequel. A probability space is the mathematical model of a physical experiment whose outcome, impossible to predict with certainty, is one of a number of possible events. To carry out the experiment is to make a trial and to observe the event realized. Knowledge of this event is the element of information conveyed to the experimenter by the trial carried out.

The second part of the chapter introduces the notion of a scalar random variable. This is the mathematical model of a physical quantity, a voltage for example, which depends on chance. When a trial is made, the random variable takes a numerical value which characterizes the event realized. The moments of the random variable, in particular its mean value and its dispersion, are obtained by taking account of the values taken over all possible trials. They characterize statistically the phenomenon represented by the random variable in question.

The notion of conditioning is essential but difficult. It is explained in detail in this chapter. It enables us to exploit the information given by the realization of such and such an event. The conditional expectation expresses in particular the statistical link which can exist between two random variables. We emphasize the importance of two examples which are elementary but of general application. One has as its subject noisy transmission and optimal reception of discrete items of information (Example 2.16). The other is devoted to tests of hypotheses and to the likelihood ratio, which is an essential tool in the theory of detection (Example 2.10).

1. Probability space

1.1. DEFINITIONS

1.1.1 Trials and events

A **probability space** (Marle, 1974, p. 429; Neveu, 1964; Pfeiffer and Schum, 1973, p. 37) is a measured space usually denoted by $(\Omega, \mathscr{A}, P)$, P being a

2 Random variables

positive normed measure (cf. Chapter 6, Section 3.1.1). The set Ω is the set of **trials**; the set $\mathscr{A}$ is the Borel field of **events**. The measure P enables us to evaluate the probabilities of realization of the events. For this reason it is called a **probability measure**.

The space $(\Omega, \mathscr{A}, P)$ is the mathematical model of a concrete experiment whose result, initially unknown, depends on chance. To conduct the experiment consists in making a trial $\omega \in \Omega$. The event A is realized when the trial made belongs to the set $A \in \mathscr{A}$. Several events can be realized simultaneously. In fact all the events containing the trial made are realized (cf. Examples 1.2 and 1.3).

The set Ω contains, as the case may be, a finite or an infinite, denumerable or not, number of trials. The singletons $\{\omega\}$ do not necessarily belong to the Borel field $\mathscr{A}$. When they do belong to it, they define events which are called **elementary**. This is particularly the case when the Borel field $\mathscr{A}$ is the Borel field of subsets $\mathscr{P}(\Omega)$ (cf. Chapter 6, Example 1.1). The elementary events are then the **atoms** of the Borel field $\mathscr{A}$ (cf. Chapter 6, Exercise 1.10). It is important to distinguish carefully between the trials and the events. The trials characterize the way in which the experiment works, whereas the events characterize the way in which its results are interpreted. An event A can be defined as a set of trials possessing a specific property Π_A: this may be written formally

$$A = \{\omega \mid \Pi_A \text{ is true}\}$$

1.1.2. Realization of events

The events obey certain laws which represent intuitive necessities. Thus, if A is an event, its realization excludes that of the **contrary event** constituted by the set $\bar{A}$ complementary to the set A in Ω. The event Ω is always realized since every trial belongs to the set Ω. This is the **certain event**. The empty event $\varnothing$ is never realized since the experiment consists in making a trial, whatever it may be. This is the **impossible event**.

Let A and B be two events. The trials belonging to the event A and to the event B constitute the event $A \cap B$. When the two sets A and B are disjoint $A \cap B = \varnothing$ and the event $A \cap B$ is impossible. In this case the events A and B are never realized together: they are **incompatible**. Two distinct atoms are incompatible events (cf. Chapter 6, Exercise 1.10).

The trials belonging to the event A or to the event B constitute the event $A \cup B$. It should be noted that the realization of the event A or the event B does not rule out that of the event A *and* the event B. The or is thus not exclusive (cf. Fig. 1.1).

When the set A is included in the set B, the realization of the event A implies that of the event B. The two events are identical if the realization of one of them implies that of the other and conversely.

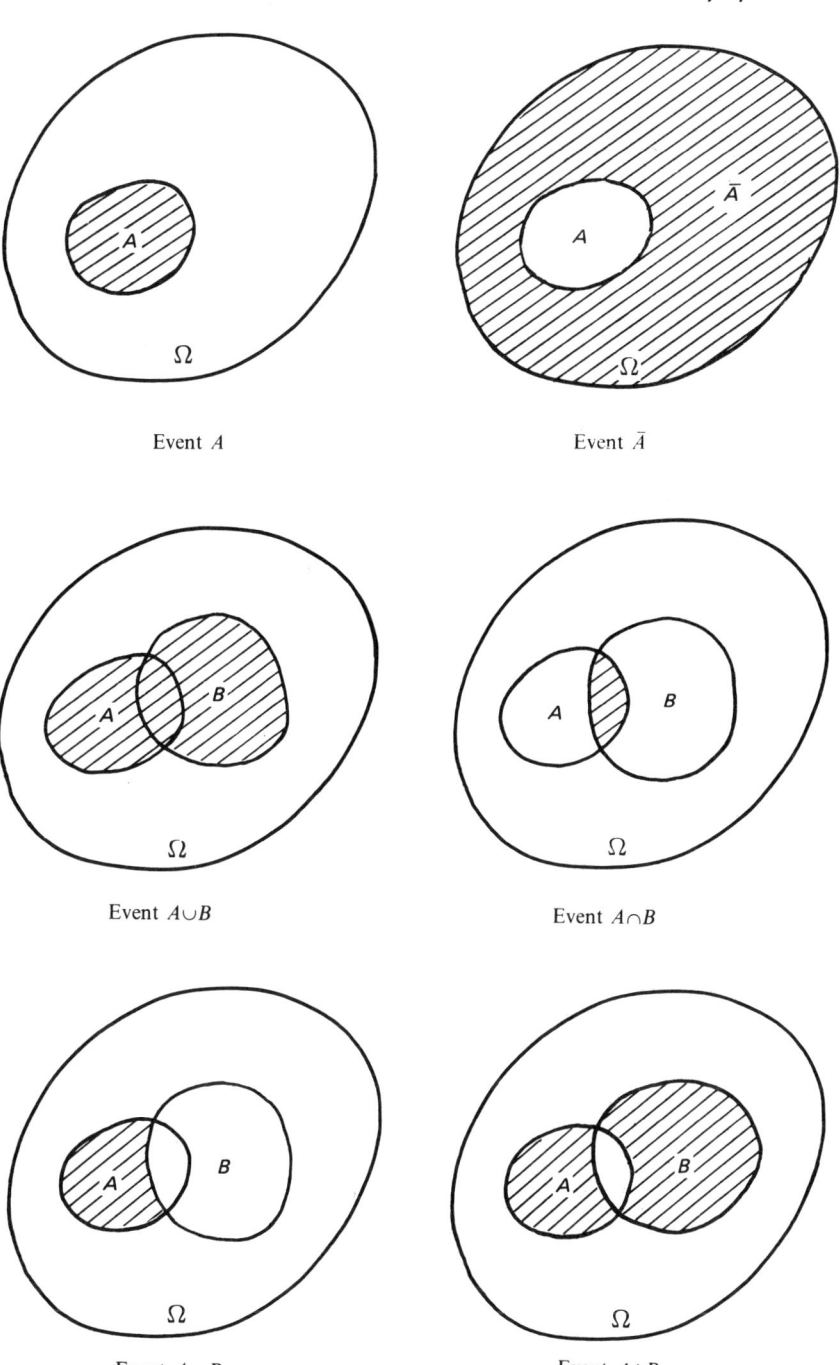

Fig. 1.1

The set of trials belonging to the event A and not belonging to the event B is the event $A \cap \bar{B} = A - B$. This is the **difference** of the event A and the event B. The set of trials belonging to one of the two events to the exclusion of the other is the event

$$A \triangle B = (A - B) \cup (B - A) \tag{1.1}$$

This is the **symmetric difference** (exclusive or) of the two events.

All these definitions are illustrated in Fig. 1.1. It is very useful to support reasoning about events by the visual evidence provided by figures of this type.

It is natural to consider that the union and the intersection of a finite number of events are events. The events then constitute an algebra of subsets of the set Ω (cf. Chapter 6, Definition 1.1.1.2). This model is in fact insufficient and it is necessary to use a more general model in which the notion of an event is conserved under denumerable union and intersection. The events then constitute a Borel field of subsets of the set Ω (cf. Chapter 6, Definition 1.1.1.1).

1.1.3. Probabilities of events

The positive normed measure P enables us to evaluate *a priori* the chances of realization of the events of the Borel field $\mathscr{A}$. To each event A is assigned a probability $P(A)$. It is a number between 0 and 1 since the measure P is normed.

The certain event has probability equal to 1 and the impossible event has zero probability (cf. Chapter 6, Section 3.1.3.1). Certain events other than the certain event can have a probability equal to 1. They are **almost surely** (we write a.s.) **certain**. Certain events other than the impossible event can have zero probability. They are **a.s. impossible**.

We shall suppose in all that follows that every subset of an event of zero probability belongs to the Borel field $\mathscr{A}$. It is then an event of zero probability and the Borel field $\mathscr{A}$ is said to be complete (cf. Chapter 6, Definition 3.1.5.3). In general, a property which is satisfied by all the trials except those which belong to an event of zero probability is said to be almost sure or **a.s. true**.

1.1.4. Relative frequency of events

How, in practice, can we define the probability of an event? This is a difficult question which we cannot answer except by mentioning the link between probability and relative frequency. Consider an experiment as a result of which an event A is either realized or not. Suppose that the conditions of the experiment are known with all required accuracy so that we can repeat it an indefinite number of times, in a strictly identical manner, and without interaction between successive experiments. If N experiments are carried out and the event A is realized N_A times, the **relative frequency** of realization of the event is N_A/N. Long experience has led us to think that when N becomes sufficiently large the relative frequency approaches closer and closer to a constant value P_A which is called the probability of the event A. It must be noted that, when two events A and B are incompatible, the relative frequency of realization of one or the other (exclusively) is the sum of the relative frequencies. In fact

$$\frac{N_{A\cup B}}{N} = \frac{N_A}{N} + \frac{N_B}{N}$$

which implies that $P_{A\cup B} = P_A + P_B$. The property of additivity is thus satisfied by the relative frequency.

Example 1.1 Uniform probability measure

Let $(\Omega, \mathcal{P}(\Omega), P)$ be a probability space. If the set Ω is finite and contains n trials, the Borel field of subsets $\mathcal{P}(\Omega)$ contains 2^n events (cf. Chapter 6, Example 1.1). Each elementary event $\{\omega_i\}$ has a probability $p_i = P(\{\omega_i\})$ and the numbers p_i satisfy

$$\sum_{i=1}^{n} p_i = P\left(\bigcup_{i=1}^{n} \{\omega_i\}\right) = P(\Omega) = 1 \tag{1.2}$$

since the measure P is normed. The measure P is discrete and can be written in the form of a combination of **Dirac measures** (cf. Chapter 6, Example 3.1):

$$P = \sum_{i=1}^{n} p_i \, \delta(\omega - \omega_i) \tag{1.3}$$

When the probabilities p_i are equal, the elementary events $\{\omega_i\}$ are equiprobable and it follows from Eqn (1.2) that

$$\forall i \in \{1, \ldots, n\} \qquad p_i = \frac{1}{\text{card } \Omega} \tag{1.4}$$

The probability of an arbitrary event A is obtained from Eqns (1.3) and (6.21). Then it follows that

$$P(A) = \frac{\text{card } A}{\text{card } \Omega} \tag{1.5}$$

The measure P is therefore proportional to the **counting measure** (cf. Chapter 6, Example 3.2). It is called **uniform**. Equation (1.5) expresses the classical definition of the probability that an event A belongs to the Borel field of subsets of a finite set of equiprobable trials. It is the ratio of the number of favourable cases (the number of trials belonging to the set A) to the number of possible trials (the number of trials contained in Ω). Calculations of probability are then a matter of enumeration (Pfeiffer and Schum, 1973, p. 10).

6 Random variables

Example 1.2 Drawing of balls from an urn

(a) Let an urn contain two black and three white balls. Drawing a single ball from the urn is an experiment which can be described by a probability space in which the set Ω contains five trials, the Borel field $\mathscr{A}$ is the Borel field of subsets $\mathscr{P}(\Omega)$ and the probability measure P is a uniform measure. Each ball has then probability 1/5 of being drawn:

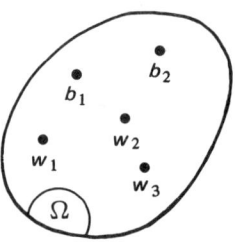

Let us define the following events:

$$A_1 = (b_1, w_1, w_2) \qquad A_2 = (b_1, w_2, w_3) \qquad A_3 = (b_1, w_3, w_1)$$
$$A_4 = (b_2, w_1, w_2) \qquad A_5 = (b_2, w_2, w_3) \qquad A_6 = (b_2, w_3, w_1)$$

These are all the sets containing one black ball and two white balls. They all have probability 3/5, by virtue of Eqn (1.5), but they are not incompatible. Let P_m, $m \in [1, 6]$, be the probability that exactly m events A_i are realized (when one and only one ball is drawn). It is easy to see, by enumeration, that

$$P_1 = P_2 = P_5 = P_6 = 0 \qquad P_3 = 2/5 \qquad P_4 = 3/5$$

The reader can verify that application of Eqn (1.32) (cf. Exercise 1.4) leads by calculation to the same results. It is thus impossible to realize exactly one event A_i by drawing only a single ball. This is obvious and shows clearly the difference between a trial and an event.

(b) Suppose now that three balls are drawn at one time. What is then the probability of drawing one black ball and two white balls? It is clear that we must change the probability space since making a trial consists now in drawing three balls. Let $(\Omega', \mathscr{A}', P')$ be the space describing the new experiment:

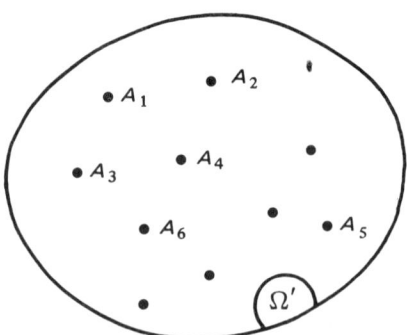

The set Ω' contains $\binom{5}{3} = 10$ trials since there are $\binom{5}{3}$ ways of choosing three balls from five. Six of these contain one black ball and two white balls and can be described in the same way as the events A_i defined above, to which they correspond. The Borel field $\mathscr{A}'$ is the Borel field of subsets. As for the probability P', it is a uniform measure. It is in fact clear, from the very conditions of the experiment, that the trials are equiprobable elementary events. Each trial thus has probability $1/10 = 0.1$. The required probability is therefore $6 \times 0.1 = 0.6$, by virtue of Eqn (1.5).

1.2. GENERAL PROPERTIES

A probability measure is a positive normed measure. The general properties of measures of this type are studied in Chapter 6, Section 3. They are given here in the language appropriate to probabilities.

1.2.1. Probability of the contrary event

Let A be an event and $\bar{A}$ the contrary event. Then $\Omega = A \cup \bar{A}$ and $\emptyset = A \cap \bar{A}$, whence there results by additivity (cf. Eqn (6.20))

$$P(\bar{A}) = 1 - P(A) \tag{1.6}$$

Since the certain event Ω has probability equal to 1, the impossible event $\emptyset$, which is its complement, has zero probability. Let A be an event other than Ω having probability equal to 1. The non-empty event $\bar{A}$ has zero probability and consequently the event A is equal to the event Ω apart from an event ($\bar{A}$) of probability zero. It is therefore a.s. certain. In the same way, an event of probability zero, other than $\emptyset$, is a.s. impossible. This does not mean that it cannot be realized. For example, a singleton can have probability zero and correspond to the value taken by a continuous random variable (cf. Section 2.2.3).

1.2.2. The formula of total probability

Let A and B be two arbitrary events. The events $A \cup B$ and $A \cap B$ are such that, from the additivity of the measure P (cf. Eqn (6.20)),

$$A \cup B = A \cup (\bar{A} \cap B) \Rightarrow P(A \cup B) = P(A) + P(\bar{A} \cap B)$$

$$B = (A \cap B) \cup (\bar{A} \cap B) \Rightarrow P(B) = P(A \cap B) + P(\bar{A} \cap B)$$

Hence we have the fundamental formula called the formula of total probability:

$$P(A \cup B) = P(A) + P(B) - P(A \cap B) \leq P(A) + P(B) \tag{1.7}$$

8 Random variables

The probability of the event $A \cup B$ is less than or equal to the sum of the probabilities of the events A and B. The equality holds if and only if the events A and B are incompatible.

1.2.3. Sequences of events

1.2.3.1. Arbitrary sequence
Let $(A_n, n \in \mathbb{N})$ be an arbitrary sequence of events. From the subadditivity of the measure P (cf. Eqn (6.25)) we have the result

$$P\left(\bigcup_{n \in \mathbb{N}} A_n\right) \leq \sum_{n \in \mathbb{N}} P(A_n) \tag{1.8}$$

The equality holds when the events are pairwise mutually incompatible.

1.2.3.2. Increasing sequence
Suppose that the sequence (A_n) is increasing with respect to the inclusion relation. This means that the realization of the event A_n implies, for all n, that of the event A_{n+1}. The property of the measure P of continuity under increasing union then enables us to write (cf. Eqn (6.26))

$$\lim_{n \to \infty} P(A_n) = P\left(\bigcup_{n \in \mathbb{N}} A_n\right) \tag{1.9}$$

1.2.3.3. Decreasing sequence
When the sequence (A_n) is decreasing with respect to the inclusion relation, the property of continuity under decreasing intersection enables us to write (cf. Eqn (6.27))

$$\lim_{n \to \infty} P(A_n) = P\left(\bigcap_{n \in \mathbb{N}} A_n\right) \tag{1.10}$$

1.2.4. Partition of an event

Let A be an event and $(B_i, i \in \mathbb{I})$ be a family, at most denumerable, of pairwise incompatible events such that

$$A \subset \left(\bigcup_{i \in \mathbb{I}} B_i\right)$$

Then

$$A = \bigcup_{i \in \mathbb{I}} (A \cap B_i)$$
$$\forall j \neq i \quad (A \cap B_i) \cap (A \cap B_j) = \emptyset$$

Probability space 9

The additivity of the measure P (cf. Eqn (6.20)) enables us to write

$$P(A) = \sum_{i \in I} P(A \cap B_i) \quad (1.11)$$

When the family (B_i) is a partition, at most denumerable, of the set Ω, Eqn (1.11) applies for an arbitrary set A.

1.3. CONDITIONING ON AN EVENT

1.3.1. *Conditional probability*

1.3.1.1. Introduction

Let $(\Omega, \mathcal{A}, P)$ be a probability space. The measure P attaches to each event $A \in \mathcal{A}$ a probability $P(A)$ whose value is determined by *a priori* knowledge of the chances of realization of the event A. It is possible, however, that better knowledge of the experiment may enable us to change the probabilities of the events and therefore to change the measure. This is the case in particular when we know before making a trial that an event C, of non-zero probability, will be realized. We then define a new measure, called conditional, which takes account of the conditions of the experiment resulting from the realization of the event C.

1.3.1.2. Definition

The **conditional probability** $P(A|C)$ of an event A, given that the event C of non-zero probability has been realized, is defined by

$$P(A|C) = \frac{P(A \cap C)}{P(B)} \quad (1.12)$$

It is easy to prove that $P(.|C)$ is a positive normed measure over the measurable space $(\Omega, \mathcal{A})$. The triplet $(\Omega, \mathcal{A}, P(.|C))$ is therefore a probability space.

Thus the probability of events belonging to the Borel field $\mathcal{A}$ can be evaluated either *a priori* from the probability P or *a posteriori* from the conditional probability $P(.|C)$, when we know that the event C, of non-zero probability, has been realized.

1.3.1.3. Elementary properties

When the event C is incompatible with the event A, $P(A \cap C) = P(\emptyset) = 0$ and the conditional probability $P(A|C)$ is zero. The event A is conditionally a.s. impossible. When realization of the event C implies that of the event A, $P(A \cap C) = P(C)$ and $P(A|C) = 1$. The event A is conditionally a.s. certain.

This allows us to write

$$A \cap C = \emptyset \Rightarrow P(A|C) = 0$$
$$C \subset A \Rightarrow P(A|C) = 1 \quad (1.13)$$

Thus conditioning can change the probabilities of events completely.

1.3.2. Realization of several events

Let $(A_1, \ldots, A_n)$ be a finite sequence of events. The event $A_1 \cap \ldots \cap A_n$ is realized when the events $A_1, \ldots, A_n$ are realized simultaneously. Suppose that $P(A_1 \cap \ldots \cap A_{n-1}) \neq 0$. Then from Eqn (1.12)

$$P(A_1 \cap \ldots \cap A_n) = P(A_1 \cap \ldots \cap A_{n-1}) \times P(A_n | A_1 \cap \ldots \cap A_{n-1})$$

Then

$$(A_1 \cap \ldots \cap A_{n-1}) \subset (A_1 \cap \ldots \cap A_{n-2}) \Rightarrow P(A_1 \cap \ldots \cap A_{n-2}) \neq 0$$

and

$$P(A_1 \cap \ldots \cap A_{n-1}) = P(A_1 \cap \ldots \cap A_{n-2}) \times P(A_{n-1} | A_1 \cap \ldots \cap A_{n-2})$$

This leads us, on applying the same reasoning several times, to write

$$P(A_1 \cap \ldots \cap A_n) = P(A_1) P(A_2 | A_1) \ldots P(A_n | A_1 \cap \ldots \cap A_{n-1}) \quad (1.14)$$

Now suppose that $P(A_1 \cap \ldots \cap A_{n-1}) = 0$. Since the numerical sequence $P(A_1), \ldots, P(A_1 \cap \ldots \cap A_n)$ is positive and decreasing, the probability $P(A_1 \cap \ldots \cap A_n)$ is zero and there exists an integer $k < n$ such that

$$P(A_1 \cap \ldots \cap A_{k-1}) > P(A_1 \cap \ldots \cap A_k) = 0$$

Hence, from Eqn (1.12) we have the result

$$\frac{P(A_1 \cap \ldots \cap A_k)}{P(A_1 \cap \ldots \cap A_{k-1})} = P(A_k | A_1 \cap \ldots \cap A_{k-1}) = 0$$

If we agree to put the right-hand side of Eqn (1.14) equal to zero when one of its terms is zero, even if certain other terms are not defined because they involve conditioning events of zero probability, we see that both sides of Eqn (1.14) are zero. The formula (1.14) is thus always valid and generalizes Eqn (1.12). This is an extremely important result.

1.3.3. Bayes' formula

Let A be an event and $(B_i, i \in I)$ be a family, at most denumerable, of events such that

$$A \subset \left(\bigcup_{i \in I} B_i \right)$$

Suppose that the events B_i are pairwise incompatible and of non-zero probability. Then from Eqns (1.11) and (1.12)

$$P(A) = \sum_{i \in I} P(A|B_i)P(B_i)$$

Suppose moreover that $P(A) \neq 0$. From Eqn (1.12)

$$\forall i \in I \quad P(B_i|A) = \frac{P(A \cap B_i)}{P(A)} = \frac{P(A|B_i)P(B_i)}{P(A)}$$

Hence the important Bayes formula results:

$$\forall i \in I \quad P(B_i|A) = P(A|B_i)P(B_i) \bigg/ \sum_{i \in I} P(A|B_i)P(B_i) \qquad (1.15)$$

Example 1.3 Tests of hypotheses and the likelihood ratio

Introduction

A probability space $(\Omega, \mathscr{A}, P)$ is a mathematical model describing a concrete experiment. It can happen that the experiment in question is insufficiently well known to be described by a single model. Thus two probability spaces $(\Omega, \mathscr{A}, P_0)$ and $(\Omega, \mathscr{A}, P_1)$ are used to describe an experiment in which the same events are produced either under a hypothesis h_0, with probability measure P_0, or under a hypothesis h_1, with probability measure P_1, the two hypotheses being mutually exclusive. This is the case, for example, when the experiment consists in observing and interpreting the output signal from a radar receiver. Under the hypothesis h_0, there is no target and therefore no radar echo and only noise is observed. Under the hypothesis h_1, there is a target and a noisy echo is observed. We must therefore decide whether what is observed is noise or a noisy echo. The object of what follows is to show what to do to minimize the probability of making an erroneous decision.

Suppose the experiment is carried out. It may be interpreted as the selection and observation of a trial $\omega \in \Omega$. Is it possible to deduce from this that the experiment has been conducted under hypothesis h_0, for example? To answer this question, let us define a partition of the set Ω into two events Ω_0 and $\Omega_1 = \bar{\Omega}_0$ such that hypothesis h_0 is chosen when the event Ω_0 is realized and hypothesis h_1 in the opposite case. It is then possible to make an erroneous choice by deciding either that hypothesis h_0 applies whereas it is in fact h_1 or that hypothesis h_1 applies whereas it is in fact h_0.

Mathematical model of the test

Let $H = (h_0, h_1)$ be the hypothesis space and $D = (d_0, d_1)$ the decision space, d_0 signifying the choice of hypothesis h_0 and d_1 that of h_1. Suppose that, by any means whatever, we can establish that the hypotheses h_0 and h_1 have *a priori* probabilities p_0 and p_1 respectively of being realized and let α and β be the

12 Random variables

conditional probabilities of making an erroneous decision, given that the hypotheses h_0 and h_1, respectively, apply.

The mathematical model of the test is then the probability space $(H \times D, \mathcal{P}(H \times D), P)$. Let us define the events:

$$H_i = \{h_i\} \times D = \{h_i, d_0\} \cup \{h_i, d_1\} \qquad i \in \{0, 1\}$$

$$D_j = H \times \{d_j\} = \{h_0, d_j\} \cup \{h_1, d_j\} \qquad j \in \{0, 1\}$$

The definition of the measure P follows from the hypotheses that we have made. We obtain

$$P(H_0) = p_0 \qquad P(H_1) = p_1 \qquad p_0 + p_1 = 1$$

$$P(D_1 | H_0) = P_0(\Omega_1) = \alpha \qquad P(D_0 | H_1) = P_1(\Omega_0) = \beta$$

The conditional probabilities α and β depend on the way in which the complementary events Ω_0 and Ω_1 are defined in the space $(\Omega, \mathcal{A})$. In the case of radar, α is the probability of a false alarm, β is that of non-detection. The probabilities of the events D_0 and D_1 may be written

$$P(D_0) = p_0(1 - \alpha) + p_1 \beta$$

$$P(D_1) = p_0 \alpha + p_1(1 - \beta)$$

The conditional probabilities α and β are *a priori* arbitrary. When they satisfy $\alpha + \beta = 1$, the events H_i and D_j are independent. This follows immediately from application of Eqn (1.20).

Optimal test
Consider the event

$$\mathscr{E} = (H_0 \cap D_1) \cup (H_1 \cap D_0)$$

It is clear that the choice of the hypothesis which applies is erroneous when the event $\mathscr{E}$ is realized. The probability of error may be written, by application of Eqn (1.12),

$$P(\mathscr{E}) = P(H_0)P(D_1 | H_0) + P(H_1)P(D_0 | H_1)$$

$$= p_0 \alpha + p_1 \beta \qquad (1.16)$$

It follows from the definition of the conditional probabilities α and β, by application of Eqn (6.39), that

$$P(\mathscr{E}) = p_0 \int_{\Omega_1} dP_0(\omega) + p_1 \int_{\Omega_0} dP_1(\omega)$$

$$= p_1 + \int_{\Omega_1} \{p_0 \, dP_0(\omega) - p_1 \, dP_1(\omega)\}$$

Suppose, in addition, that the measure P_0 is absolutely continuous with respect to the measure P_1, and let $\psi = dP_0/dP_1$ be the density (Radon–

Nikodym derivative) of the measure P_0 with respect to the measure P_1. This is a positive measurable function defined a.s. on the probability space $(\Omega, \mathcal{A}, P_1)$ (cf. Chapter 6, Section 4.1.5) and we have

$$P(\mathcal{E}) = p_1 + p_0 \int_{\Omega_1} \left\{ \psi(\omega) - \frac{p_1}{p_0} \right\} dP_1(\omega)$$

The probability of error depends on the way in which the events Ω_0 and Ω_1 are defined. An elementary variational calculation shows that the probability of error is minimal when the event Ω_1 is such that

$$\Omega_1 = \{ \omega \,|\, \psi(\omega) \leqslant p_1/p_0 \}$$

We can then answer the question posed. The criterion used to choose the hypothesis presumed to be realized at the time of the experiment consists in comparing with the ratio p_1/p_0 the value taken by the function ψ over the selected trial and making the decision:

$$\psi(\omega) = \frac{dP_0(\omega)}{dP_1(\omega)} \underset{H_1}{\overset{H_0}{\gtrless}} \frac{p_1}{p_0} \tag{1.17}$$

The direction of the inequality determines the choice of hypothesis, as the preceding relation shows. The definition of the event Ω_1 implies that the probability of error is minimal.

The function ψ is also called the **likelihood ratio** in this kind of application. The likelihood ratio is widely used in carrying out tests of hypotheses, particularly in detection problems (Levine, 1973). More specific applications are the subject of Example 2.10 and Chapter 2, Example 1.3.

1.4. NOTION OF INDEPENDENCE

1.4.1. Independence of two events

1.4.1.1. Definitions

Let $(\Omega, \mathcal{A}, P)$ be a probability space. Two events A and B are **independent** with respect to the measure P when

$$P(A \cap B) = P(A) \times P(B) \tag{1.18}$$

Let C be an event of non-zero probability and let the conditional probability measure be $P(.|C)$. The events A and B are **conditionally independent** with respect to the event C when they are independent with respect to the probability measure $P(.|C)$:

$$P(A \cap B \,|\, C) = P(A \,|\, C) \times P(B \,|\, C) \tag{1.19}$$

The results which follow relate to independence with respect to the measure P. They can be extended immediately to conditional independence by replacing the measure P by the measure $P(.|C)$.

1.4.1.2. Property

If one of the pairs $\{A, B\}$, $\{A, \bar{B}\}$, $\{\bar{A}, B\}$ and $\{\bar{A}, \bar{B}\}$ is composed of independent events, so also are the others.

Suppose, for example, that the events A and B are independent and let us show that the events A and $\bar{B}$ are independent also. Let us write

$$A = (A \cap B) \cup (A \cap \bar{B}) \qquad \emptyset = (A \cap B) \cap (A \cap \bar{B})$$

It results from the additivity of the measure P (cf. Eqn (6.20)) that

$$P(A) = P(A \cap B) + P(A \cap \bar{B})$$
$$\Rightarrow P(A \cap \bar{B}) = P(A) - P(A)P(B) = P(A)P(\bar{B})$$

1.4.1.3. Property

Let an event be a.s. impossible or a.s. certain. It is independent of every other event.

In fact let A be an arbitrary event and N an a.s. impossible event. Since $(A \cap N) \subset N$ it follows that

$$0 \leqslant P(A \cap N) \leqslant P(N) = 0 \Rightarrow P(A \cap N) = P(A) \times P(N) = 0$$

The events A and N are therefore independent. It results from the preceding property that the a.s. certain event $\bar{N}$ and the event A are also independent.

1.4.1.4. Property

Let A and B be two events of non-zero probability. They are independent if and only if one of the following equalities is satisfied:

$$P(A|B) = P(A) \qquad P(B|A) = P(B) \tag{1.20}$$

The proof follows immediately from Eqn (1.12).

1.4.1.5. Property

Let two events be incompatible. They are independent only if at least one of them has zero probability.

This follows immediately from the fact that, if two events A and B are incompatible, $P(A \cap B) = P(\emptyset) = 0$. Thus we must not confuse the notions of incompatibility and independence. Incompatibility depends on the way in which events are defined, independently of their probabilities, whereas independence depends on these probabilities. This is why it is sometimes a mistake to try to prove that two events are independent other than by verifying formula (1.18).

1.4.2. Independence of a family of events

1.4.2.1. Definition

Let $(A_i, i \in \mathbb{I})$ be an arbitrary family of events, finite or infinite. It is independent with respect to the probability measure P when the equality

$$P(A_{i_1} \cap \ldots \cap A_{i_n}) = P(A_{i_1}) \ldots P(A_{i_n}) \qquad (1.21)$$

is satisfied for every finite sequence $i_1, \ldots, i_n$ extracted from the set $\mathbb{I}$.

1.4.2.2. Property
Let $(A_i, i \in \mathbb{I})$ be an independent family of events and let the family $(\hat{A}_i, i \in \mathbb{I})$ be such that $\hat{A}_i$ is an event of the family $\{\emptyset, \Omega, A_i, \bar{A}_i\}$. The family $(\hat{A}_i)$ is independent.

The proof is left to the reader.

1.4.2.3. Property
Let $(A_n, n \in \mathbb{N})$ be a sequence of events. It is independent if and only if

$$\forall n \in \mathbb{N} \qquad P(A_0 \cap \ldots \cap A_n) = P(A_0) \ldots P(A_n) \qquad (1.22)$$

It is immediate that Eqn (1.22) is satisfied if the sequence (A_n) is independent. Conversely, let $i_1, \ldots, i_k$ be a finite sequence contained in $\mathbb{N}$. It is easy to prove that Eqn (1.21) can be obtained from Eqn (1.22) by taking $n = \sup i_j$ and by replacing the events not belonging to the sequence $A_{i_1}, \ldots, A_{i_n}$ by the event Ω.

1.4.2.4. Property
Let $(A_n, n \in \mathbb{N})$ be an independent sequence of events. Then

$$P\left(\bigcap_{n \in \mathbb{N}} A_n\right) = \prod_{n \in \mathbb{N}} P(A_n) \qquad (1.23)$$

Put $B_n = A_0 \cap \ldots \cap A_n$. It follows from Eqn (1.22) that

$$P(B_n) = P(A_0) \ldots P(A_n)$$

Moreover, the sequence $(B_n, n \in \mathbb{N})$ is decreasing and has as limit the event (cf. Chapter 6, Exercise 1.1)

$$B = \bigcap_{n \in \mathbb{N}} B_n = \bigcap_{n \in \mathbb{N}} A_n$$

The result stated follows immediately from application of Eqn (1.10).

1.4.2.5. Important remark
A family of pairwise independent events is not necessarily an independent family. This is illustrated by the following example.

Example 1.4
Let $\Omega = \{\omega_1, \omega_2, \omega_3, \omega_4\}$ be a set and $(\Omega, \mathscr{P}(\Omega), P)$ be the probability space, P being a uniform measure (cf. Example 1.1). Consider the events

$$A = \{\omega_1, \omega_2\} \qquad B = \{\omega_2, \omega_3\} \qquad C = \{\omega_1, \omega_3\}$$

16 Random variables

It follows that

$$P(A) = P(B) = P(C) = 1/2$$
$$P(A \cap B) = P(B \cap C) = P(C \cap A) = 1/4$$
$$P(A \cap B \cap C) = P(\emptyset) = 0 \neq P(A) \times P(B) \times P(C)$$

The events A, B and C are pairwise independent but event A, although independent of B and of C, is not independent of the event $B \cap C$.

Example 1.5 The Borel–Cantelli lemma

Let $(\Omega, \mathscr{A}, P)$ be a probability space and $(A_n, n \in \mathbb{N})$ a sequence of events. Consider the events

$$\forall n \in \mathbb{N} \quad C_n = \bigcup_{k \geq n} A_k \quad A = \limsup A_n = \bigcap_{n \in \mathbb{N}} C_n$$

The event A is realized if and only if an infinity of events A_n is realized (cf. Chapter 6, Exercise 1.1).

(a) Suppose that we have

$$\sum_{n \in \mathbb{N}} P(A_n) < +\infty$$

the sequence (A_n) being arbitrary, independent or not. It follows then from Eqn (1.8) that

$$\forall \varepsilon \in \mathbb{R}_+^* \quad \exists n \in \mathbb{N} \quad 0 \leq P(A) \leq P(C_n) \leq \sum_{k \geq n} P(A_k) \leq \varepsilon$$

Hence it results that $P(A) = 0$. The realization of an infinity of events A_n is therefore a.s. impossible.

(b) Now suppose that we have

$$\sum_{n \in \mathbb{N}} P(A_n) = +\infty$$

and that the sequence (A_n) is independent. It follows from Eqn (1.20) that

$$P(\bar{C}_n) = \prod_{k \geq n} P(\bar{A}_k) = \prod_{k \geq n} \{1 - P(A_k)\}$$

The inequality $1 - P(A_k) \leq \exp\{-P(A_k)\}$ then enables us to write

$$\forall n \in \mathbb{N} \quad 0 \leq P(\bar{C}_n) \leq \exp\left\{-\sum_{k \geq n} P(A_k)\right\} = 0$$

whence it results that the events C_n are a.s. certain. Moreover, the sequence (C_n) is decreasing and has as limit the event A (cf. Chapter 6, Exercise 1.1). It follows then from Eqn (1.10) that

$$P(A) = \lim_{n \to \infty} P(C_n) = 1$$

The realization of an infinity of events A_n is therefore a.s. certain.

Example 1.6 Independent binary random sequences

Binary random sequences composed of the symbols 0 and 1 play a very important role in modern telecommunications. This is due to the penetration of data processing into this field and to the generalization of the process of presenting analogue data numerically.

Thus, pieces of information carried by telecommunication channels are binary sequences, possessing structures which make them intelligible and protect them against transmission errors.

The aim of what follows is to study the properties of the simplest binary sequences, those whose symbols are independent. In practice, things are much more complicated since the structures introduce links between the symbols.

Introduction

Let $(\Omega_0, \mathcal{A}_0, P_0)$ be a probability space. It is the mathematical model of an experiment realized in practice by the possible occurrence of events belonging to the Borel field $\mathcal{A}_0$. Let E be an event having probability $P_0(E) = p$. Consider the probability space $(\Omega, \mathcal{P}(\Omega), \Pi)$, $\Omega = \{E, \bar{E}\}$, the probability measure Π being defined by $\Pi(E) = P_0(E) = p$. This is the mathematical model of the same experiment as before except that we are interested only in the realization or non-realization of the event E.

Finite sequences

Consider the probability space $(\Omega^n, \mathcal{P}(\Omega^n), P)$, $n \in \mathbb{N}^*$, P being the product measure $\otimes_n \Pi$ (cf. Chapter 6, Section 3.2). This is the mathematical model of the composite experiment realized by the repetition, in conditions which are strictly identical (successive repetitions being independent of each other), of n experiments whose model is the probability space $(\Omega, \mathcal{P}(\Omega), \Pi)$. The composite experiment is also called a Bernoulli experiment. The set Ω^n contains 2^n trials of the form $E_1 \times E_2 \times \ldots \times E_n$ with $E_i = E$ or $\bar{E}$. The probability of such a trial is obtained from Eqn (6.33) since P is the product measure. Thus

$$P(E_1 \times \ldots \times E_n) = P(E_1) \times \ldots \times P(E_n) \tag{1.24}$$

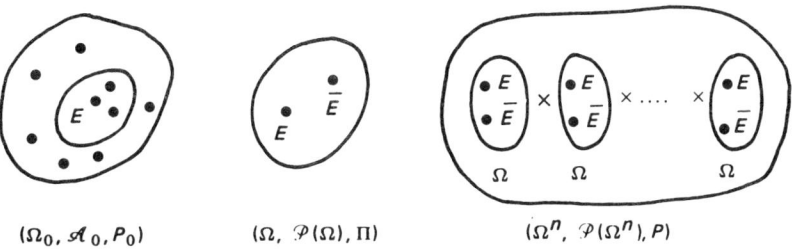

$(\Omega_0, \mathcal{A}_0, P_0)$ $(\Omega, \mathcal{P}(\Omega), \Pi)$ $(\Omega^n, \mathcal{P}(\Omega^n), P)$

18 Random variables

Let us agree to represent the trial E of the set Ω by the symbol 1 and the trial $\bar{E}$ by 0. We can then consider a trial of the set Ω^n as a sequence of n binary symbols belonging to the alphabet $\{0, 1\}$. We also say that such a sequence is a **binary word** of length n. An event of the Borel field $\mathscr{P}(\Omega'')$ is a set of binary words.

Let a well-defined word contain k symbols 1, $k \in [0, n]$. Its probability is $p^k(1-p)^{n-k}$ by virtue of Eqn (1.24). Let A_k be the event consisting of the $\binom{n}{k}$ distinct words containing exactly k symbols 1. It follows, from the additivity of the measure P, that

$$P(A_k) = \binom{n}{k} p^k(1-p)^{n-k} \tag{1.25}$$

This is the probability of obtaining a word containing exactly k symbols 1.

Infinite sequences

Now consider the probability space $(\Omega^{\mathbb{N}^*}, \mathscr{P}(\Omega^{\mathbb{N}^*}), P)$, P being the product measure $\otimes_n \Pi$. The set $\Omega^{\mathbb{N}^*}$ contains a non-denumerable infinity of trials, a trial being an infinite sequence of symbols 0 and 1.

Let B_n be the event consisting of the set of sequences whose nth symbol is a 1. Using an obvious notation, it can be written

$$B_n = \Omega_1 \times \ldots \times \Omega_{n-1} \times \{1\} \times \Omega_{n+1} \times \ldots$$

Its probability is obtained from Eqn (6.32) and is

$$\forall n \in \mathbb{N}^* \qquad P(B_n) = p \tag{1.26}$$

Consider the sequence $(B_n, n \in \mathbb{N}^*)$ and a finite subsequence $B_{i_1}, \ldots, B_{i_n}$. Let us write

$$B_{i_1} \cap \ldots \cap B_{i_n} = \Omega_1 \times \ldots \times \Omega_{i_1-1} \times \{1\} \times \Omega_{i_1+1}$$
$$\times \ldots \times \Omega_{i_n-1} \times \{1\} \times \Omega_{i_n+1} \times \ldots$$

It then follows from Eqn (6.32) that

$$P(B_{i_1} \cap \ldots \cap B_{i_n}) = P(B_{i_1}) \times \ldots \times P(B_{i_n}) = p^n \tag{1.27}$$

The sequence (B_n) is therefore independent. Clearly this follows from the fact that P is the product measure. It is immediate moreover that

$$\sum_{n \in \mathbb{N}^*} P(B_n) = +\infty$$

The Borel–Cantelli lemma then enables us to state that the event consisting of the sequences containing an infinity of symbols 1 is a.s. certain (cf. Example 1.5(b)).

Let C_n be the event consisting of the sequences each of whose first n symbols is 1. It can be written

$$C_n = \{1\} \times \ldots \times \{1\} \times \Omega_{n+1} \times \ldots$$

and its probability is $P(C_n) = p^n$. It is easy to prove that the sequence $(C_n, n \in \mathbb{N}^*)$ is decreasing. It therefore has as its limit the event (cf. Chapter 6, Exercise 1.1)

$$C = \bigcap_{n \in \mathbb{N}} C_n$$

This is the elementary event consisting of the sequence composed entirely of symbols 1. It follows from Eqn (1.10) that

$$P(C) = \lim_{n \to \infty} P(C_n) = 0$$

The event C is therefore a.s. impossible.

To sum up, it is a.s. impossible to have a sequence composed entirely of symbols 1 whereas it is a.s. certain that the sequence will contain an infinity of symbols 1.

1.4.3. Independence of a family of sub-Borel fields

1.4.3.1. Definition

Let $(\Omega, \mathcal{A}, P)$ be a probability space and let $(\mathcal{B}_i, i \in \mathbb{I})$ be an arbitrary family, finite or infinite, of sub-Borel fields. The family $(\mathcal{B}_i)$ is independent when every family $(B_i, B_i \in \mathcal{B}_i, i \in \mathbb{I})$ of events belonging to each sub-Borel field $\mathcal{B}_i$ is independent.

1.4.3.2. Property

Let $(B_i, i \in \mathbb{I})$ be a family of events and $(\mathcal{B}_i, i \in \mathbb{I})$ the family of sub-Borel fields generated by the events B_i. The family $(\mathcal{B}_i)$ is independent if and only if the family (B_i) is independent.

The Borel field $\mathcal{B}_i$ generated by the event B_i is the family $(\emptyset, \Omega, B_i, \bar{B}_i)$ (cf. Chapter 6, Example 1.3). It is then immediate that the independence of the family $(\mathcal{B}_i)$ implies that of the family (B_i).

Conversely, if the family (B_i) is independent, every family which can be derived from it by replacing an arbitrary number of events B_i by one of the events $\emptyset, \Omega$ or $\bar{B}_i$ is still independent (cf. Section 1.4.1). The family $(\mathcal{B}_i)$ is therefore indeed independent.

Exercises

Exercise 1.1 Discrete probability measures
Let $\Omega = \{\omega_1, \ldots, \omega_n\}$ be a finite set and let P and Q be two probability

measures over the measurable space $(\Omega, \mathscr{P}(\Omega))$. Put

$$\forall i \in [1, n] \qquad P(\{\omega_i\}) = p_i \qquad Q(\{\omega_i\}) = q_i$$

Show that

$$\sum_{i=1}^{n} p_i \log\left(\frac{q_i}{p_i}\right) \leq 0 \tag{1.28}$$

(Start from the inequality, $\forall x \geq 0$, $\log x \leq x - 1$.)

Exercise 1.2 Symmetric difference

Let $(\Omega, \mathscr{A}, P)$ be a probability space and let A and B be two events and $A \triangle B$ be the symmetric difference event defined by Eqn (1.1). Show that

$$|P(A) - P(B)| \leq P(A \triangle B) = P(A) + P(B) - 2P(A \cap B) \tag{1.29}$$

Exercise 1.3 Incompatible events

Let $(\Omega, \mathscr{A}, P)$ be a probability space and $(A_1, \ldots, A_n)$ be a finite sequence of arbitrary events. Show that

$$P\left(\bigcup_{i=1}^{n} A_i\right) = 1 - P\left(\bigcap_{i=1}^{n} \bar{A}_i\right) \tag{1.30}$$

The formula (1.30) is particularly interesting when the sequence (A_i) is independent. Hint: Use the incompatibility of the events

$$\bigcup_{i=1}^{n} A_i \quad \text{and} \quad \bigcap_{i=1}^{n} \bar{A}_i$$

Exercise 1.4 Poincaré's formula

Let $(\Omega, \mathscr{A}, P)$ be a probability space and $(A_1, \ldots, A_n)$ be a finite sequence of arbitrary events. Put

$$S_k = \sum_{i_1, \ldots, i_k} P(A_{i_1} \cap \ldots \cap A_{i_k})$$

where the sum ranges over the $\binom{n}{k}$ combinations of k different indices belonging to the set $\{1, 2, \ldots, n\}$, $1 \leq k \leq n$.

(1) Reasoning by recurrence starting from Eqn (1.7), prove Poincaré's formula:

$$P\left(\bigcup_{i=1}^{n} A_i\right) = S_1 - S_2 + S_3 - \ldots + (-1)^{n+1} S_n \tag{1.31}$$

This gives the probability that at least one of the events A_i is realized.

(2) Hence deduce that the probability P_m that the number of events realized is exactly m ($1 \leq m \leq n$) is

$$P_m = S_m - \binom{m+1}{1} S_{m+1} + \binom{m+2}{2} S_{m+2} - \ldots + (-1)^{n-m} \binom{n}{n-m} S_n$$

$$\tag{1.32}$$

Verify that we have

$$\sum_{m=1}^{n} P_m = P\left(\bigcup_{i=1}^{n} A_i\right) = 1 - P\left(\bigcap_{i=1}^{n} \bar{A}_i\right)$$

(3) Hence deduce that the probability Q_m that the number of events realized is at least equal to m ($1 \leq m \leq n$) is

$$Q_m = S_m - \binom{m}{1} S_{m+1} + \binom{m+1}{2} S_{m+2} + \ldots + (-1)^{n-m} \binom{n-1}{n-m} S_n \quad (1.33)$$

(4) Show, reasoning by recurrence, that

$$P\left(\bigcup_{i=1}^{n} A_i\right) \geq S_1 - S_2$$

Hint: suppose the property is true for $n-1$ and prove that it is true for n by applying Eqn (1.8) to the event

$$\bigcup_{i=1}^{n} (A_i \cap A_n)$$

Exercise 1.5 Permutations

Let $\{1, 2, \ldots, n\}$ be a finite ordered sequence and Ω the set of its permutations. Let $(\Omega, \mathscr{P}(\Omega), P)$ be a probability space, P being a uniform measure. Prove that the probability P_0 of having a permutation σ such that, for all $i \in [1, n]$, $\sigma(i) \neq i$ is

$$P_0 = \sum_{k=0}^{n} \frac{(-1)^k}{k!}$$

(Use the results of the preceding exercise and Eqn (1.31).)

Exercise 1.6 Binary random sequences

Let $(\Omega^n, \mathscr{P}(\Omega^n), P)$ be the probability space defined in Example 1.6. Prove that the probability that, in a sequence of n symbols containing exactly k symbols 1 ($k \geq 1$), the symbol in position $i \leq n$ is a 1 is k/n.

Exercise 1.7 Independent events

Let $(\Omega, \mathscr{A}, P)$ be a probability space.

(1) Let A and B be two events of non-zero probability. Show that they are independent if and only if

$$P(A|B) = P(A|\bar{B})$$

(2) Let A, B and C be three events of non-zero probability such that $P(B|C) = 1$. Prove that

$$P(A \cap B|C) = P(A|C)$$

22 Random variables

Exercise 1.8 Multisymbol random sequences

Let $(E, \mathcal{P}(E), P)$ be the probability space such that

$$E = \{a_1, \ldots, a_m\} \qquad P = \sum_{i=1}^{m} p_i \delta_{a_i}$$

Let the product probability space be $(\Omega = E^{\mathbb{N}^*}, \otimes_{\mathbb{N}^*} \mathcal{P}(E), \otimes_{\mathbb{N}^*} P)$.

A trial of the set Ω is an infinite sequence of symbols belonging to the set E. Let A_{ij} be the event consisting of the set of sequences whose symbol in position $i \in \mathbb{N}^*$ is the symbol a_j.

(1) Show that $P(A_{ij}) = p_j$.
(2) Show that the family $\{A_{ij}, i \in \mathbb{N}^*, j \in [1, m]\}$ is independent.

Exercise 1.9 Probabilistic judgment

An individual charged with an offence goes forward for judgment. The judge estimates that his *a priori* chance of being guilty is 90%. Two eye-witnesses are heard. Witness a is a friend of the defendant. The probability that he will give false testimony if the accused is guilty is 0.6 whereas he will tell the truth if the accused is innocent. Witness b does not like the accused. With probability 0.3, he will give false testimony if the accused is innocent but he will tell the truth if the accused is guilty. Consider the following events:

C   the accused is guilty
A   witness a gives false testimony
B   witness b gives false testimony
D   the two testimonies are contradictory

(1) Calculate the probabilities $P(A)$ and $P(B)$. ($P(A) = 0.54$, $P(B) = 0.03$.)
(2) Calculate the probability $P(D)$. ($P(D) = 0.57$.)

Exercise 1.10 Independent binary sequences

Consider an independent binary sequence of n symbols 0 or 1, p being the probability of the symbol 1 and $q = 1 - p$ being that of the symbol 0. Such a sequence is a trial in the probability space defined in Example 1.6. Let $P_{n,m}$ be the probability that a sequence contains no block of m, or more than m, consecutive symbols 1, with $0 \leq m \leq n$.

(1) The probability $P_{n,m}$ can be obtained using Poincaré's formula (1.31).
  (1.1) Prove

$$P_{n,m} = 1 - S_1 + S_2 - \ldots + (-1)^{n-m+1} S_{n-m+1}$$

by putting

$$S_k = \sum_{i_1, \ldots, i_k} P(A_{i_1} \cap \ldots \cap A_{i_k})$$

the sum ranging over the $\binom{n-m+1}{k}$ combinations of k different indices belonging to the set $\{1, 2, \ldots, n - m + 1\}$ and A_j being the event consisting of the set of

sequences whose symbols in positions $j, j+1, \ldots, j+m-1$ are symbols 1 with $j \in \{1, 2, \ldots, n-m+1\}$, the other symbols of the sequence being arbitrary.
(1.2) Verify that $P_{8,4} = 1 - 5p^4 + 4p^5$.

(2) Use of Eqn (1.31) is very laborious when n is large. Also it is interesting to establish recurrence formulae.
(2.1) Prove directly that

$$P_{n,m} = qP_{n-1,m} + qpP_{n-2,m} + \ldots + qp^{m-1}P_{n-m,m}$$

with, $\forall i \leq m$, $P_{i,m} = 1$.
(2.2) Hence deduce

$$\forall m \quad \lim_{n \to \infty} P_{n,m} = 0$$

2. Random variables

2.1. DEFINITIONS AND GENERAL PROPERTIES

2.1.1. Real or complex random variables

2.1.1.1. Definitions
Let $(\Omega, \mathscr{A}, P)$ be a probability space. A real **random variable** is a measurable mapping ξ from the space $(\Omega, \mathscr{A})$ into the space $(\mathbb{R}, \mathscr{R})$ (cf. Chapter 6, Section 2.1) (Neveu, 1964; Pfeiffer and Schum, 1973, p. 159).

A complex random variable is a mapping ξ from the space $(\Omega, \mathscr{A})$ into $\mathbb{C}$ such that the mappings Re ξ and Im ξ are real random variables.

In what follows we shall write r.v. systematically for random variable.

2.1.1.2. Sub-Borel field generated by a random variable
The measurable space $(\mathbb{R}, \mathscr{R})$ is defined in Chapter 6, Example 1.4. The elements of the Borel field $\mathscr{R}$ are the **Borel sets** of the set $\mathbb{R}$. The most common example of a Borel set is any interval, whether open, semi-open or closed. Measurable functions are defined and studied in Chapter 6, Section 2.

Let ξ be a real r.v. Since the function ξ is measurable with respect to the Borel fields $\mathscr{A}$ and $\mathscr{R}$, the inverse image $\xi^{-1}(\beta)$ of a Borel set $\beta \in \mathscr{R}$ is an event belonging to the Borel field $\mathscr{A}$. It is also denoted $\{\xi \in \beta\}$. It contains all the trials ω such that $\xi(\omega) \in \beta$. This is illustrated by Fig. 1.2 in the case where the Borel set β is an interval.

The set of events corresponding to the Borel sets of $\mathbb{R}$ is the Borel field $\mathscr{B}_\xi$ generated by the mapping ξ (cf. Chapter 6, Section 1.1.2). It is the inverse image $\xi^{-1}(\mathscr{R})$. It is contained in the Borel field $\mathscr{A}$ (cf. Eqn (6.16)):

$$\mathscr{B}_\xi = \xi^{-1}(\mathscr{R}) \subset \mathscr{A} \tag{1.34}$$

24 Random variables

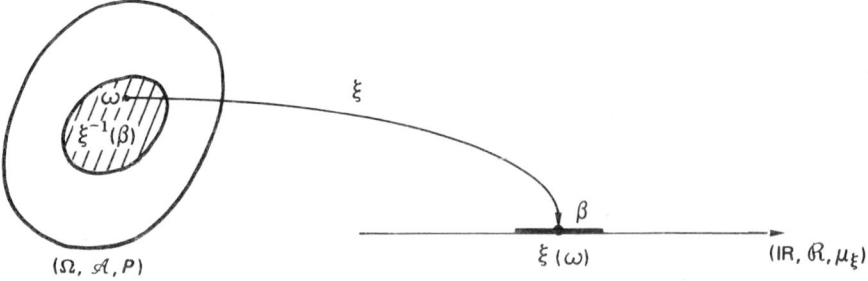

Fig. 1.2

Hence we say that $\mathscr{B}_\xi$ is a sub-Borel field of the Borel field $\mathscr{A}$. Thus there can exist, in the Borel field $\mathscr{A}$, events which do not belong to $\mathscr{B}_\xi$ and do not depend on the r.v. ξ.

2.1.1.3. Integration of a random variable

Let ξ be a complex r.v. The function ξ can be integrable with respect to the measure P. Integration with respect to a positive measure is defined and studied in Section 6.4. It is traditional to denote by $E[\,.\,]$ the linear integration operator with respect to a probability measure P. The integral of the r.v. ξ is then written

$$E[\xi] = P(\xi) = \int_\Omega \xi(\omega)\,\mathrm{d}P(\omega) \tag{1.35}$$

The integrable complex r.v.s defined over the space $(\Omega, \mathscr{A}, P)$ constitute the vector space $\mathscr{L}^1(\Omega, \mathscr{A}, P)$ (cf. Chapter 6, Section 4.1.6.1).

2.1.1.4. Interpretation

The ensemble of the space $(\Omega, \mathscr{A}, P)$ and of the real r.v. ξ is the mathematical model of an experiment in which making a trial $\omega \in \Omega$ consists of actual observation of the value $\xi(\omega)$ taken in $\mathbb{R}$ by the r.v. ξ. The event $\{\xi \in \beta\}$ is realized when $\xi(\omega) \in \beta$. From the above it follows that the only events observable are those which depend on the r.v. ξ and hence belong to the sub-Borel field $\mathscr{B}_\xi$.

Example 2.1 Indicator random variable of an event

Let $A \in \mathscr{A}$ be an event and I_A its **indicator** function (cf. Chapter 6, Example 2.4). This is a real r.v. since it is a measurable mapping from $(\Omega, \mathscr{A})$ into $(\mathbb{R}, \mathscr{R})$. By definition

$$\omega \in A \Rightarrow I_A(\omega) = 1 \qquad \omega \in \bar{A} \Rightarrow I_A(\omega) = 0$$

The r.v. I_A takes the value 1 when event A is realized and the value 0 in the opposite case. Hence we say that the r.v. I_A is **binary**.

The sub-Borel field generated by the r.v. I_A is the family $\{\emptyset, \Omega, A, \bar{A}\}$. This is the sub-Borel field generated by the event A (cf. Chapter 6, Example 1.3). In general, properties relating to events can be transposed to the indicator r.v.s of these events. For this we must use the following obvious relations (cf. Fig. 1.1):

$$I_{\bar{A}} = 1 - I_A \qquad I_{A \cap B} = I_A I_B \qquad I_{A \cup B} = I_A + I_B - I_A I_B \qquad (1.36)$$

The r.v. I_A is integrable and has as its integral the probability of event A. It follows in fact from Eqn (6.39) that

$$E[I_A] = P(A) \qquad (1.37)$$

When the event considered is the certain event Ω, the r.v. I_Ω is constant and equal to 1. It follows from Eqn (1.37) that its integral is equal to 1 since the measure P is normed.

The complex r.v. $cI_\Omega, c \in \mathbb{C}$, is a constant r.v. It takes the value c over all trials $\omega \in \Omega$. It is integrable and its integral is equal to c. Usually, we omit the I_Ω and simply speak of the constant r.v. c.

Example 2.2 Discrete random variables

A complex r.v. ξ is **discrete** when the set $\xi(\Omega)$ of values which it takes is at most denumerable. It can always be written

$$\xi = \sum_{n \in \mathbb{N}} x_n I_{A_n} \qquad A_n = \xi^{-1}(\{x_n\}) = \{\xi = x_n\} \qquad (1.38)$$

When the event A_n is realized, the r.v. ξ takes the value x_n.

The family $(A_n, n \in \mathbb{N})$ is a denumerable partition of the set Ω. The sub-Borel field $\mathscr{B}_\xi$ generated by the r.v. ξ is the Borel field generated by the family (A_n). It is made up of the finite or denumerable unions of events A_n (cf. Chapter 6, Exercises 1.7 and 1.8):

$$\mathscr{B}_\xi = \left\{ \bigcup_{n \in \mathbb{I}} A_n \,\middle|\, \mathbb{I} \subset \mathbb{N} \right\}$$

2.1.2. Random function of a random variable

Let ξ be an r.v. and g be a measurable function, real or complex, defined over $\mathbb{C}$ if the r.v. ξ is complex and over $\mathbb{R}$ if ξ is real. The composite function $\eta = g \circ \xi$ is measurable (cf. Chapter 6, Section 2.2.1). It is therefore a complex or a real r.v., as the case may be, which is defined as a function of the r.v. ξ. The sub-Borel fields $\mathscr{B}_\xi$ and $\mathscr{B}_\eta$ generated respectively by the r.v.s ξ and η satisfy $\mathscr{B}_\eta \subset \mathscr{B}_\xi$ (cf. Chapter 6, Exercise 2.4). This means that all events of the Borel field $\mathscr{A}$ which depend on the r.v. η depend also on the r.v. ξ, which is intuitively obvious.

Conversely, if an r.v. η is such that the sub-Borel field which it generates is contained in the sub-Borel field generated by an r.v. ξ, there exists a measurable function g such that $\eta = g \circ \xi$ (cf. Chapter 6, Example 4.3). The r.v. η is then a function of the r.v. ξ.

Example 2.3
Let ξ be a real r.v. The complex function $\exp\{ia(.)\}$, $a \in \mathbb{R}$, is measurable since it is continuous (cf. Chapter 6, Example 2.3). It enables us to define the complex r.v.

$$\eta = \exp(ia\xi) = \cos(a\xi) + i\sin(a\xi)$$

where the real part is the real r.v. $\cos(a\xi)$ and the imaginary part is the real r.v. $\sin(a\xi)$.

2.1.3. Almost surely equal random variables

Consider the set of complex r.v.s defined over the probability space $(\Omega, \mathcal{A}, P)$. Almost certain equality (almost everywhere with respect to the measure P) establishes an equivalence relation over this set (cf. Chapter 6, Section 4.1.4.3).

Two complex r.v.s ξ and η are equivalent when they belong to the same equivalence class. The set of trials ω for which the r.v.s take different values is the event $\{\omega \,|\, \xi(\omega) \neq \eta(\omega)\}$ which we can write more simply $\{\xi \neq \eta\}$. Its probability is zero, by definition. Two equivalent r.v.s ξ and η are therefore a.s. equal and we write $\xi \underset{\text{a.s.}}{=} \eta$.

The two r.v.s ξ and η are a.s. equal if and only if the r.v. $|\xi - \eta|^2$ is a.s. zero, which is the case if and only if the positive function $|\xi - \eta|^2$ has zero integral with respect to the measure P (cf. Chapter 6, Section 4.1.4.2). This allows us to write

$$\xi \underset{\text{a.s.}}{=} \eta \Leftrightarrow E[|\xi - \eta|^2] = 0 \tag{1.39}$$

This property is used notably to show that an r.v. is a.s. constant and equal to its mean value (cf. Eqn (1.70)).

2.2. PROBABILITY LAW OF A REAL RANDOM VARIABLE

2.2.1. Definitions and general properties

2.2.1.1. Definition
Let ξ be a real r.v. defined over a probability space $(\Omega, \mathcal{A}, P)$. Its **probability law** μ_ξ is the image of the measure P through the mapping ξ (cf. Chapter 6, Section 3.1.4). It is a measure over the measurable space $(\mathbb{R}, \mathcal{R})$ such that (cf. Eqn (6.28))

$$\forall \beta \in \mathcal{R} \qquad \mu_\xi(\beta) = P\{\xi^{-1}(\beta)\} = P\{\xi \in \beta\} \tag{1.40}$$

2.2.1.2. Properties
It is immediate that μ_ξ is a positive normed measure. It is therefore a probability measure. The number $\mu_\xi(\beta)$ is the probability of the event

$\xi^{-1}(\beta) = \{\xi \in \beta\}$. It is the probability that the r.v. ξ takes a value belonging to the Borel set β.

The r.v. ξ therefore makes the probability space $(\mathbb{R}, \mathscr{R}, \mu_\xi)$ correspond to the original probability space $(\Omega, \mathscr{A}, P)$ and the Borel sets of the Borel field $\mathscr{R}$ correspond to the events of the Borel field $\mathscr{A}$. Most frequently the r.v. ξ is defined directly by its probability law and the space $(\Omega, \mathscr{A}, P)$ is not made explicit. We work then in the space $(\mathbb{R}, \mathscr{R}, \mu_\xi)$, observing the values taken in $\mathbb{R}$ by the r.v. All this is illustrated by Fig. 1.2 in the case where the Borel set β is an interval.

Example 2.4 Almost surely equal random variables

Let ξ and η be two r.v.s defined over one and the same probability space. Suppose that they are a.s. equal; we shall show that they have the same probability law which may be expressed by

$$\forall \beta \in \mathscr{R} \qquad \mu_\xi(\beta) = \mu_\eta(\beta)$$

Consider the events $A = \{\xi \in \beta\}$ and $B = \{\eta \in \beta\}$. We have to prove that their probabilities are equal. Let $A \triangle B$ be the symmetric difference event (cf. Eqn (1.1)). It satisfies

$$A \triangle B \subset \{\xi \neq \eta\}$$

since it contains the trials for which only one of the r.v.s ξ and η (but not the other) takes a value belonging to the Borel set β, which implies that the two r.v.s take different values. The stated result then follows from Eqns (1.37) and (1.29).

It follows from the above that two a.s. equal real r.v.s have the same probability law. It is impossible to distinguish them by observing the values which they take in $\mathbb{R}$. This is why it is common to assign an r.v. to the class to which it belongs.

A probability law characterizes a class of a.s. equal real r.v.s but the converse is not true. Two real r.v.s can have the same probability laws without being a.s. equal. Exercise 2.13 gives an example of this.

2.2.2. Discrete probability laws

2.2.2.1. Definition
A probability law is **discrete** if it is the probability law of a discrete r.v. (cf. Example 2.2).

2.2.2.2. Property
Let ξ be the discrete real r.v. defined in Example 2.2. Its probability law is obtained by applying the definition of the Dirac measure (cf. Chapter 6,

Example 3.1). It follows that

$$\mu_\xi = \sum_{n\in\mathbb{N}} P\{\xi = x_n\} \delta(x - x_n) \quad (1.41)$$

A discrete probability law is therefore a combination of Dirac measures.

Example 2.5 The Poisson law
The **Poisson law** of parameter $\lambda \in \mathbb{R}$ is the probability law of the Poisson r.v. defined by

$$\xi = \sum_{n\in\mathbb{N}} n I_{A_n} \qquad A_n = \{\xi = n\} \qquad P(A_n) = \exp(-\lambda)\frac{\lambda^n}{n!}$$

It follows from Eqn (1.41) that

$$\mu_\xi(x) = \sum_{n\in\mathbb{N}} \exp(-\lambda) \frac{\lambda^n}{n!} \delta(x - n) \quad (1.42)$$

The Poisson law is very important. In fact, a large number of physical quantities may be represented, to a good approximation, by a Poisson r.v., e.g. the number of calls received per unit time by a telephone exchange and the number of photons received per unit time by a photodetector exposed to monochromatic light.

Exercise 2.7 is devoted to the Poisson law.

Example 2.6 The binomial or Bernoulli law
Let $(\Omega^n, \mathcal{P}(\Omega^n), P)$, $n\in\mathbb{N}^*$, be the probability space defined in Example 1.6. The set Ω^n contains 2^n trials, each trial being a binary word of n symbols 0 and 1. Let B_i, $i \in [1, n]$, be the event consisting of the set of words whose symbol in position i is a 1, and let A_k, $k \in [0, n]$, be the event consisting of the set of $\binom{n}{k}$ words containing exactly k symbols 1. We associate with each word the number of symbols 1 which it contains. This is a discrete random variable ξ, defined over the space $(\Omega^n, \mathcal{P}(\Omega^n), P)$, which takes its values in the set $[0, n]$. It can be written in two equivalent ways (cf. Example 2.2):

$$\xi = \sum_{i=1}^n I_{B_i} = \sum_{k=0}^n k I_{A_k}$$

It is obvious that the events B_i are not incompatible since a binary word can contain many symbols 1. In contrast the events A_k are incompatible, and they constitute a partition of the set Ω^n. It follows from Eqns (1.41) and (1.24) that the probability law of the r.v. ξ can be written

$$\mu_\xi(x) = \sum_{k=0}^n \binom{n}{k} p^k (1-p)^{n-k} \delta(x - k) \quad (1.43)$$

This is the **binomial law** of parameters (n, p). It is also called the **Bernoulli law**.
The binomial law is much used in problems of enumeration. We can show

that the law of parameters (n, p) tends towards a Poisson law of parameter λ when n tends towards infinity and p tends towards zero in such a way that the product np tends towards λ (cf. Chapter 2, Example 2.5).

2.2.3. Continuous probability laws

2.2.3.1. Definition
A probability law is **continuous** if it is an absolutely continuous measure with respect to the Lebesgue measure.

2.2.3.2. Properties
The Lebesgue measure over the space $(\mathbb{R}, \mathscr{R})$ is defined in Chapter 6, Example 3.3. Absolutely continuous measures are the subject of Chapter 6, Section 4.1.5.

Let μ_ξ be a continuous probability law. It has a density p_ξ which is a positive real measurable function which is integrable (with respect to the Lebesgue measure). Its integral is equal to 1 since the measure μ_ξ is normed. The function p_ξ is the **probability density** of the r.v. ξ. It must be noted that it is defined only almost everywhere over the space $(\mathbb{R}, \mathscr{R}, \mu_0)$, μ_0 being the Lebesgue measure.

The probability that the r.v. ξ takes a value belonging to a Borel set β is given by the integral (cf. Eqn (6.60))

$$P\{\xi \in \beta\} = \mu_\xi(\beta) = \int_\beta p_\xi(x)\,dx \qquad (1.44)$$

The probability that the r.v. ξ takes the value a is

$$P\{\xi = a\} = \int_{\{a\}} p_\xi(x)\,dx = 0 \qquad (1.45)$$

This results from the fact that the function $p_\xi I_{\{a\}}$ is negligible with respect to the Lebesgue measure (cf. Chapter 6, Section 4.1.4 and Example 3.5). The event $\{\xi = a\}$ thus has zero probability.

Example 2.7 **The gaussian or normal law**

The **gaussian law** of parameters m and σ or the normal law $N(m, \sigma)$ is defined by the density

$$p_\xi(x) = \frac{1}{(2\pi\sigma^2)^{1/2}} \exp\left\{-\frac{(x-m)^2}{2\sigma^2}\right\} \qquad m \in \mathbb{R} \quad \sigma^2 \in \mathbb{R}_+^* \qquad (1.46)$$

A real r.v. ξ is gaussian when its probability law is a gaussian law. The parameters m and σ^2 are respectively the mean value and the variance of the r.v. (cf. Section 2.4.1 and Exercise 2.4). The graph of the function p_ξ is represented in Fig. 1.3 for different values of the standard deviation σ.

It is interesting to calculate the probability that the r.v. ξ takes a value

30 Random variables

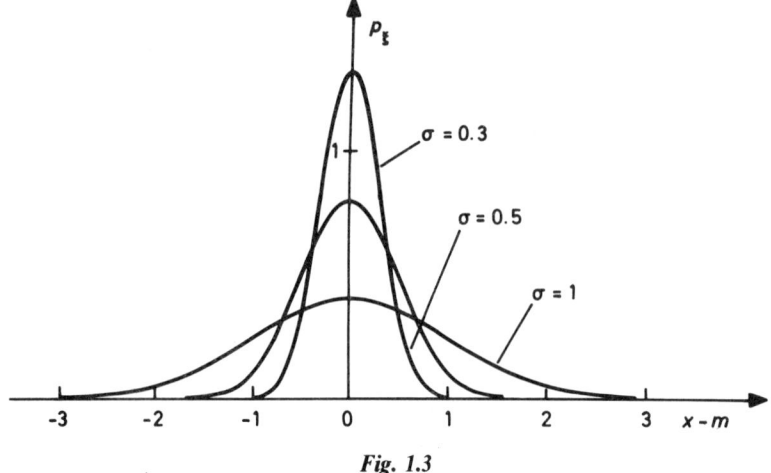

Fig. 1.3

belonging to the interval $[m - k\sigma, m + k\sigma]$. This is the probability of the event $\{|\xi - m| \leq k\sigma\}$. Applying Eqn (1.63) and using the table of numerical values in Appendix 4, we obtain

$$P\{|\xi - m| \leq \sigma\} = 0.682\,68$$
$$P\{|\xi - m| \leq 2\sigma\} = 0.954\,50$$
$$P\{|\xi - m| \leq 3\sigma\} = 0.997\,30$$
$$P\{|\xi - m| \leq 4\sigma\} = 0.999\,94$$

The probability that the r.v. deviates from its mean value m by more than three times its standard deviation σ is equal to 2.7×10^{-3}. It is thus extremely small. This enables us to accept, in many applications, that a gaussian r.v. takes values within the interval $[m - 3\sigma, m + 3\sigma]$.

The gaussian law is very widely used. Its importance follows from the central limit theorem (cf. Chapter 2, Section 2.5). Noise which affects the accuracy of measurement of physical quantities is almost always represented by a gaussian r.v.

Example 2.8   The exponential law

The **exponential law** of parameter a is defined by the density

$$p_\xi(x) = Y(x)a \exp(-ax) \qquad a \in \mathbb{R}_+^* \qquad (1.47)$$

The function p_ξ has domain $\mathbb{R}_+$, which means that the r.v. ξ takes values in $\mathbb{R}_+$ and is, in fact, positive. The graph of the function p_ξ is represented in Fig. 1.4 for different values of the parameter a.

The exponential law plays a very important role in numerous fields, notably the theory of reliability and queueing theory (Kleinrock, 1975). We may mention among physical quantities that are represented to a good

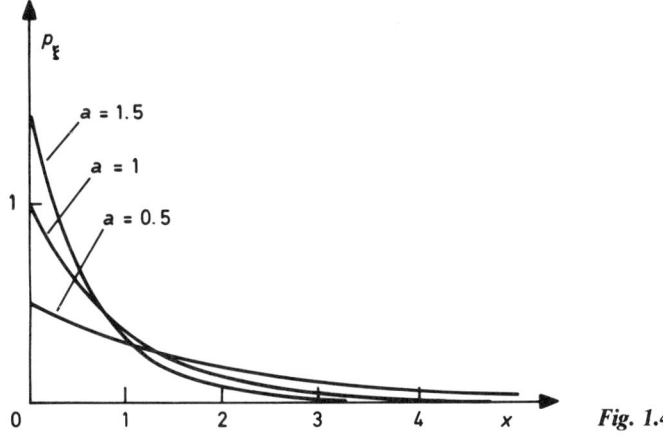

Fig. 1.4

approximation by an r.v. having the exponential probability law the following: the duration of a telephone call; the length of life of an electronic component; the interval of time separating the arrivals of two successive photons at a photodetector exposed to monochromatic light.

We prove moreover in Exercise 2.16 that the exponential law is the only law which possesses the interesting property of having no memory.

Example 2.9 The Rayleigh law

The **Rayleigh law** of parameter a is defined by the density

$$p_\xi(x) = Y(x) \frac{x}{a^2} \exp\left(-\frac{x^2}{2a^2}\right) \qquad a \in \mathbb{R}_+^* \qquad (1.48)$$

The function p_ξ has domain $\mathbb{R}_+$, which means that the r.v. ξ is positive. The graph of the function p_ξ is represented in Fig. 1.5 for different values of the parameter a.

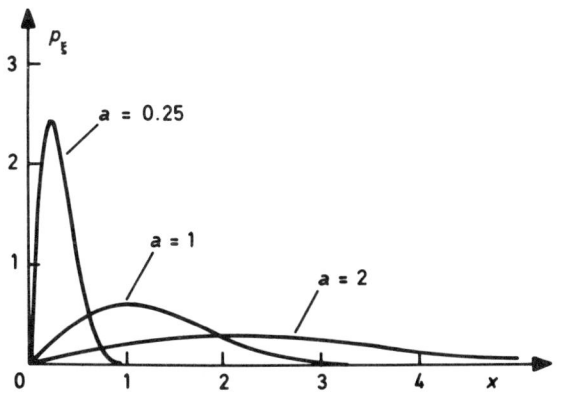

Fig. 1.5

32 Random variables

The Rayleigh law is used in the theory of radar and in the theory of propagation of electromagnetic waves to represent the phenomenon of 'fading'.

Example 2.10   Hypothesis test and the likelihood ratio

General case

Let $(\Omega, \mathscr{A}, P)$ be a probability space and ξ_0 and ξ_1 be two real r.v.s having probability densities p_{ξ_0} and p_{ξ_1}. Suppose an experiment is carried out in which either the r.v. ξ_0 or the r.v. ξ_1 is involved but it is impossible to know *a priori* which of the two is actually the case.

When the experiment is carried out, a trial ω is made and the value $a \in \mathbb{R}$ taken by the r.v. concerned is observed. The question is to know whether $a = \xi_0(\omega)$ or $a = \xi_1(\omega)$. We are thus concerned with a hypothesis test, as formulated in Example 1.3. The two probability spaces are here $(\mathbb{R}, \mathscr{R}, \mu_{\xi_0})$ and $(\mathbb{R}, \mathscr{R}, \mu_{\xi_1})$.

The answer is provided by formula (1.17) which gives the criterion enabling us to choose one of the two hypotheses with minimal probability of making an error. This criterion uses the likelihood ratio $d\mu_{\xi_0}/d\mu_{\xi_1}$. It may be written here, taking account of the existence of the probability densities p_{ξ_0} and p_{ξ_1} (cf. Chapter 6, Example 4.7)

$$\frac{d\mu_{\xi_0}}{d\mu_{\xi_1}} = \frac{p_{\xi_0}}{p_{\xi_1}}$$

The criterion leading to our choice is therefore

$$\frac{p_{\xi_0}(a)}{p_{\xi_1}(a)} \underset{H_1}{\overset{H_0}{\gtrless}} \frac{p_1}{p_0} \tag{1.49}$$

The situation is represented by the graph in Fig. 1.6(a). This figure shows the

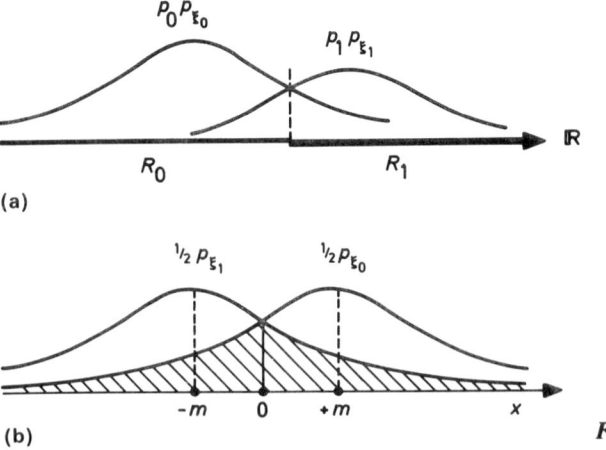

Fig. 1.6

partition of ℝ into two complementary events R_0 and R_1. When the event R_0 is realized, i.e. when $a \in R_0$, we make the decision that the hypothesis h_0 is realized and that $a = \xi_0(\omega)$. In the other case we make the decision that the hypothesis h_1 is realized. The probability of making a wrong choice is then written (cf. Eqn (1.16))

$$P(\mathscr{E}) = p_0 \int_{R_1} p_{\xi_0}(x)\,dx + p_1 \int_{R_0} p_{\xi_1}(x)\,dx \qquad (1.50)$$

Application to the detection of a binary symbol

Consider, by way of illustration, a telecommunications receiver which receives one or other of two binary symbols 0 or 1 with equal probability. In the absence of noise, the output voltage from the receiver is equal to $+m$ when the symbol received is 0. It is equal to $-m$ when the symbol received is 1. In the presence of gaussian centred noise, the output voltage from the receiver is the value taken by the r.v. ξ_0 when the symbol received is 0 and the value taken by the r.v. ξ_1 when the symbol received is 1. The probability densities of the two r.v.s are respectively (cf. Eqn (1.46))

$$p_{\xi_0}(x) = \frac{1}{(2\pi\sigma^2)^{1/2}} \exp\left\{ -\frac{(x-m)^2}{2\sigma^2} \right\}$$

and

$$p_{\xi_1}(x) = \frac{1}{(2\pi\sigma^2)^{1/2}} \exp\left\{ -\frac{(x+m)^2}{2\sigma^2} \right\}$$

The value a of the output voltage enables us to decide whether the receiver has received a 0 or a 1. The situation is illustrated by Fig. 1.6(b). It is immediate that the choice depends on the sign of a and that the criterion is written

$$a \underset{H_1}{\overset{H_0}{\gtrless}} 0$$

The probability of error may be obtained from Eqn (1.50). It is represented by the shaded area in Fig. 1.6(b). It follows (cf. Eqn (1.63)) that

$$P(\mathscr{E}) = \frac{1}{2} P\{\xi_0 < 0\} + \frac{1}{2} P\{\xi_1 > 0\}$$

$$= \frac{1}{2} - \Phi\left(\frac{m}{\sigma}\right) \qquad (1.51)$$

The function Φ takes the value 0 at the origin and tends to $\tfrac{1}{2}$ at infinity (cf. Appendix 4). The probability of error is thus between 0 and $\tfrac{1}{2}$. It tends to 0 when the ratio m/σ tends to infinity and to $\tfrac{1}{2}$ when the ratio tends to 0.

2.2.4. Probability law of a random variable function of a random variable

2.2.4.1. General case

Let $g: (\mathbb{R}, \mathcal{R}) \to (\mathbb{R}, \mathcal{R})$ be a measurable function and ξ a real r.v. The probability law of the real r.v. $\eta = g \circ \xi$ is the image of the measure P through the measurable function η. Consequently (cf. Eqn (1.40))

$$\forall \beta \in \mathcal{R} \qquad \mu_\eta(\beta) = P\{\eta^{-1}(\beta)\}$$

Now we have

$$\eta^{-1}(\beta) = (\xi^{-1} \circ g^{-1})(\beta) = \xi^{-1}\{g^{-1}(\beta)\}$$

Hence we have the result

$$\mu_\eta(\beta) = P[\xi^{-1}\{g^{-1}(\beta)\}] = \mu_\xi\{g^{-1}(\beta)\} \tag{1.52}$$

The probability law of the r.v. $\eta = g \circ \xi$ is thus the image of the probability law of the r.v. ξ through the measurable function g (cf. Eqn (6.28)).

Example 2.11 Translation of a random variable

Let ξ be a real r.v. and η the r.v. $\eta = a + \xi$, $a \in \mathbb{R}$. We have from Eqn (1.52)

$$\mu_\eta(\beta) = \mu_\xi\{\omega \mid \omega + a \in \beta\} = \mu_\xi(\beta - a)$$

Hence we have, from Eqn (6.78),

$$\mu_\eta = \mu_\xi * \delta_a \tag{1.53}$$

The probability law of the translated r.v. η is the convolution of the probability law of the r.v. ξ with the Dirac measure δ_a.

2.2.4.2. Case of a continuous random variable

Suppose that the r.v. ξ is continuous and let p_ξ be its probability density. From Eqn (1.44) we have

$$\mu_\eta(\beta) = \mu_\xi\{g^{-1}(\beta)\} = \int_\mathbb{R} I_{g^{-1}(\beta)}(x) p_\xi(x) \, dx$$

Put $A = \xi(\Omega)$ and $A' = \eta(\Omega) = g(A)$. Then the r.v. ξ takes its values in A and the r.v. η takes its values in A'. Therefore

$$\mu_\eta(\beta) = \int_A I_{g^{-1}(\beta)}(x) p_\xi(x) \, dx \tag{1.54}$$

Suppose the function g defines a continuously differentiable homomorphism from $\mathring{A}$ onto $\mathring{A}'$. The theorem for a change of variable in integrals enables us to write, taking account of the fact that $I_{g^{-1}(\beta)} = I_\beta \circ g$,

$$\mu_\eta(\beta) = \int_{A'} I_\beta(y) p_\xi\{g^{-1}(y)\} \left|\frac{dg^{-1}(y)}{dy}\right| dy$$

$$= \int_{\mathbb{R}} I_\beta(y) p_\beta\{g^{-1}(y)\} \left|\frac{dg^{-1}(y)}{dy}\right| dy$$

Since the preceding equality holds for every Borel set β, the r.v. η has for density the function (cf. Chapter 6, Exercise 4.2)

$$p_\eta(y) = p_\xi\{g^{-1}(y)\} \left|\frac{dg^{-1}(y)}{dy}\right| \tag{1.55}$$

It has as domain the set A' since $\eta(\Omega) = A'$.

Example 2.12 **The exponential and uniform laws**

Let ξ be a real r.v. uniformly distributed over the interval $[0, 1]$. Its probability density p_ξ is defined by

$$p_\xi(x) = 1 \quad x \in [0, 1]$$
$$p_\xi(x) = 0 \quad x \notin [0, 1]$$

Consider the positive r.v. $\eta = -(1/a)\log \xi$, $a \in \mathbb{R}_+^*$. It is clear that the conditions for applications of Eqn (1.55) are fulfilled with

$$\left|\frac{dg^{-1}(y)}{dy}\right| = a \exp(-ay)$$

Hence

$$p_\eta(y) = Y(y) a \exp(-ay)$$

The probability law of the r.v. η is therefore exponential. This property is used in practice to obtain sample drawings from the exponential law starting from drawings from the uniform law.

2.3. DISTRIBUTION FUNCTION OF A REAL RANDOM VARIABLE

2.3.1. *Definitions and general properties*

2.3.1.1. Definition

Let ξ be a real r.v. Its **distribution function** is the positive real function F_ξ defined by

$$\forall x \in \mathbb{R} \quad F_\xi(x) = P(\xi < x) = \mu_\xi(]-\infty, x[) \tag{1.56}$$

This is the convolution of the measure μ_ξ with the Heaviside function Y, which may be written $F_\xi = \mu_\xi * Y$ (cf. Eqn (6.80)).

The value of the function F_ξ at the point x is the probability that the r.v. ξ takes a value strictly less than x.

2.3.1.2. Properties

The distribution function is monotonic increasing. In fact we can write

$$]-\infty, x_2[=]-\infty, x_1[\cup [x_1, x_2[\quad x_1 \leqslant x_2$$

Hence, by additivity of the measure P

$$P\{x_1 \leqslant \xi < x_2\} = F_\xi(x_2) - F_\xi(x_1) \geqslant 0 \tag{1.57}$$

Its minimum value is $F_\xi(-\infty) = \mu_\xi(\emptyset) = 0$. Its maximum value is $F_\xi(+\infty) = \mu_\xi(\mathbb{R}) = 1$.

Like every monotonic function, the distribution function has at most a denumerable number of discontinuities of the first kind. Moreover, it is continuous to the left. In fact

$$]-\infty, x[= \bigcup_{n \in \mathbb{N}} \left( ]-\infty, x - \frac{1}{n}[\right)$$

and it follows from Eqn (1.9) that

$$F_\xi(x) = \lim_{n \to \infty} F_\xi\left(x - \frac{1}{n}\right) = F_\xi(x - 0)$$

Next we write

$$]-\infty, x[= \bigcap_{n \in \mathbb{N}} \left( ]-\infty, x + \frac{1}{n}[\right) = (]-\infty, x[) \cup (\{x\})$$

It follows from Eqn (1.10) that

$$P\{\xi \leqslant x\} = F_\xi(x) + P(\xi = x) = \lim_{n \to \infty} F_\xi\left(x + \frac{1}{n}\right) = F_\xi(x + 0) \tag{1.58}$$

Thus, when the distribution function is discontinuous at x, the jump is equal to the probability that the r.v. ξ takes the value x.

2.3.2. Distribution function of a discrete random variable

Let ξ be the discrete r.v. whose probability law is given by Eqn (1.41). Its distribution function is obtained from Eqn (1.56) and the definitions of the Dirac measure and of the Heaviside function (cf. Chapter 6, Example 3.1). It follows that

$$F_\xi(x) = \sum_{n \in \mathbb{N}} P\{\xi = x_n\} Y(x - x_n) \tag{1.59}$$

This is a step function. The jump at the point x_n is equal to the probability that the r.v. takes the value x_n.

The discontinuous function F_ξ is bounded. Thus it defines a tempered distribution which is differentiable in the space $\mathscr{S}'$ and has as derivative the tempered distribution (Dupraz, 1977, p. 18)

$$F'_\xi = \sum_{n \in \mathbb{N}} P\{\xi = x_n\} \delta(x - x_n) = \mu_\xi \tag{1.60}$$

This is the distribution defined by the measure μ_ξ, i.e. by the probability law of the r.v. ξ (cf. Eqn (1.41)).

2.3.3. Distribution function of a continuous random variable

Let ξ be a continuous real r.v. having as probability density the function p_ξ. Its distribution function is obtained from Eqns (1.56) and (1.44). We obtain

$$\forall x \in \mathbb{R} \quad F_\xi(x) = \int_{-\infty}^{x} p_\xi(u)\, du \tag{1.61}$$

Then it follows from Eqn (1.58), since $P\{\xi = x\} = 0$,

$$\forall x \in \mathbb{R} \quad F_\xi(x) = F_\xi(x + 0)$$

The distribution function of a continuous r.v. is thus continuous.

The function p_ξ is integrable. It follows then from Eqn (1.61) that the function F_ξ is almost everywhere (a.e.) differentiable and has as derivative the function p_ξ which can be discontinuous:

$$F'_\xi \underset{\text{a.e.}}{=} p_\xi \tag{1.62}$$

When the function p_ξ is continuous, the above equality holds everywhere in $\mathbb{R}$.

Example 2.13 Gaussian random variable

The probability density of a gaussian r.v. is given by Eqn (1.46). Its distribution function is thus defined by the integral

$$F_\xi(x) = \frac{1}{(2\pi\sigma^2)^{1/2}} \int_{-\infty}^{x} \exp\left\{-\frac{(u - m)^2}{2\sigma^2}\right\} du$$

This may be written

$$F_\xi(x) = \frac{1}{2} + \Phi\left(\frac{x - m}{\sigma}\right) = \frac{1}{2}\left[1 + \text{erf}\left\{\frac{x - m}{(2\sigma)^{1/2}}\right\}\right] \tag{1.63}$$

with

$$\Phi(x) = \frac{1}{(2\pi)^{1/2}} \int_{0}^{x} \exp\left(-\frac{u^2}{2}\right) du \quad \text{erf}(x) = \frac{2}{\pi^{1/2}} \int_{0}^{x} \exp(-u^2)\, du \tag{1.64}$$

The probability that the r.v. ξ deviates from its mean value by less than k times

38 Random variables

its standard deviation is then (cf. Example 2.7)
$$P\{|\xi - m| \leqslant k\sigma\} = 2\Phi(k)$$

We are often required in practice to calculate probabilities involving numerical values of the function Φ. We then use the table in Appendix 4.

Example 2.14 Probability density of the square of a random variable

Let ξ be a real r.v. and let the r.v. $\eta = \xi^2$. The distribution function F_η is obtained by applying Eqn (1.56). It is clearly zero for all $y < 0$ and for $y \geqslant 0$ we have

$$F_\eta(y) = P\{-y^{1/2} < \xi < y^{1/2}\} = P\{\xi < y^{1/2}\} - P\{\xi \leqslant -y^{1/2}\}$$
$$= F_\xi(y^{1/2}) - F_\xi(-y^{1/2}) - P\{\xi = -y^{1/2}\}$$

When the r.v. ξ is continuous, $P\{\xi = -y^{1/2}\} = 0$ and

$$F_\eta(y) = Y(y)\{F_\xi(y^{1/2}) - F_\xi(-y^{1/2})\}$$

The probability density p_η cannot be obtained from Eqn (1.55) since the inverse mapping does not exist here. It can, however, be obtained by application of Eqn (1.62) to differentiate the function F_η. We obtain

$$p_\eta(y) = Y(y) \frac{1}{2y^{1/2}} \{p_\xi(y^{1/2}) + p_\xi(-y^{1/2})\} \qquad (1.65)$$

2.4. MOMENTS OF A RANDOM VARIABLE

2.4.1. Definitions and general properties

2.4.1.1. Definitions

Let $(\Omega, \mathscr{A}, P)$ be a probability space and ξ a real or complex r.v. Its **moment of order** $n, n \in \mathbb{N}^*$, is the integral $E[\xi^n]$ of the r.v. ξ^n (with respect to the measure P):

$$E[\xi^n] = \int_\Omega \xi^n(\omega) \, dP(\omega) \qquad (1.66)$$

We say also that the moment of order n is the **mathematical expectation** of the r.v. ξ^n.

The moment of order 1 is the **mean value** $E[\xi]$ of the r.v. ξ. We generally denote this value by m_ξ. An r.v. whose mean value is zero is called **centred**.

The moment of order 2 is the **mean square value** $E[\xi^2]$ of the r.v. ξ.

2.4.1.2. Properties

We note first that, if ξ is an r.v., ξ^n is also an r.v. (cf. Chapter 6, Example 2.3). The r.v. ξ^n is not necessarily integrable, so that it is possible that the r.v. ξ does not have a moment of order n. In contrast, it is sufficient that the r.v. $|\xi|$ has a moment of order n for the r.v. ξ to have a moment of the same order (cf. Eqn (6.51)). When it exists, the moment of order n of an r.v. ξ is the complex number (cf. Eqn (6.48))

$$E[\xi^n] = E[\operatorname{Re} \xi^n] + iE[\operatorname{Im} \xi^n] \qquad (1.67)$$

Let $\overset{\circ}{\xi}$ be the r.v. defined by

$$\overset{\circ}{\xi} = \xi - E[\xi] \qquad (1.68)$$

The linearity of the operator E allows us to write (cf. Chapter 6, Section 4.1.3.3)

$$E[\overset{\circ}{\xi}] = E[\xi] - E[\xi] = 0$$

The r.v. $\overset{\circ}{\xi}$ has zero mean value. It is therefore centred.

It must be remarked finally that if two r.v.s ξ and η are a.s. equal, the r.v.s ξ^n and η^n are also a.s. equal. Therefore they have equal integrals (cf. Chapter 6, Section 4.1.4.3). Two a.s. equal r.v.s consequently have identical moments of all orders, which is equivalent to saying that they have identical laws of probability (cf. Example 2.4).

2.4.2. Square integrable random variables

2.4.2.1. Definitions

A complex r.v. ξ is **square integrable** or **of order 2** when the mean square value $E[|\xi|^2]$ of the real r.v. $|\xi|$ is finite.

The variance $V[\xi]$ of the r.v. ξ is the mean square value $E[|\overset{\circ}{\xi}|^2]$. The positive square root σ_ξ of the variance is the **dispersion** or **standard deviation** of the r.v. ξ.

2.4.2.2. Properties

The square integrable r.v.s defined over a probability space $(\Omega, \mathscr{A}, P)$ constitute the vector space $\mathscr{L}^2(\Omega, \mathscr{A}, P)$. The equivalence classes of a.s. equal square integrable r.v.s constitute the Hilbert vector space $L^2(\Omega, \mathscr{A}, P)$ (cf. Chapter 6, Section 4.1.7). The norm of a square integrable r.v. ξ is defined by

$$\|\xi\| = (E[|\xi|^2])^{1/2} \qquad (1.69)$$

In practice, we often do not distinguish between a square integrable r.v. and the equivalence class to which it belongs. This means that we do not distinguish between r.v.s which are a.s. equal.

A square integrable r.v. has finite mean and mean square values (cf. Chapter 6, Section 4.1.7.3). The linearity of the operator E allows us to write

$$V[\xi] = \sigma_\xi^2 = E[|\overset{\circ}{\xi}|^2] = E[|\xi|^2] - |E[\xi]|^2 \qquad (1.70)$$

The variance of a square integrable r.v. is therefore finite.

Square integrable r.v.s play a considerable part in practical work. They represent random physical quantities of finite mean value, a current or a

40 Random variables

voltage for example. They are sometimes called of order 2 to signify that their moments of order 2 are finite.

Example 2.15 Approximation to a random variable by a constant

Let ξ be a complex r.v. having a mean value $E[\xi] = m$ and let a be a complex constant. Using the linearity of the operator E we can write

$$E[|\xi - a|^2] = E[|\xi - m + m - a|^2]$$
$$= E[|\xi - m|^2 + |m - a|^2 + (m^* - a^*)(\xi - m) + (m - a)(\xi^* - m^*)]$$
$$= V[\xi] + |m - a|^2$$
$$\geq V[\xi]$$

The mean square value of the absolute deviation $|\xi - a|$ is minimal, and equal to the variance $V[\xi]$, when $m = a$.

The mean value of an r.v. is thus the best approximation, in quadratic mean, to this r.v. by a constant. The mean square error is equal to the variance $V[\xi]$.

2.4.3. Moments of a real random variable

(a) Let $(\Omega, \mathcal{A}, P)$ be a probability space and ξ a real r.v. having a probability law μ_ξ. Let $g: (\mathbb{R}, \mathcal{R}) \to (\mathbb{R}, \mathcal{R})$ be a measurable function and let η be the real r.v. $\eta = g \circ \xi$.

The moments of order n, $n \in \mathbb{N}^*$, of the r.v. η are obtained from the general formula (1.66). Now the integral of the r.v. $(g \circ \xi)^n$ with respect to the measure P is equal to the integral of the function g^n with respect to the measure μ_ξ (cf. Chapter 6, Example 4.6). This enables us to write

$$E[\eta^n] = \int_\Omega \eta^n(\omega) \, dP(\omega) = \int_\mathbb{R} g^n(x) \, d\mu_\xi(x) \qquad (1.71)$$

In the case where g is the identity function, we obtain

$$E[\xi^n] = \int_\Omega \xi^n(\omega) \, dP(\omega) = \int_\mathbb{R} x^n \, d\mu_\xi(x) \qquad (1.72)$$

The two preceding formulae are very important. Formula (1.72) shows that the moment of order n of a real r.v. is obtained either by integration with respect to the measure P or by integration with respect to the measure μ_ξ. In practice it is the second form which is used. In fact, a real r.v. is defined by its probability law, whereas the measure P is not in general known explicitly. Formula (1.71) shows that it is not necessary to know explicitly the probability law of the r.v. η in order to calculate its moments.

(b) Let ξ be a discrete r.v. whose probability law is given by Eqn (1.41). Its moment of order n is written (cf. Eqn (6.53))

$$E[\xi^n] = \sum_{i \in \mathbb{N}} x_i^n P\{\xi = x_i\} \tag{1.73}$$

Let ξ be a real continuous r.v. having probability density p_ξ. Its moment of order n is written (cf. Eqn (6.60))

$$E[\xi^n] = \int_\mathbb{R} x^n p_\xi(x) \, dx \tag{1.74}$$

2.4.4. The Markov and Tchebycheff inequalities

2.4.4.1. The Markov inequality

Let $]-a, a[$ be an interval and $\mathring{I}_{2a}, a \in \mathbb{R}_+^*$, its indicator function. It is easy to verify that

$$\forall a \in \mathbb{R}_+^* \quad \forall n \in \mathbb{N}^* \quad \forall x \in \mathbb{R} \quad 0 \leqslant 1 - \mathring{I}_{2a}(x) \leqslant \frac{1}{a^n} |x|^n$$

Let ξ be a real r.v. having a finite absolute moment of order k: $E[|\xi|^k] < +\infty$, $k \in \mathbb{N}^*$. The preceding inequality enables us to write, by applying the increasing property of the integral with respect to the measure μ_ξ (cf. Eqn (6.50)),

$$\mu_\xi(1 - \mathring{I}_{2a}) \leqslant \frac{1}{a^k} \mu_\xi(|x|^k)$$

Hence, taking account of Eqn (1.37), we have

$$\forall a \in \mathbb{R}_+^* \quad P\{|\xi| \geqslant a\} \leqslant \frac{1}{a^k} E[|\xi|^k] \tag{1.75}$$

This is the **Markov inequality**.

2.4.4.2. The Tchebycheff inequality

Suppose that the r.v. ξ is square integrable and put

$$m_\xi = E[\xi] \quad \sigma_\xi^2 = V[\xi]$$

Apply the inequality (1.75) to the centred r.v. $\mathring{\xi}$ with $a = \varepsilon$ and $k = 2$. We obtain

$$\forall \varepsilon \in \mathbb{R}_+^* \quad P\{|\xi - m_\xi| < \varepsilon\} \geqslant 1 - \frac{\sigma_\xi^2}{\varepsilon^2} \tag{1.76}$$

This is the **Tchebycheff inequality**. It justifies the name **dispersion** given to the positive square root of the variance. It shows in fact that a real r.v. takes values that are the more widely dispersed on both sides of its mean value, the greater its dispersion is. In the limit, when the variance σ_ξ^2 is zero, the r.v. ξ is a.s. constant and equal to its mean value m_ξ. This also follows from Eqn (1.39).

Example 2.16 Noisy transmission of discrete information

(a) Let $A = \{a_1, \ldots, a_m\}$ and $B = \{b_1, \ldots, b_n\}$ be two finite sets of symbols, $\Omega = A \times B$ be the product set and $(\Omega, \mathcal{P}(\Omega), P)$ the probability space defined as follows. The trials are the $m \times n$ pairs (a_i, b_j) and the measure P is defined by the $m \times n$ probabilities p_{ij} such that

$$P\{(a_i, b_j)\} = p_{ij} \qquad \sum_{i=1}^{m} \sum_{j=1}^{n} p_{ij} = 1$$

Let us define the events

$$\forall i \in \{1, 2, \ldots, m\} \qquad A_i = \bigcup_{j=1}^{n} (a_i, b_j)$$

$$\forall j \in \{1, 2, \ldots, n\} \qquad B_j = \bigcup_{i=1}^{m} (a_i, b_j)$$

The event A_i is composed of the trials whose first element is the symbol a_i and the event B_j is composed of the trials whose second element is the symbol b_j. It is immediate that $A_i \cap B_j = (a_i, b_j)$, that the families (A_i) and (B_j) are partitions of the set Ω and that the probabilities of the events A_i and B_j are respectively

$$P(A_i) = \sum_{j=1}^{n} p_{ij} \qquad P(B_j) = \sum_{i=1}^{n} p_{ij}$$

The families (A_i) and (B_j) are not mutually independent in the general case because

$$P(A_i \cap B_j) \neq P(A_i) P(B_j)$$

(b) Let ξ and η be real discrete r.v.s defined over the trials $\omega \in \Omega$ by

$$\begin{aligned} \xi(\omega) &= -\log_2 P(A_i) & \omega &\in A_i \\ \eta(\omega) &= -\log_2 P(A_i | B_j) & \omega &\in A_i \cap B_j \end{aligned} \qquad (1.77)$$

The positive quantity $-\log_2 P(A_i)$ is called the **uncertainty** of the event A_i. It is in fact greater according as the event A_i is less probable. It is zero when the event is a.s. certain and equal to unity when the event is a.s. impossible. The quantity $-\log_2 P(A_i | B_j)$ is the conditional uncertainty of the event A_i, given that the event B_j is realized.

The mean value of the r.v. ξ, by virtue of Eqn (1.73), is

$$E[\xi] = -\sum_{i=1}^{m} P(A_i) \log_2 P(A_i) = H(A) \qquad (1.78)$$

This is the **entropy** of the family (A_i). It is a measure of the mean uncertainty of the events A_i. This is equivalent to saying, taking account of the nature of the events A_i, that it is the entropy of the set A. In the same way

$$E[\eta] = - \sum_{i=1}^{m} \sum_{j=1}^{n} p_{ij} \log_2 P(A_i|B_j)$$

$$= - \sum_{i=1}^{m} \sum_{j=1}^{n} p_{ij} \log_2 \left\{\frac{p_{ij}}{P(B_j)}\right\} = H(A|B) \qquad (1.79)$$

This is the **conditional entropy** of the set A, conditional on the set B.

It is interesting to compare the two entropies. We can write

$$H(A|B) - H(A) = - \sum_{i=1}^{m} \sum_{j=1}^{n} p_{ij} \log_2 \left(\frac{p_{ij}}{q_{ij}}\right)$$

by putting $Q\{a_i, b_j\} = P(A_i)P(B_j) = q_{ij}$. It is easy to verify that Q is a probability measure over the space $(\Omega, P(\Omega))$ with respect to which the families (A_i) and (B_j) are mutually independent. It follows then (cf. Exercise 1.1) that

$$H(A|B) \leq H(A) \qquad (1.80)$$

which signifies that the conditioning of the set A on the set B cannot increase the entropy of the set A.

Let us examine the two extreme cases. In the first, the families (A_i) and (B_j) are mutually independent with respect to the measure P. The measures P and Q are then identical, i.e. $p_{ij} = q_{ij}$, and hence it follows that $H(A|B) = H(A)$. The entropy of the set A is not affected by the conditioning on the set B when the two sets are mutually independent. In the other extreme case, the two sets A and B are completely connected; they are identical and the measure p_{ij} is such that $p_{ij} = p_i \delta_{ij}$, δ_{ij} being the Kronecker delta. Then

$$P(A_i) = p_i \qquad P(B_j) = p_j \qquad P(A_i|B_j) = \frac{p_i \delta_{ij}}{p_j} = \delta_{ij}$$

Hence $H(A|B) = 0$. The entropy of the set A is reduced to zero by the conditioning on the set B when the two sets are completely connected.

(c) The space $(\Omega, \mathcal{A}, P)$ describes an experiment involving *noisy* transmission of discrete information. It is represented by the diagram in Fig. 1.7. The sets A and B are then called **alphabets**. A source of 'information' emits a symbol a_i belonging to the alphabet A with probability $P(A_i)$. The presence of 'noise' in the 'transmission medium' has the effect that the receiver receives a symbol b_j belonging to the alphabet B with probability $P(B_j)$. The transmission is represented by the trial (a_i, b_j). Before the transmission, the receiver does not know what the symbol emitted by the source will be. Its uncertainty is measured *a priori* by the entropy $H(A)$. When the transmission has taken place, the receiver has received a symbol b_j but still does not know what the symbol a_i is which was emitted. Its uncertainty is then measured *a posteriori* by the conditional entropy $H(A|B)$. Formula (1.80) expresses a fundamental result: the transmission can in no case increase the uncertainty of the receiver.

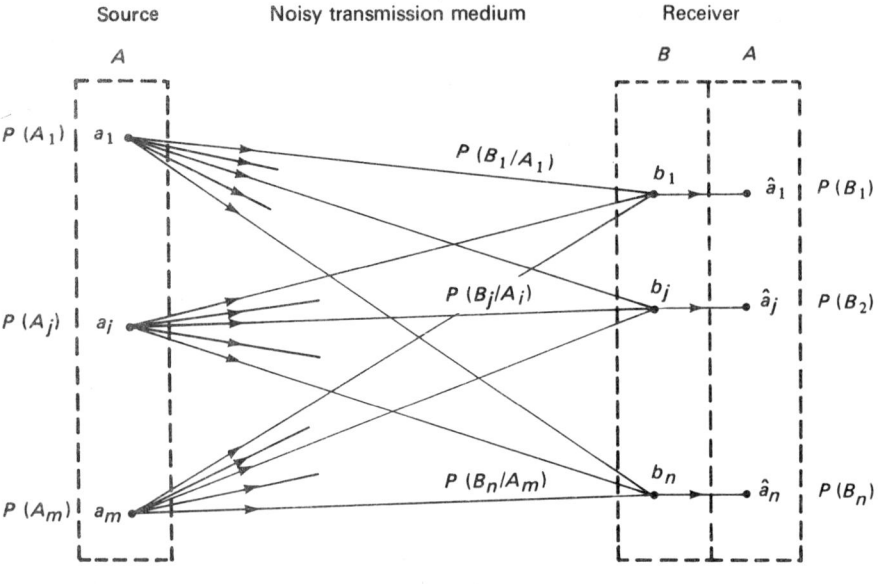

Fig. 1.7

We say, equivalently, that the transmission contributes to the receiver a **quantity of information** $I(A, B)$, positive or zero, measured by the eventual reduction in entropy

$$I(A, B) = H(A) - H(A|B) \geq 0$$

When there is a high level of 'noise' in the 'transmission medium', the two alphabets are independent, i.e. $H(A|B) = H(A)$, and the quantity of information received by the receiver is zero. The transmission is of no help. In contrast, when there is no 'noise', the two alphabets are completely connected, i.e. $H(A|B) = 0$ and $I(A, B) = H(A)$. The quantity of information received by the receiver is maximal and equal to the entropy $H(A)$. The transmission reduces to zero the uncertainty of the receiver.

(d) In the general case, knowledge by the receiver of the symbol b_j received is not sufficient to enable him to identify with certainty the symbol a_i emitted by the source. He must then make a choice and this choice can obviously be a bad one. Suppose that the receiver decides that the source has emitted the symbol $\hat{a}_j$ when he receives the symbol b_j. The probability that his decision is correct is the conditional probability $P(\hat{A}_j|B_j)$. The probability $P(C)$, averaged over the alphabet B, of making a correct decision is

$$P(C) = \sum_{j=1}^{n} P(B_j) P(\hat{A}_j|B_j) \tag{1.81}$$

It is natural that the criterion used by the receiver to make his decision should be such that the probability $P(C)$ is a maximum. Now, this probability is a sum of positive terms in which the factors $P(B_j)$ are independent of the choices made since they depend only on the transmission. The probability $P(C)$ is therefore maximal if each of the factors $P(\hat{A}_j | B_j)$ is maximal. The criterion for optimal choice results from this. When he receives the symbol b_j the receiver chooses from the symbols of the alphabet A the symbol $\hat{a}_j$ such that the conditional probability $P(\hat{A}_j | B_j)$ is a maximum. We note that the choice can be ambiguous and furthermore that the receiver must know the $m \times n$ conditional probabilities $P(A_i | B_j)$ to be able to work at all.

2.5. CHARACTERISTIC FUNCTION AND GENERATING FUNCTION

2.5.1. *Characteristic function of a real random variable*

2.5.1.1. Definition

Let ξ be a real r.v. having the probability law μ_ξ. Its **characteristic function** Θ_ξ is defined by the integral

$$\Theta_\xi(u) = \int_\mathbb{R} \exp(iux)\, d\mu_\xi(x) \qquad u \in \mathbb{R} \tag{1.82}$$

2.5.1.2. Properties

It is immediate that the function $\Theta_\xi(2\pi y)$ is the inverse Fourier transform of the measure μ_ξ (cf. Eqn (6.81)). It is therefore bounded and uniformly continuous (cf. Chapter 6, Section 4.3.2). It follows that

$$|\Theta_\xi(u)| \leq \int_\mathbb{R} d\mu_\xi(x) = \Theta_\xi(0) = 1$$

More generally, the function $\Theta_\xi(2\pi y)$ and the measure μ_ξ define tempered distributions, connected by the Fourier transformation. We can write, defining *ipso facto* the function φ_ξ,

$$\varphi_\xi(y) = \Theta_\xi(2\pi y) \qquad \mu_\xi \underset{\mathscr{F}}{\overset{\mathscr{F}}{\rightleftarrows}} \varphi_\xi \tag{1.83}$$

The class of real r.v.s a.s. equal to the r.v. ξ is completely characterized either by the probability law μ_ξ or by the characteristic function Θ_ξ. The table in Appendix 3 enables us to apply Eqn (1.83) to pass from the former to the latter and conversely.

When the r.v. ξ has a probability density p_ξ, the integral in Eqn (1.82) can be written

$$\Theta_\xi(u) = \int_\mathbb{R} \exp(iux) p_\xi(x)\, dx \tag{1.84}$$

Suppose that the absolute moment of order n of the r.v. ξ is finite; this may be

46 Random variables

written $E[|\xi|^n] < +\infty$. The theorem for differentiation under the summation sign applied to the integral in Eqn (1.82) enables us to write

$$\Theta_\xi^{(n)}(0) = i^n E[\xi^n] \tag{1.85}$$

When all the absolute moments up to order n inclusive are finite, the Taylor–Young formula enables us to write the expansion

$$\Theta_\xi(u) = \sum_{k=0}^{n} \frac{(iu)^k}{k!} E[\xi^k] + o(u^n) \tag{1.86}$$

Example 2.17 Characteristic function of a gaussian random variable

Let ξ be the gaussian r.v. whose probability density is defined by Eqn (1.46). Its characteristic function may be obtained by use of the table in Appendix 3. We obtain

$$\Theta_\xi(u) = \exp\left(imu - \frac{\sigma^2 u^2}{2}\right) \tag{1.87}$$

The characteristic function of the centred gaussian r.v. $\overset{\circ}{\xi}$ can be written

$$\Theta_{\overset{\circ}{\xi}}(u) = \exp\left(-\frac{\sigma^2 u^2}{2}\right) = 1 - \frac{\sigma^2 u^2}{2} + \ldots + (-1)^n \frac{\sigma^{2n}}{2^n n!} u^{2n} + \ldots$$

Hence we have the result, by identification with Eqn (1.86),

$$E[\overset{\circ}{\xi}^{2n}] = \frac{\sigma^{2n}(2n)!}{2^n n!} \qquad E[\overset{\circ}{\xi}^{2n+1}] = 0 \tag{1.88}$$

2.5.2. Generating function of a real random variable

2.5.2.1. Definition

Let ξ be a real r.v. having probability law μ_ξ. Its generating function G_ξ is defined by the integral

$$G_\xi(p) = \int_{\mathbb{R}} \exp(-px) \, d\mu_\xi(x) \qquad p \in \mathbb{C} \tag{1.89}$$

2.5.2.2. Properties

The function G_ξ is the two-sided Laplace transform of the measure μ_ξ. In contrast with the characteristic function, the generating function is not always defined. When it exists, it is a holomorphic function within a strip of the complex plane defined by $\text{Re } p \in \overset{\circ}{I}$, I being an interval which always contains the origin since $G_\xi(0) = 1$ (Dupraz, 1977, p. 115). When the variable p is real, the generating function has special properties of convexity which are studied in Exercise 2.19.

Suppose that the absolute moment of order n of the r.v. ξ is finite, which may be written $E[|\xi|^n] < +\infty$. The theorem for differentiation under the summation sign applied to the integral in Eqn (1.89) enables us to write

$$E[\xi^n] = (-1)^n G_\xi^{(n)}(0) \qquad (1.90)$$

The moment of order n of the r.v. ξ is given by the value at the origin of the derivative of order n of the function G_ξ. Hence the name (moment) generating function is given to the function G_ξ.

Generating functions are used principally when we are concerned with positive r.v.s for which the measure μ_ξ has domain $\mathbb{R}_+$. These are then holomorphic functions in a half-plane defined by $\operatorname{Re} p \in \,]a, +\infty[$ with $a < 0$ (Dupraz, 1977, p. 117). The use of generating functions enables us, in addition, to solve certain problems involving Wiener–Hopf convolution equations (Dupraz, 1977, p. 18). A brief table of two-sided Laplace transforms is given in Appendix 3.

Example 2.18 Square of a centred gaussian random variable

Let ξ be a centred gaussian r.v. and let η be the positive r.v. $\eta = \xi^2$. The probability density p_η may be obtained from Eqns (1.65) and (1.46):

$$p_\eta(y) = Y(y) \frac{1}{(2\pi\sigma^2)^{1/2}} \frac{1}{y^{1/2}} \exp\left(-\frac{y}{2\sigma^2}\right) \qquad (1.91)$$

The generating function G_η is given by the table in Appendix 3. We obtain

$$G_\eta(p) = \frac{1}{(1 + 2\sigma^2 p)^{1/2}} \qquad \operatorname{Re} p > \frac{1}{2\sigma^2} \qquad (1.92)$$

The function G_η is holomorphic in the half-plane $\operatorname{Re} p > 1/2\sigma^2$. The moments $E[\xi^{2n}]$ are obtained by a series expansion of G_η. We thus find again the result given by Eqn (1.88).

2.6. CONDITIONING OF A REAL RANDOM VARIABLE BY A SUB-BOREL FIELD

2.6.1. Introduction

(a) Let $(\Omega, \mathcal{A}, P)$ be a probability space. The measure P enables us to evaluate *a priori* the probabilities of events of the Borel field $\mathcal{A}$. When the event C, of non-zero probability, is realized, the probabilities of events are evaluated *a posteriori* by the conditional probability measure $P(.\,|\,C)$. This has the effect of replacing the space $(\Omega, \mathcal{A}, P)$ by the space $(\Omega, \mathcal{A}, P(.\,|\,C))$ (cf. Section 1.3).

It is much more fruitful to consider that the conditioning is effected by the events belonging to a sub-Borel field $\mathcal{B} \subset \mathcal{A}$. The object of this section is to study more precisely the case when the events generated by a real r.v. ξ are

48 Random variables

conditioned by events belonging to a sub-Borel field $\mathscr{B}$. We say then, more simply, that the r.v. ξ is conditioned by the sub-Borel field $\mathscr{B}$.

The notion of conditioning is extremely important in signal theory. It is rather intricate, however. Accordingly, it is useful to start by recalling a certain number of results. Everything which follows uses the properties proved in Chapter 6, Section 4.1.8.

(b) Let ξ be a real r.v. defined over the probability space $(\Omega, \mathscr{A}, P)$. It is measurable with respect to the Borel field $\mathscr{A}$ and generates a sub-Borel field $\mathscr{B}_\xi \subset \mathscr{A}$. When the r.v. ξ has a finite mean value it is integrable with respect to the measure P and belongs to the vector space $\mathscr{L}^1(\Omega, \mathscr{A}, P)$. The classes of integrable real r.v.s constitute the vector space $L^1(\Omega, \mathscr{A}, P)$ (cf. Chapter 6, Section 4.1.6.1).

Now let $\mathscr{B}$ be an arbitrary sub-Borel field $\mathscr{B} \subset \mathscr{A}$. We denote by $P_\mathscr{B}$ the restriction of the measure P to the sub-Borel field $\mathscr{B}$. This means simply that $(\Omega, \mathscr{B}, P_\mathscr{B})$ is a probability space for which

$$\forall B \in \mathscr{B} \qquad P_\mathscr{B}(B) = P(B)$$

A real r.v. ξ defined over the space $(\Omega, \mathscr{B}, P_\mathscr{B})$ is measurable with respect to the sub-Borel field $\mathscr{B}$. It generates a sub-Borel field $\mathscr{B}_\xi \in \mathscr{B}$. When it has a finite mean value, the r.v. ξ belongs to the vector space $\mathscr{L}^1(\Omega, \mathscr{B}, P_\mathscr{B})$ whose equivalence classes constitute the vector space $L^1(\Omega, \mathscr{B}, P_\mathscr{B})$.

2.6.2. Definitions and general properties

2.6.2.1. Definition

Let $(\Omega, \mathscr{A}, P)$ be a probability space, $\mathscr{B} \in \mathscr{A}$ be a sub-Borel field, $(\Omega, \mathscr{B}, P_\mathscr{B})$ be a probability space and let $\xi \in \mathscr{L}^1(\Omega, \mathscr{A}, P)$ be a real r.v. The **conditional expectation** of the r.v. ξ with respect to the sub-Borel field $\mathscr{B}$ is a real r.v., denoted $E[\xi | \mathscr{B}]$, defined on the space $(\Omega, \mathscr{B}, P_\mathscr{B})$, such that

$$\forall B \in \mathscr{B} \qquad E[\xi I_B] = E[E[\xi | \mathscr{B}] I_B] \qquad (1.93)$$

or, making the operator E explicit,

$$\forall B \in \mathscr{B} \qquad \int_B \xi(\omega) \, \mathrm{d}P(\omega) = \int_B E[\xi | \mathscr{B}](\omega) \, \mathrm{d}P(\omega)$$

2.6.2.2. Properties

The r.v. $E[\xi | \mathscr{B}]$ is defined only a.s. over the space $(\Omega, \mathscr{B}, P_\mathscr{B})$. In fact, all the r.v.s belonging to the equivalence class satisfy Eqn (1.93). In practice, we do not distinguish between the conditional expectation r.v. and its equivalence class.

The r.v. $E[\xi | \mathscr{B}]$ is measurable with respect to the sub-Borel field $\mathscr{B}$. The sub-Borel field which it generates is consequently contained within the sub-Borel field $\mathscr{B}$ (cf. Chapter 6, Section 2.1). This means that the events which depend on the conditional expectation r.v. belong to the conditioning sub-Borel field.

2.6.2.3. Definition

Usually the conditioning sub-Borel field is the sub-Borel field $\mathscr{B}_\eta$ generated by a real r.v. η or, more generally, by a real random vector (r.v.) $\vec{\eta}$ of finite or infinite dimension (cf. Chapter 2, Section 1.1). We say then that the r.v. ξ is conditioned by the r.v. η or r.v. $\vec{\eta}$ and we change the notation to define the r.v. $E[\xi|\eta]$.

The conditional expectation $E[\xi|\eta]$ of the r.v. ξ with respect to the r.v. η or to the r.v. $\vec{\eta}$ is the r.v. $E[\xi|\mathscr{B}_\eta]$.

2.6.2.4. Properties

The r.v. $E[\xi|\eta]$ satisfies the general relation (1.93). We can show, moreover, that it is a function of the r.v. η or the r.v. $\vec{\eta}$ (cf. Chapter 6, Exercise 4.11).

Suppose that η is a real r.v., of dimension n. There therefore exists a real measurable function ψ, defined almost everywhere over the space $(\mathbb{R}^n, \mathscr{R}^n, \mu_\eta)$, such that

$$E[\xi|\mathscr{B}_\eta] = E[\xi|\eta] = \psi \circ \eta \quad (1.94)$$

This is illustrated by Fig. 1.8. For a.s. each trial ω, the r.v. η takes the value $\eta(\omega) \in \mathbb{R}^n$ and the r.v. $E[\xi|\eta]$ takes the value $\psi[\eta(\omega)] \in \mathbb{R}$.

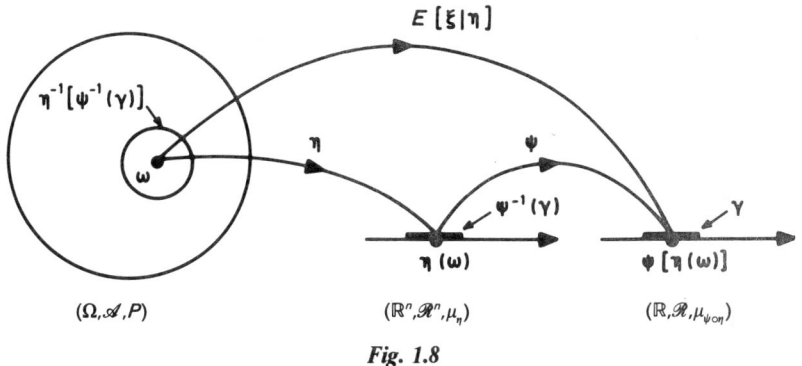

Fig. 1.8

The function ψ is defined almost everywhere by (cf. Eqn (6.93))

$$\forall \beta \in \mathscr{R}^n \quad \mu_\eta(\psi I_\beta) = E[\xi(I_\beta \circ \eta)] \quad (1.95)$$

It is integrable with respect to the measure μ_η. We find in fact from Eqn (1.95), by taking $\beta = \mathbb{R}^n$ and noting that $I_{\mathbb{R}^n} \circ \eta = I_\Omega$,

$$\mu_\eta(\psi) = \int_{\mathbb{R}^n} \psi(y) \, d\mu_\eta(y) = E[\xi] \quad (1.96)$$

The integral is equal to the mean value of the r.v. ξ.

Example 2.19 Conditioning on a discrete random variable

Suppose that the conditioning sub-Borel field $\mathscr{B}_\eta$ is generated by the discrete r.v.

$$\eta = \sum_{n \in \mathbb{N}} x_n I_{B_n} \qquad B_n = \{\eta = x_n\}$$

The family $(B_n, n \in \mathbb{N})$ is a partition of the set Ω which generates the sub-Borel field $\mathscr{B}_\eta$ (cf. Example 2.2).

Let ξ be a real r.v. with finite mean value. The conditional expectation r.v. $E(\xi|\eta)$ is the function $\Omega \to \mathbb{R}$ defined by

$$\omega \to \frac{E[\xi I_{B_n}]}{E[I_{B_n}]} \qquad \omega \in B_n$$

In fact, the linearity of the operator E allows us to write

$$E\left[\frac{E[\xi I_{B_n}]}{E[I_{B_n}]} I_{B_n}\right] = \frac{E[\xi I_{B_n}]}{E[I_{B_n}]} E[I_{B_n}] = E[\xi I_{B_n}]$$

Formula (1.93) is thus verified for all the sets B_n. We can then show (cf. Chapter 6, Exercise 1.8) that

$$\forall B \in \mathscr{B}_\eta \qquad B = \bigcup_{n \in \mathbb{I}} B_n \Leftrightarrow I_B = \sum_{n \in \mathbb{I}} I_{B_n} \qquad \mathbb{I} \subset \mathbb{N}$$

It follows from this, by linearity, that Eqn (1.93) is verified for all the sets of the sub-Borel field $\mathscr{B}_\eta$. The function defined above is thus indeed the conditional expectation r.v. $E[\xi|\eta]$.

When the r.v. η takes the value x_n, the event B_n is realized and the r.v. $E[\xi|\eta]$ takes the value $E[\xi I_{B_n}]/E[I_{B_n}]$. The r.v. $E[\xi|\eta]$ is therefore discrete and the sub-Borel field which it generates is the sub-Borel field $\mathscr{B}_\eta$ itself. It may be written

$$E[\xi|\eta] = \sum_{n \in \mathbb{N}} \frac{E[\xi I_{B_n}]}{E[I_{B_n}]} I_{B_n} \qquad (1.97)$$

It is easy to deduce from this that the function $\psi: \mathbb{R} \to \mathbb{R}$ defined by Eqn (1.95) may be written

$$\psi = \sum_{n \in \mathbb{N}} \frac{E[\xi I_{B_n}]}{E[I_{B_n}]} I_{\{x_n\}} \qquad (1.98)$$

$I_{\{x_n\}}$ being the indicator function of the singleton $\{x_n\} \in \mathscr{R}$.

2.6.3. Properties of conditional expectation random variables

The properties of conditional expectation r.v.s $E[\xi|\mathscr{B}]$ are much used in signal theory. They are the subject of this section.

2.6.3.1. Conditioning of a positive or zero random variable

When the r.v. ξ is positive or zero, the r.v. $E[\xi|\mathscr{B}]$ is also positive or zero:

$$\xi \geq 0 \Rightarrow E[\xi|\mathscr{B}] \geq 0 \quad (1.99)$$

More precisely, if the r.v. ξ is a.s. positive or zero, the same is true of the r.v. $E[\xi|\mathscr{B}]$ (cf. Exercise 2.22).

2.6.3.2. Linearity of the conditional expectation

Let ξ_1 and ξ_2 be two real r.v.s with finite mean value and let ξ be the r.v. $\xi = a_1\xi_1 + a_2\xi_2$, a_1 and a_2 being two real numbers. It follows immediately from Eqn (1.93) that

$$E[\xi|\mathscr{B}] = a_1 E[\xi_1|\mathscr{B}] + a_2 E[\xi_1|\mathscr{B}] \quad (1.100)$$

2.6.3.3. Conditioning of a random variable $\xi \in \mathscr{L}^1(\Omega, \mathscr{B}, P_\mathscr{B})$

When the real r.v. ξ belongs to the space $\mathscr{L}^1(\Omega, \mathscr{B}, P_\mathscr{B})$, the sub-Borel field $\mathscr{B}_\xi$ which it generates is contained within the sub-Borel field $\mathscr{B}$. The r.v.s ξ and $E[\xi|\mathscr{B}]$ are then a.s. equal (cf. Eqn (6.74)):

$$\xi \in \mathscr{L}^1(\Omega, \mathscr{B}, P_\mathscr{B}) \Rightarrow E[\xi|\mathscr{B}] \underset{\text{a.s.}}{=} \xi \quad (1.101)$$

This means that the conditioning of an r.v. ξ by a sub-Borel field containing the sub-Borel field $\mathscr{B}_\xi$ which it generates almost surely does not change the r.v. ξ.

Example 2.20 Conditioning of a constant random variable

Let $\xi = aI_\Omega$ be a constant r.v. It is integrable and generates the sub-Borel field $\{\varnothing, \Omega\}$ which is contained in all the sub-Borel fields $\mathscr{B} \subset \mathscr{A}$ (cf. Example 2.1). Formula (1.101) can then be applied, whatever the sub-Borel field $\mathscr{B}$ may be:

$$E[aI_\Omega|\mathscr{B}] \underset{\text{a.s.}}{=} aI_\Omega \quad (1.102)$$

Conditioning of a constant r.v. by any sub-Borel field whatever does not change its value.

2.6.3.4. Mean value of the random variable $E[\xi|\mathscr{B}]$

The r.v. $E[\xi|\mathscr{B}]$ belongs to the space $\mathscr{L}^1(\Omega, \mathscr{B}, P_\mathscr{B})$. It is integrable and it follows from Eqn (1.93) by taking $B = \Omega$ that

$$E[E[\xi|\mathscr{B}]] = E[\xi] \quad (1.103)$$

The r.v.s ξ and $E[\xi|\mathscr{B}]$ have thus the same mean value. Conditioning of an r.v. by any sub-Borel field whatever does not change its mean value.

2.6.3.5. Conditioning of a random variable $\xi \in \mathscr{L}^2(\Omega, \mathscr{A}, P)$

When the r.v. ξ is square integrable, the r.v. $E[\xi|\mathscr{B}]$ is also square integrable (cf. Chapter 6, Section 4.1.8(c)):

$$\xi \in \mathscr{L}^2(\Omega, \mathscr{A}, P) \Rightarrow E[\xi|\mathscr{B}] \in \mathscr{L}^2(\Omega, \mathscr{B}, P_\mathscr{B}) \quad (1.104)$$

The mean square values of the two r.v.s satisfy (cf. Eqn (6.75))

$$E[|E[\xi|\mathscr{B}]|^2] \leq E[|\xi|^2]$$

Hence the following result obtains, since the mean values are the same (cf. Eqn (1.70)):

$$V[E[\xi|\mathscr{B}]] \leq V[\xi] \quad (1.105)$$

Conditioning of a square integrable r.v. by any sub-Borel field whatever does not increase its variance and therefore the same is true of the dispersion of the values which it takes.

Let $\mathscr{L}_\mathscr{B}^2(\Omega, \mathscr{A}, P)$ be the vector space of square integrable r.v.s a.s. equal to an r.v. belonging to the vector space $\mathscr{L}^2(\Omega, \mathscr{B}, P_\mathscr{B})$. The r.v. $E[\xi|\mathscr{B}]$ belongs to the space $\mathscr{L}_\mathscr{B}^2(\Omega, \mathscr{A}, P)$. This is the orthogonal projection of the r.v. ξ onto the space $\mathscr{L}_\mathscr{B}^2(\Omega, \mathscr{A}, P)$ which may be written

$$\forall \zeta \in \mathscr{L}_\mathscr{B}^2(\Omega, \mathscr{A}, P) \quad E[|\xi - E[\xi|\mathscr{B}]|^2] \leq E[|\xi - \zeta|^2] \quad (1.106)$$

This means that the r.v. $E[\xi|\mathscr{B}]$ is the best approximation, in quadratic mean, to the r.v. ξ by an r.v. belonging to the space $\mathscr{L}_\mathscr{B}^2(\Omega, \mathscr{A}, P)$. Figure 1.9 illustrates formula (1.106).

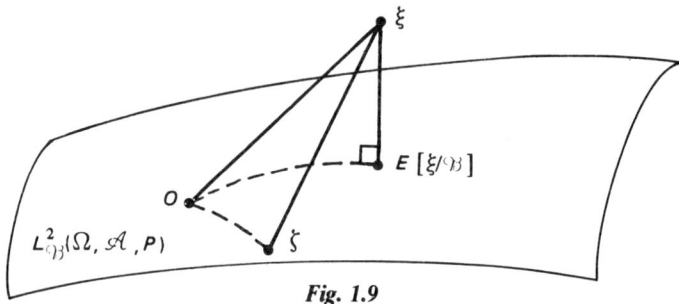

Fig. 1.9

When the sub-Borel field $\mathscr{B}$ is generated by an r.v. or an $\vec{r.v.}$ η, the r.v. $E[\xi|\mathscr{B}]$ is a function of η (cf. Eqn (1.94)). It is the best approximation, in quadratic mean, to the r.v. ξ by an r.v. which is a function of η.

Example 2.21 Estimation of a probability

Let $(\Omega^n, \mathscr{P}(\Omega^n), P)$ be the probability space defined in Example 1.6. It describes a composite Bernoulli experiment. This is a sequence of independent experiments resulting in the realization or non-realization of an event E of probability p. Let η be the real discrete r.v. representing the number of times that event E is realized. Its probability law is a binomial law such that (cf. Eqn (1.25))

$$P\{\eta = k\} = \binom{n}{k} p^k (1-p)^{n-k} \qquad 0 \leqslant k \leqslant n$$

Suppose now that the probability p is unknown and let us accept that it is the value taken by an r.v. ξ uniformly distributed over the interval $[0, 1]$. The object of the Bernoulli experiment is to enable us to evaluate p as a function of the number of times that the event E is realized. Thus we wish to estimate the square integrable r.v. ξ as a function of the discrete r.v. η. We know that the best estimate, in quadratic mean, is the conditional expectation r.v. $\hat{\xi} = E[\xi | \eta]$. The result obtained in Example 2.19 may be applied and it follows from Eqn (1.97) that

$$\hat{\xi} = \sum_{k=0}^{n} \frac{E[\xi I_k]}{E[I_k]} I_k$$

where we denote by I_k the indicator r.v. of the event $A_k = \{\eta = k\}$. When the r.v. ξ takes the value p, we must write

$$E[I_k | \xi = p] = P(A_k | \xi = p) = \binom{n}{k} p^k (1-p)^{n-k}$$

Hence the following result obtains, p_ξ being the probability density of the r.v. ξ:

$$E[I_k] = \int_{\mathbb{R}} p_\xi(x) \binom{n}{k} x^k (1-x)^{n-k} \, dx$$

$$= \binom{n}{k} \int_0^1 x^k (1-x)^{n-k} \, dx$$

$$= \binom{n}{k} \frac{k!(n-k)!}{(n+1)!}$$

$$E[\xi I_k] = \int_{\mathbb{R}} p_\xi(x) \binom{n}{k} x^{k+1} (1-x)^{n-k} \, dx$$

$$= \binom{n}{k} \frac{(k+1)!(n-k)!}{(n+2)!}$$

Finally, the r.v. $\hat{\xi}$ is given by

$$\hat{\xi} = \sum_{k=0}^{n} \frac{k+1}{n+2} I_k \qquad (1.107)$$

When the event E is realized k times, the r.v. I_k takes the value 1 and the r.v. $\hat{\xi}$ takes the value $(k+1)/(n+2)$. In these conditions this is the best estimate of the value of the probability p.

Example 2.22 Repeated conditioning

Let $(\Omega, \mathscr{A}, P)$ be a probability space, ξ a real r.v. having finite mean value, and $\mathscr{B}$ and $\mathscr{C}$ two conditioning sub-Borel fields such that $\mathscr{C} \subset \mathscr{B}$.

Consider the conditional expectation r.v. $E[\xi|\mathscr{C}]$; we seek its conditional expectation with respect to the sub-Borel field $\mathscr{B}$. This is the r.v. $E[E[\xi|\mathscr{C}]|\mathscr{B}]$. It follows from Eqn (1.93) that

$$\forall B \in \mathscr{B} \quad E[E[E[\xi|\mathscr{C}]|\mathscr{B}]I_B] = E[E[\xi|\mathscr{C}]I_B]$$

The r.v. $E[\xi|\mathscr{C}]$ is measurable with respect to the sub-Borel field $\mathscr{C}$. It is therefore measurable with respect to the sub-Borel field $\mathscr{B}$ which contains $\mathscr{C}$. The r.v.s $E[\xi|\mathscr{C}]$ and $E[E[\xi|\mathscr{C}]|\mathscr{B}]$ are therefore measurable with respect to the sub-Borel field $\mathscr{B}$. It results then from the preceding equality that they are a.s. equal and from this very fact they define the same conditional expectation r.v. (cf. Chapter 6, Exercise 4.2):

$$E[\xi|\mathscr{C}] \underset{\text{a.s.}}{=} E[E[\xi|\mathscr{C}]|\mathscr{B}]$$

Let us now consider the r.v. $E[\xi|\mathscr{B}]$; we seek its conditional expectation with respect to the sub-Borel field $\mathscr{C}$. This is the r.v. $E[E[\xi|\mathscr{B}]|\mathscr{C}]$. It follows from Eqn (1.93) that

$$\forall C \in \mathscr{C} \quad E[E[E[\xi|\mathscr{B}]|\mathscr{C}]I_C] = E[E[\xi|\mathscr{B}]I_C]$$

The r.v. $E[\xi|\mathscr{B}]$ is measurable with respect to the sub-Borel field $\mathscr{B}$. It follows from Eqn (1.93), since $\mathscr{C} \subset \mathscr{B}$, that

$$\forall C \in \mathscr{C} \quad E[E[\xi|\mathscr{B}]I_C] = E[\xi I_C]$$

It then follows from Eqn (1.93), by definition of the r.v. $E[\xi|\mathscr{C}]$, that

$$\forall C \in \mathscr{C} \quad E[E[\xi|\mathscr{C}]I_C] = E[\xi I_C]$$

Hence, from the above,

$$\forall C \in \mathscr{C} \quad E[E[E[\xi|\mathscr{C}]|\mathscr{B}]I_C] = E[E[\xi|\mathscr{C}]I_C]$$

Application of Eqn (6.88) enables us to conclude that the r.v.s $E[\xi|\mathscr{C}]$ and $E[E[\xi|\mathscr{B}]|\mathscr{C}]$ are a.s. equal and from this very fact they define the same conditional expectation.

Summarizing, we can write

$$\mathscr{C} \subset \mathscr{B} \Rightarrow E[\xi|\mathscr{C}] \underset{\text{a.s.}}{=} E[E[\xi|\mathscr{C}]|\mathscr{B}] \underset{\text{a.s.}}{=} E[E[\xi|\mathscr{B}]|\mathscr{C}] \quad (1.108)$$

Repeated conditioning of an r.v. by embedded sub-Borel fields reduces to conditioning by the smallest of the sub-Borel fields. This is a very important property in signal theory (cf. Exercises 2.11, 2.12 and Chapter 5, Exercise 2.4).

Exercises

Exercise 2.1 The Rayleigh law

Let ξ be a real r.v. uniformly distributed over the interval $[0, 1]$ and let η be the real r.v. given by $\eta = (-2a^2 \log \xi)^{1/2}$. Show that the probability law of the r.v. η is a Rayleigh law of parameter a (cf. Example 2.9).

Exercise 2.2 Random functions of a random variable

Let ξ be a real r.v. having probability density p_ξ and η and ζ the real, positive, r.v.s

$$\eta = Y(\xi)\xi \qquad \zeta = |\xi|$$

obtained respectively by truncation and taking the absolute value of the r.v. ξ. Show that

$$F_\eta'(y) = F_\xi(0)\delta(y) + Y(y)p_\xi(y) \qquad (1.109)$$
$$p_\zeta(z) = Y(z)\{p_\xi(z) + p_\xi(-z)\}$$

The r.v. η is neither discrete nor continuous. Its probability law is the sum of a Dirac measure at the origin, of mass $F_\xi(0)$, and an absolutely continuous measure (with respect to the Lebesgue measure), having density function $Y(y)p_\xi(y)$. (Apply the results of Section 2.3.3.)

Exercise 2.3 Sinusoidal function of a random variable

Let ϑ be a real r.v. uniformly distributed over the interval $[-\pi, \pi]$ and let η be the real r.v. $\eta = A \cos \vartheta$. Show that the probability density of the r.v. η is the function

$$p_\eta(y) = \frac{1}{\pi A}\left(1 - \frac{y^2}{A^2}\right)^{1/2} \qquad y \in [-A, A]$$
$$p_\eta(y) = 0 \qquad y \notin [-A, A] \qquad (1.110)$$

This is represented in Fig. 1.10. (Note that p_η is an even function and apply Eqn (1.55), working over the interval $[0, \pi]$.)

Exercise 2.4 Gaussian random variable

Let ξ be the gaussian r.v. whose probability density is given by Eqn (1.46). Prove that

$$E[\xi] = m \qquad V[\xi] = \sigma^2 \qquad E[|\xi - m|] = \left(\frac{2}{\pi}\right)^{1/2}\sigma \qquad (1.111)$$

(Apply Eqn (1.85).)

56 Random variables

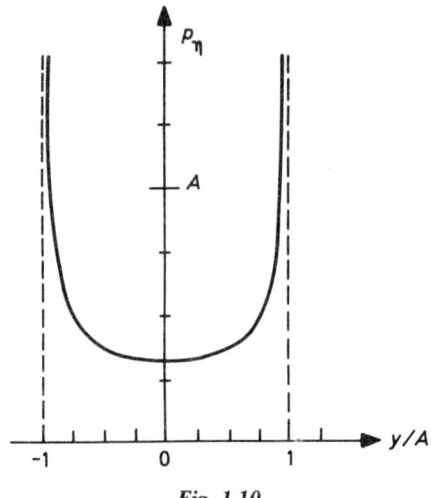

Fig. 1.10

Exercise 2.5 Rayleigh random variables

Let ξ be a Rayleigh r.v. of parameter a (cf. Example 2.9). Show that

$$E[\xi] = a\left(\frac{\pi}{2}\right)^{1/2} \qquad V[\xi] = \frac{4-\pi}{2}a^2 \qquad (1.112)$$

Exercise 2.6 Binomial law

Let ξ be an r.v. having a binomial probability law of parameters (n, p) (cf. Example 2.6). Show, putting $q = 1 - p$, that its characteristic function is

$$\Theta_\xi(u) = \{p\exp(iu) + q\}^n \qquad (1.113)$$

Hence deduce (apply Eqn (1.85)) that

$$E[\xi] = np \qquad V[\xi] = npq \qquad (1.114)$$

Exercise 2.7 Poisson random variable

Let ξ be a Poisson r.v. of parameter λ (cf. Example 2.5). Show that its characteristic function is written

$$\Theta_\xi(u) = \exp[\lambda\{\exp(iu) - 1\}] \qquad (1.115)$$

Hence deduce that

$$E[\lambda] = V[\lambda] = \lambda \qquad (1.116)$$

It should be noted that the mean value of a Poisson r.v. is equal to its variance.

Exercise 2.8 Rayleigh random variable

Let ξ be a Rayleigh r.v. of parameter a (cf. Example 2.9) and let η be the r.v. $\eta = \xi^2$. Prove that the r.v. η has an exponential probability law of parameter $\frac{1}{2}a^2$ (cf. Example 2.8). (Use Example 2.14.)

Exercise 2.9 Gaussian random variable

Let ξ be a centred gaussian r.v., a a positive number, and $Q(a)$ the probability $Q(a) = P(\xi > a)$. Show, using successive integrations by parts, that

$$\frac{1}{(2\pi)^{1/2}}\frac{\sigma}{a}\left(1-\frac{\sigma^2}{a^2}\right)\exp\left(-\frac{a^2}{2\sigma^2}\right) < Q(a) < \frac{1}{(2\pi)^{1/2}}\frac{\sigma}{a}\exp\left(-\frac{a^2}{2\sigma^2}\right) \qquad a \in \mathbb{R}_+^*$$

Hence deduce that, when a tends to infinity, we have

$$Q(a) \approx \frac{1}{(2\pi)^{1/2}}\frac{\sigma}{a}\exp\left(-\frac{a^2}{2\sigma^2}\right)$$

Exercise 2.10 Exponential law

Let ξ be an r.v. having an exponential probability law of parameter a (cf. Example 2.8).

(1) Verify that its characteristic function may be written

$$\Theta_\xi(u) = \frac{a}{a-iu} \qquad (1.117)$$

Hence deduce that

$$E[\xi] = \frac{1}{a} \qquad V[\xi] = \frac{1}{a^2} \qquad (1.118)$$

(Apply Eqns (1.83) and (1.85), using the table in Appendix 3.)

(2) Prove that

$$\forall x \geqslant 0 \qquad P\{\xi \geqslant x\} = \exp(-ax) \qquad (1.119)$$

(Apply Eqn (1.61).)

Comment: when the r.v. ξ represents the lifetime of an electronic component, the parameter a is the failure rate of the component and the probability $P\{\xi \geqslant x\}$ that its lifetime is greater than or equal to x seconds is its reliability over this lifetime.

Exercise 2.11 Conditioning of a product of two random variables

Let $(\Omega, \mathcal{A}, P)$ be a probability space, $\mathcal{B}$ and $\mathcal{C}$ be two sub-Borel fields such that $\mathcal{C} \subset \mathcal{B}$, and ξ be a real r.v. with finite mean value.

(1) Let B be an event $B \in \mathcal{B}$ and let I_B be the indicator r.v. Prove that

$$E[\xi I_B | \mathcal{B}] \underset{\text{a.s.}}{=} I_B E[\xi | \mathcal{B}]$$

(Apply the result obtained in Chapter 6, Exercise 4.2.)

(2) Let η be a real r.v., measurable with respect to the Borel field $\mathcal{B}$. Deduce from the above, taking the limit of an increasing sequence of step functions (cf. Chapter 6, Section 4.1.2), the important relation

$$E[\xi\eta | \mathcal{B}] \underset{\text{a.s.}}{=} \eta E[\xi | \mathcal{B}] \qquad (1.120)$$

58 Random variables

Hence deduce, under the same conditions, the other important relation

$$E[\xi\eta] = E[E[\xi|\mathscr{B}]\eta] \qquad (1.121)$$

(Take the expectation of both sides of Eqn (1.120).)

(3) Let η be a real r.v. measurable with respect to the Borel field $\mathscr{C}$. Show that

$$E[\xi\eta|\mathscr{B}] \underset{\text{a.s.}}{=} E[E[\xi|\mathscr{C}]\eta|\mathscr{B}]$$

(Apply Eqns (1.108) and (1.41).)

Exercise 2.12 Conditional expectation random variable

Let $(\Omega, \mathscr{A}, P)$ be a probability space, ξ and η be real r.v.s, and g be a Borel function $g: \mathbb{R} \to \mathbb{R}$. Show that

$$E[g \circ \eta | \eta] \underset{\text{a.s.}}{=} g \circ \eta$$

$$E[E[\xi|\eta]|g \circ \eta] \underset{\text{a.s.}}{=} E[E[\xi|g \circ \eta]|\eta] \underset{\text{a.s.}}{=} E[\xi|g \circ \eta] \qquad (1.122)$$

(For the first relation apply Eqn (6.19) and then Eqn (1.101); for the second, apply Eqn (6.19) and then Eqn (1.108).)

Exercise 2.13 Almost surely equal random variables

Let $(\Omega, \mathscr{A}, P)$ be a probability space and A and B two events such that $P(A) = P(B)$. Verify that the indicator r.v.s I_A and I_B have the same probability law but that they are a.s. equal only if the event $A \triangle B$ has zero probability. (Apply Eqns (2.41) and (1.1).)

Exercise 2.14

Let η be the r.v. $\eta = a\xi, a \in \mathbb{R}^*$, ξ being a real r.v. having probability density p_ξ. Show that the probability density of the r.v. η is the function

$$p_\eta(y) = \frac{1}{|a|} p_\xi\left(\frac{y}{a}\right) \qquad (1.123)$$

Exercise 2.15 Conditional probability density

Let ξ be a real r.v. having probability density p_ξ and distribution function F_ξ.
(1) Show that

$$\forall x \in \mathbb{R} \quad \forall y \in \mathbb{R}_+ \quad P(\{\xi \geq x+y\}|\{\xi \geq x\}) = \frac{1 - F_\xi(x+y)}{1 - F_\xi(x)} \qquad (1.124)$$

Hence deduce that

$$\forall x \in \mathbb{R} \quad \forall y \in \mathbb{R}_+ \quad P(\{\xi < x+y\}|\{\xi \geq x\}) = \frac{F_\xi(x+y) - F_\xi(x)}{1 - F_\xi(x)} \qquad (1.125)$$

(2) Let η be the real r.v. $\eta = \xi - x$, $x \in \mathbb{R}$. Show that the conditional probability density of the r.v. η, given that the r.v. ξ takes a value greater than or equal to x, is written

$$\forall x \in \mathbb{R} \qquad p_{\eta|\{\xi \geq x\}}(y) = Y(y) \frac{p_\xi(x+y)}{1 - F_\xi(x)} \qquad (1.126)$$

(Obtain from Eqn (1.125) the conditional distribution function of the r.v. η and then apply Eqn (1.62).)

Exercise 2.16 Exponential law

This exercise is an extension of Exercise 2.15. Let us suppose now that the r.v. ξ has an exponential probability law of parameter a (cf. Example 2.8).

(1) Verify that

$$\forall x \in \mathbb{R}_+ \quad \forall y \in \mathbb{R}_+ \quad P(\{\xi \geq x + y\} | \{\xi \geq x\}) = \exp(-ay) \qquad (1.127)$$

(Apply Eqn (1.124).)

(2) Let η be the r.v. $\eta = \xi - x$, $x \geq 0$. Verify that

$$p_{\eta|\{\xi \geq x\}}(y) = Y(y) a \exp(-ay) \qquad (1.128)$$

The conditional probability law of the r.v. η, given that the event $\{\xi \geq x\}$ is realized, is thus identical with the probability law of the r.v. ξ itself. This property is extremely important in applications and the exponential law is the only one with this property, as the following question shows. (Apply Eqn (1.126).)

(3) Show that the r.v. ξ has an exponential probability law of parameter a if and only if

$$\forall x > 0 \qquad P(\{\xi < x + \Delta x\} | \{\xi \geq x\}) = a\,\Delta x + o(\Delta x) \qquad (1.129)$$

Comment: an r.v. with an exponential probability law of parameter a is the mathematical model of a telephone call commencing at the instant 0 and of mean duration $1/a$ seconds (cf. Eqn (1.118)). The preceding results show that the probability law of the residual duration of a call which has lasted x seconds is independent of x and is identical with the probability law of the total duration of the call. We say, from this fact, that the exponential law has no memory.

Exercise 2.17 Poisson law

Let ξ_1 and ξ_2 be two independent r.v.s having Poisson probability laws of parameters λ_1 and λ_2. Show that the conditional expectation r.v. of the r.v. ξ_1 with respect to the r.v. $\xi_1 + \xi_2$ may be written

$$E[\xi_1 | \xi_1 + \xi_2] = \sum_{n \in \mathbb{N}} \frac{n\lambda_1}{\lambda_1 + \lambda_2} I_{B_n} \qquad (1.130)$$

60 Random variables

where I_{B_n} is the indicator r.v. of the event $B_n = \{\xi_1 + \xi_2 = n\}$. (Use Example 2.19.)

Exercise 2.18 Jensen inequality
Let g be a convex Borel function $g: \mathbb{R} \to \mathbb{R}$ and ξ an r.v. with finite mean value. Show that

$$g(E[\xi]) = E[g \circ \xi] \tag{1.131}$$

(Apply the increasing property of the measure μ_ξ to the function g and to its tangent at the point of abscissa $E[\xi]$, cf. Eqn (6.50).)

Exercise 2.19 Study of the random variable $\exp(\lambda \xi)$
Let ξ be a real r.v. having a mean finite value m_ξ. Consider the real r.v. $\exp(\lambda \xi)$, $\lambda \in \mathbb{R}$.

(1) Let I be the set of values of λ such that $E[\exp(\lambda \xi)] < +\infty$.
  (1.1) Show that I is an interval which contains the origin. (Use

$$\forall \lambda \in [\lambda_1, \lambda_2] \quad \forall x \in \mathbb{R} \quad 0 \leqslant \exp(\lambda x) \leqslant \exp(\lambda_1 x) + \exp(\lambda_2 x)$$

and apply the increasing property of the expectation (cf. Eqn (6.50)).)
  (1.2) Verify that $I = \mathbb{R}$ for discrete r.v.s (cf. Eqn (1.41)) and for gaussian r.v.s (cf. Eqn (1.46)).

(2) Consider the function $g_\xi(\lambda) = E[\exp(\lambda \xi)]$.
  (2.1) Show that it is differentiable up to order n in the interval $\mathring{I}$ and that

$$\forall n \in \mathbb{N}^* \quad g_\xi^{(n)}(\lambda) = E\{\xi^n \exp(\lambda \xi)\}$$

(Use the same inequality as for problem (1) and apply the theorem for differentiation under the summation sign.)
  (2.2) Hence deduce that the function g_ξ is convex in the interval $\mathring{I}$. (Verify that its second derivative is positive.)

(3) Consider the function

$$h_\xi(\lambda) = \log(E[\exp(\lambda \xi)]) \tag{1.132}$$

  (3.1) Verify that $h_\xi'(0) = m_\xi$.
  (3.2) Prove that

$$\forall \lambda_1, \lambda_2 \in I \quad h_\xi\left(\frac{\lambda_1 + \lambda_2}{2}\right) \leqslant \tfrac{1}{2} h_\xi(\lambda_1) + \tfrac{1}{2} h_\xi(\lambda_2)$$

which implies, since the function h_ξ is continuous in the interval $\mathring{I}$, that it is convex in the same interval. (Write

$$\exp\left(\frac{\lambda_1 + \lambda_2}{2} x\right) = \exp\left(\frac{\lambda_1 x}{2}\right) \exp\left(\frac{\lambda_2 x}{2}\right)$$

and apply Schwarz's inequality (Eqn (6.70)).)

Exercise 2.20 Chernoff's inequalities

Let ξ be a real r.v. having finite mean value m_ξ.

(1) Prove the inequalities

$$\forall a \in \mathbb{R} \quad \forall \lambda \geq 0 \quad P\{\xi \geq a\} \leq E[\exp\{\lambda(\xi - a)\}]$$

$$\forall a \in \mathbb{R} \quad \forall \lambda \leq 0 \quad P\{\xi \leq a\} \leq E[\exp\{\lambda(\xi - a)\}]$$

(Apply Eqn (6.50) to the indicator functions of the intervals $]-\infty, a]$ and $[a, +\infty[$ and to the function $\exp\{\lambda(\xi - a)\}$.)

(2) Suppose that the interval I defined in problem (1) of Exercise 2.19 is not empty.

(2.1) Let a be a number $a > m_\xi$ such that the equation $h_\xi(\lambda) = a\lambda$ has a root which is positive and belongs to the interval I, h_ξ being the function defined in problem (3) of Exercise 2.19. Prove that

$$\forall \lambda \in I \cap \mathbb{R}_+$$

$$P\{\xi \geq a\} \leq \exp[-\{a\lambda_0 - h_\xi(\lambda_0)\}] \leq \exp[-\{a\lambda - h_\xi(\lambda)\}] \quad (1.133)$$

λ_0 being the root of the equation $a = h_\xi'(\lambda)$. (Examine the form of the graph of the function h_ξ and use the facts that it is convex and that it passes through the origin with slope equal to m_ξ.)

(2.2) Let a be a number $a < m_\xi$ such that the equation $h_\xi(\lambda) = a\lambda$ has a root which is negative and belongs to the interval I. Prove that

$$\forall \lambda \in I \cap \mathbb{R}_-$$

$$P\{\xi \leq a\} \leq \exp[-\{a\lambda_0 - h_\xi(\lambda_0)\}] \leq \exp[-\{a\lambda - h_\xi(\lambda)\}] \quad (1.134)$$

λ_0 being again the root of the equation $a = h_\xi'(\lambda)$. (Argue as in the previous question.)

The inequalities (1.133) and (1.134) are those of **Chernoff**.

(3) Consider an independent sequence $(\xi_n, n \in \mathbb{N}^*)$ of real r.v.s having the same probability law as the r.v. ξ defined above.

Consider the r.v.

$$X_n = \frac{1}{n} \sum_{i=1}^{n} \xi_i$$

Show that Chernoff's inequalities apply to it and may be written, a satisfying the same conditions and λ_0 being the root of the same equation as in problems (2.1) and (2.2),

$$\forall \lambda \in I \cap \mathbb{R}_+ \quad P\{X_n \geq a\} \leq \exp[-n\{a\lambda_0 - h_\xi(\lambda_0)\}]$$
$$\forall \lambda \in I \cap \mathbb{R}_- \quad P\{X_n \leq a\} \leq \exp[-n\{a\lambda_0 - h_\xi(\lambda_0)\}] \quad (1.135)$$

It follows from this, since the exponents are negative, that these probabilities tend exponentially to zero when n tends to infinity. This is an example of the

62 Random variables

importance of Chernoff's inequalities which are extensively used to give strict bounds, decreasing exponentially (Wozencraft and Jacobs, 1965).

Exercise 2.21 The conditional variance random variable

Let ξ be a real r.v. with finite mean value and let $\mathscr{B}$ be a sub-Borel field such that $\mathscr{B} \subset \mathscr{A}$. The **conditional variance** of the r.v. conditioned by the sub-Borel field $\mathscr{B}$ is the real r.v. $V[\xi|\mathscr{B}]$ defined by

$$V[\xi|\mathscr{B}] = E[(\xi - E[\xi|\mathscr{B}])^2 | \mathscr{B}] \tag{1.136}$$

(1) Prove, when the r.v. ξ is measurable with respect to the Borel field $\mathscr{B}$, that the conditional variance is a.s. zero.

(2) Verify that $E[\xi E[\xi|\mathscr{B}]] = E[(E[\xi|\mathscr{B}])^2]$. (Apply Eqns (1.103) and (1.120).)

(3) Prove that

$$V[\xi] = E[V[\xi|\mathscr{B}]] + V[E[\xi|\mathscr{B}]] \tag{1.137}$$

The variance is thus the sum of the expectation of the conditional variance and the variance of the conditional expectation. (Start from $\xi - E[\xi] = \xi - E[\xi|\mathscr{B}] + E[\xi|\mathscr{B}] - E[\xi]$ and apply Eqn (1.103) as well as the result proved in problem (2).)

Exercise 2.22 The conditional expectation random variable

Let $(\Omega, \mathscr{A}, P)$ be a probability space, $\mathscr{B}$ a sub-Borel field and ξ an r.v., a.s. positive or zero, with finite mean value. Prove that the r.v. $E[\xi|\mathscr{B}]$ is a.s. positive or zero. (Apply Eqn (1.93) to the event $\{E[\xi|\mathscr{B}] < 0\} \in \mathscr{B}$ and hence deduce that its probability is zero.)

Exercise 2.23 Linear quantification of a real random variable

Let g be an even step function $g: \mathbb{R} \to \mathbb{R}$ such that

$$\forall x \in]k\Delta, (k+1)\Delta] \quad \forall k \in \{0, 1, \ldots, n-1\} \quad g(x) = \frac{\Delta}{2} + k\Delta$$

$$\forall x \in [n\Delta, +\infty[\quad g(x) = n\Delta - \frac{\Delta}{2}$$

Put $A = n\Delta$. The function g is the characteristic of a **linear quantifier** of which Δ is the quantification step.

Let ξ be a real r.v. having density function p_ξ. The r.v. $\xi_q = g \circ \xi$ results from the quantification, by the characteristic g, of the r.v. ξ and the r.v. $\varepsilon = \xi - \xi_q$ is the quantification error. Let $h(x)$ be the function $h(x) = x - g(x)$. Then we have $\varepsilon = h \circ \xi$. Figure 1.11 represents the graphs of the functions g and h in the particular case where $n = 3$.

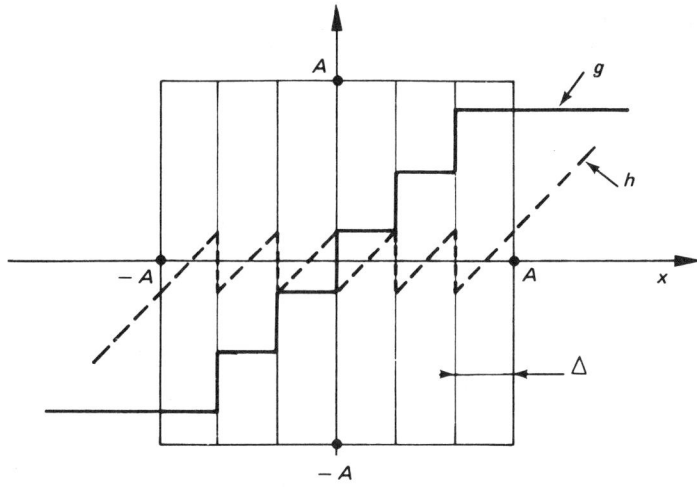

Fig. 1.11

(1) Verify that

$$E[\varepsilon] = E[h \circ \xi] = \int_{\mathbb{R}} h(x) p_\xi(x) \, dx \qquad (1.138)$$

(Use Eqn (1.71).)

(2) Prove, assuming that $\Delta \ll A$, that

$$E[\varepsilon^2] \neq \frac{\Delta^2}{12} \int_{-A}^{A} p_\xi(x) \, dx + 2 \int_{A}^{\infty} (x - A)^2 \{p_\xi(x) + p_\xi(-x)\} \, dx \qquad (1.139)$$

The first term is due to the quantification within the interval $[-A, A]$, the second to the truncation outside that interval.

(3) Hence deduce, when the r.v. ξ takes its values in the interval $[-A, A]$, that

$$E[\varepsilon^2] \approx \frac{\Delta^2}{12} \qquad (1.140)$$

The mean square value of the quantification error thus depends only on the step of quantification.

References

DUPRAZ J. (1977): *La Théorie des Distributions*, Cepadues, Toulouse.
KLEINROCK L. (1975): *Queuing Systems*, Vol. I, p. 10, Wiley, New York.
LEVINE B. (1973): *Fondements Théoriques de la Radiotechnique Statistique*, Vol. 2, Editions de Moscou, Moscow.
MARLE C.-M. (1974): *Mesures et Probabilités*, p. 429, Hermann, Paris.

NEVEU J. (1964): *Bases Mathématiques du Calcul des Probabilités*, Masson, Paris.
PFEIFFER P. and SCHUM D. (1973): *Introduction to Applied Probability*, Academic Press, New York.
WOZENCRAFT J.M. and JACOBS I.M. (1965): *Principles of Communication Engineering*, p. 97, Wiley, New York.

Chapter 2
Random vectors

Random vectors ($\vec{\text{r.v}}$.s) of finite and of infinite dimension are studied in this chapter, which is complementary to the preceding one.

The notion of independence of a family of random variables (r.v.s) or $\vec{\text{r.v}}$.s is developed in connection with that of conditioning. These notions have great practical importance. In fact, when two r.v.s are independent, it is fruitless to hope to learn anything about one of them by observing the values taken by the other. The opposite is the case when the r.v.s are not independent but are statistically linked. We can then use this link to estimate one of them as a function of the other. It must be stated that independence is a property which it is not always easy to prove. We must very often be content with establishing non-correlation.

Gaussian $\vec{\text{r.v}}$.s play a large part in practice because of the use of a gaussian random signal as a model of physical phenomena. They are the subject of numerous examples and exercises.

The chapter ends with the study of modes of convergence of sequences of r.v.s. The example of an independent binary random sequence is used to compare the weak and the strong laws of large numbers. A sequence of this type is the simplest model of a numerical message.

1. Random vectors of finite dimension

1.1. DEFINITIONS AND GENERAL PROPERTIES

1.1.1. *Real or complex random vectors*

1.1.1.1. Definition

Let $(\Omega, \mathscr{A}, P)$ be a probability space. An n-dimensional real $\vec{\text{r.v}}$. (Neveu, 1970) is a mapping ξ from the space $(\Omega, \mathscr{A})$ into the measurable product space $(\mathbb{R}^n, \mathscr{R}^n)$, $n \geq 2$.

1.1.1.2. Components of a random vector

Measurable product spaces are defined in Chapter 6, Section 1.2. The elements of the Borel field $\mathscr{R}^n$ are the Borel sets of the set $\mathbb{R}^n$. The simplest example of a Borel set is a rectangular block $\beta_1 \times \ldots \times \beta_n$, β_i being an arbitrary interval of the 'factor' set $\mathbb{R}_i$ numbered i, $i \in \{1, \ldots, n\}$ (cf. Fig. 2.1, where $n = 2$).

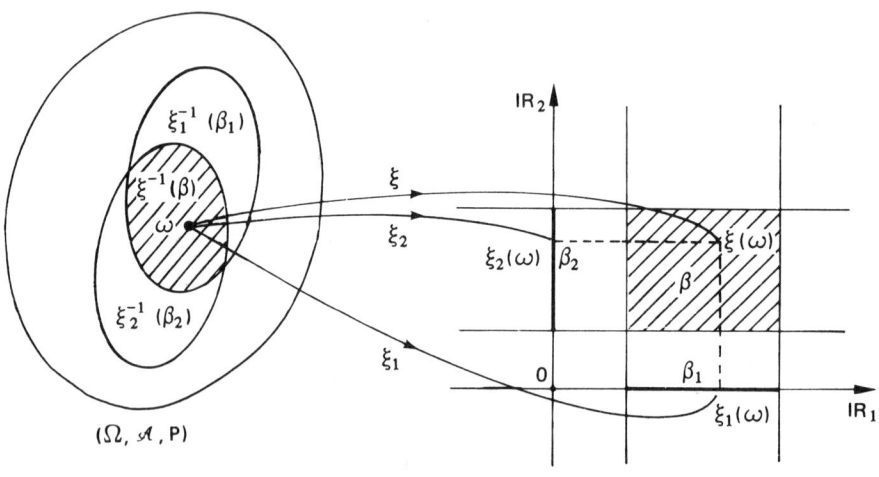

Fig. 2.1

The **projections** $\varphi_i: \mathbb{R}^n \to \mathbb{R}_i$ are measurable with respect to the Borel fields $\mathscr{R}^n$ and $\mathscr{R}_i$ (cf. Chapter 6, Section 1.2.3). Let the compound mappings then be

$$\forall i \in \{1, \ldots, n\} \qquad \xi_i = \varphi_i \circ \xi : (\Omega, \mathscr{A}) \to (\mathbb{R}_i, \mathscr{R}_i) \tag{2.1}$$

They are measurable and consequently they are real r.v.s defined over the space $(\Omega, \mathscr{A}, P)$ (cf. Chapter 6, Section 2.2.2). These are the components of the $\vec{\text{r.v.}}$ ξ. This leads to the following equivalent definition, extended to the complex case, of an $\vec{\text{r.v.}}$.

1.1.1.3. Definitions

A complex n-dimensional $\vec{\text{r.v.}}$ ξ defined over a probability space $(\Omega, \mathscr{A}, P)$ is a sequence $(\xi_1, \ldots, \xi_n)$ of n complex r.v.s defined over that space.

We can associate with the $\vec{\text{r.v.s}}$ ξ a random column matrix X whose element in the ith row is the r.v. ξ_i. According to circumstances, we shall write the $\vec{\text{r.v.}}$ ξ either in the normal form or in matrix form. Thus we shall write

$$\xi = (\xi_1, \ldots, \xi_n) \qquad X^T = [\xi_1 \ldots \xi_n] \tag{2.2}$$

It must be noted that the sequence $(\xi_1, \ldots, \xi_n)$ is not independent in the general case.

1.1.1.4. Sub-Borel field generated by a random vector

Let ξ be a real n-dimensional $\vec{\text{r.v.}}$ The inverse image $\xi^{-1}(\beta)$ of a Borel set $\beta \in \mathscr{R}^n$ is an event belonging to the Borel field $\mathscr{A}$. It is also denoted by $\{\xi \in \beta\}$. It contains all the trials ω such that $\xi(\omega) \in \beta$. This is illustrated by Fig. 2.1 in the case where $n = 2$ and where the Borel set considered is a rectangle $\beta_1 \times \beta_2$.

The inverse image of the Borel field $\mathscr{R}^n$ is the sub-Borel field generated by the $\vec{\text{r.v.}}$ ξ:

$$\mathscr{B}_\xi = \xi^{-1}(\mathscr{R}^n) \subset \mathscr{A}$$

It contains all the events which depend on the $\vec{\text{r.v.}}$ ξ. It is the smallest sub-Borel field containing the sub-Borel fields $\mathscr{B}_{\xi_i}$ generated by the r.v.s ξ_i respectively (cf. Chapter 6, Exercise 1.11).

1.1.2. Random vector function of a random vector

Let ξ be a real m-dimensional $\vec{\text{r.v.}}$ and $g: (\mathbb{R}^m, \mathscr{R}^m) \to (\mathbb{R}^n, \mathscr{R}^n)$ be a measurable function. The measurable function $\eta = g \circ \xi$ is a real n-dimensional $\vec{\text{r.v.}}$ The sub-Borel fields $\mathscr{B}_\xi$ and $\mathscr{B}_\eta$ generated by the $\vec{\text{r.v.}}$s ξ and η respectively satisfy $\mathscr{B}_\eta \subset \mathscr{B}_\xi$, which means that the events which depend on the $\vec{\text{r.v.}}$ η depend also on the $\vec{\text{r.v.}}$ ξ (cf. Eqn (6.19)).

The properties of $\vec{\text{r.v.}}$s which are functions of $\vec{\text{r.v.}}$s are formally the same as the properties of r.v.s which are functions of r.v.s (cf. Chapter 1, Section 2.1.2).

1.1.3. Almost surely equal random vectors

Let ξ and η be two complex n-dimensional $\vec{\text{r.v.}}$s. They are almost surely (a.s.) equal when the event $\{\omega \,|\, \xi(\omega) \neq \eta(\omega)\} = \{\xi \neq \eta\}$ has zero probability:

$$\xi \underset{\text{a.s.}}{=} \eta \Leftrightarrow P\{\xi \neq \eta\} = 0$$

1.2. PROBABILITY LAW OF A REAL RANDOM VECTOR

1.2.1. Definitions and general properties

1.2.1.1. Definition

Let $\xi = (\xi_1, \ldots, \xi_n)$ be a real n-dimensional $\vec{\text{r.v.}}$ Its probability law is the measure over the measurable space $(\mathbb{R}^n, \mathscr{R}^n)$ which is the image of the measure P through the mapping ξ (cf. Chapter 6, Section 3.1.4).

We shall write, according to circumstances, $\mu_{\xi_1, \ldots, \xi_n}$ or μ_X for the probability law of the $\vec{\text{r.v.}}$ ξ (cf. Eqn (2.2)).

1.2.1.2. Property

The preceding definition is formally the same as that of the probability law of a real r.v. It follows from Eqn (6.28) that

$$\forall \beta \in \mathcal{R}^n \quad \mu_X(\beta) = P\{\xi^{-1}(\beta)\} = P\{\xi \in \beta\} \quad (2.3)$$

Let us take as Borel set the rectangular block $\beta = \beta_1 \times \ldots \times \beta_n$, $\beta_i \in \mathcal{R}_i$. It follows immediately that

$$\{\xi \in \beta\} = \{\xi_1 \in \beta_1\} \cap \ldots \cap \{\xi_n \in \beta_n\}$$

This is illustrated by Fig. 2.1 in the case where $n = 2$ and the Borel sets β_1 and β_2 are intervals. It follows from the above that

$$\mu_{\xi_1 \ldots \xi_n}(\beta_1 \times \ldots \times \beta_n) = P(\{\xi_1 \in \beta_1\} \cap \ldots \cap \{\xi_n \in \beta_n\}) \quad (2.4)$$

It is always possible to apply Eqn (1.14) to write the above probability in another form but it is impossible to go further in the general case because the events $\{\xi_i \in \beta_i\}$ are not necessarily independent.

1.2.1.3. Marginal probability law

Consider the $\vec{\text{r.v.}}$ $\eta = (\xi_1, \ldots, \xi_m)$, $m < n$. This is a function $g \circ \eta$ of the $\vec{\text{r.v.}}$ η, the function g being the projection from $\mathbb{R}^n$ onto $\mathbb{R}^m$. The probability law of the $\vec{\text{r.v.}}$ η may be obtained from Eqn (2.3) by observing that

$$\forall \beta \in \mathcal{R}^m \quad \{\eta \in \beta\} = \{\eta \in \beta\} \cap \{\xi_{m+1} \in \mathbb{R}_{m+1}\} \cap \ldots \cap \{\xi_n \in \mathbb{R}_n\}$$

since $\{\xi_i \in \mathbb{R}_i\} = \Omega$. It follows that

$$\forall \beta \in \mathcal{R}^m \quad \mu_{\xi_1 \ldots \xi_m}(\beta) = \mu_{\xi_1 \ldots \xi_n}(\beta \times \mathbb{R}^{n-m}) \quad (2.5)$$

The measure $\mu_{\xi_1 \ldots \xi_m}$ is thus the projection of the measure $\mu_{\xi_1 \ldots \xi_n}$ onto the space $(\mathbb{R}^m, \mathcal{R}^m)$ (cf. Chapter 6, Section 3.2.2).

The probability laws $\mu_{\xi_1 \ldots \xi_m}$, $m < n$, are called **marginal**. They are all obtained by projection of the probability law $\mu_{\xi_1 \ldots \xi_n}$. The converse is false except in certain particular cases such as that of gaussian $\vec{\text{r.v.}}$s (cf. Example 1.6).

1.2.2. Discrete and continuous probability laws

1.2.2.1. Definitions

Discrete and continuous probability laws are defined formally for an $\vec{\text{r.v.}}$ in the same way as for an r.v. (cf. Chapter 1, Sections 2.2.2 and 2.2.3).

1.2.2.2. Property

A real n-dimensional $\vec{\text{r.v.}}$ $\xi = (\xi_1, \ldots, \xi_n)$ has a continuous probability law when the latter is defined by a probability density which we shall write, according to circumstances, $p_{\xi_1 \ldots \xi_n}$ or p_X. This is a real positive function defined over the space $\mathbb{R}^n$. It is measurable and integrable (with respect to Lebesgue measure), its integral being equal to unity.

The probability that the r.v. $\vec{\xi}$ takes a value belonging to a Borel set $\beta \in \mathbb{R}^n$ is therefore

$$P\{\xi \in \beta\} = \int_\beta p_{\xi_1 \ldots \xi_n}(x_1, \ldots, x_n) \, dx_1 \ldots dx_n \tag{2.6}$$

The marginal probability density $p_{\xi_1 \ldots \xi_m}$ of the r.v. $\vec{\eta} = (\xi_1, \ldots, \xi_m)$, $m < n$, is obtained from Eqn (2.5). We get

$$p_{\xi_1 \ldots \xi_m}(x_1, \ldots, x_m) = \int_{\mathbb{R}^{n-m}} p_{\xi_1 \ldots \xi_n}(x_1, \ldots, x_n) \, dx_{m+1} \ldots dx_n \tag{2.7}$$

This formula is very important in practice.

1.2.3. Probability law of a random vector function of a random vector

The probability law of a real r.v. $\vec{\eta} = g \circ \vec{\xi}$ which is a function of a real r.v. $\vec{\xi}$ is obtained formally in the same way as for r.v.s (cf. Chapter 1, Section 2.2.4). The probability density p_η is obtained from Eqn (1.55) by considering $dg^{-1}(y)/dy$ as the jacobian of the inverse mapping g^{-1}.

Suppose that the r.v.s $\vec{\xi}$ and η are n-dimensional and let us consider a Borel function $g: \mathbb{R}^n \to \mathbb{R}^n$. Suppose that the inverse mapping g^{-1} is defined by

$$(y_1, \ldots, y_n) \mapsto (x_1 = f_1(y_1, \ldots, y_n), \ldots, x_n = f_n(y_1, \ldots, y_n))$$

The jacobian J of the mapping g^{-1} is written (Bass, 1956)

$$J = \begin{vmatrix} \dfrac{\partial f_1}{\partial y_1} & \cdots & \dfrac{\partial f_1}{\partial y_n} \\ \cdots & \cdots & \cdots \\ \dfrac{\partial f_n}{\partial y_1} & \cdots & \dfrac{\partial f_n}{\partial y_n} \end{vmatrix} \tag{2.8}$$

Example 1.1 Transformation to polar coordinates

Let $\xi = (\xi_1, \xi_2)$ be a real two-dimensional r.v. and let the real two-dimensional r.v. $g \circ \vec{\xi} = (\rho, \vartheta)$ be such that

$$\xi_1 = \rho \cos \vartheta \qquad \xi_2 = \rho \sin \vartheta \tag{2.9}$$

This is illustrated in Fig. 2.2.

The above results apply and the jacobian of the reciprocal mapping g^{-1} is $J = r$. The probability density of the r.v. (ρ, ϑ) has as support the set $\mathbb{R}_+ \times [0, 2\pi]$. It may be written (cf. Eqn (1.55))

$$p_{\rho\vartheta}(r, \vartheta) = r p_{\xi_1 \xi_2}(r \cos \vartheta, r \sin \vartheta) \qquad r \in \mathbb{R}_+ \qquad \vartheta \in \,]0, 2\pi[\tag{2.10}$$

Random vectors

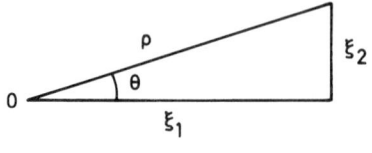

Fig. 2.2

The marginal probability densities of the r.v.s ρ and ϑ may be obtained by applying Eqn (2.7). We get

$$p_\rho(r) = \int_0^{2\pi} r p_{\xi_1 \xi_2}(r \cos \vartheta, r \sin \vartheta) \, d\vartheta \qquad r \in \mathbb{R}_+$$

$$p_\vartheta(\vartheta) = \int_0^\infty r p_{\xi_1 \xi_2}(r \cos \vartheta, r \sin \vartheta) \, dr \qquad \vartheta \in]0, 2\pi[$$

(2.11)

1.2.4. Conditional probability density

1.2.4.1. Introduction

Let $\xi = (\xi_1, \ldots, \xi_m)$ be a real r.v. such that the r.v.s ξ_i have finite mean value and let $\eta = (\eta_1, \ldots, \eta_n)$ be a real n-dimensional r.v. Suppose that the real r.v. $(\xi_1, \ldots, \xi_m, \eta_1, \ldots, \eta_n)$ has a probability density $p_{\xi\eta}(x, y)$, where we have put for simplification $(x_1, \ldots, x_m) = x$ and $(y_1, \ldots, y_n) = y$.

Consider the conditional expectation r.v. $E[\xi_i | \eta] = \psi_i \circ \eta$ and let us seek an expression for the function ψ_i. This is defined in Chapter 1, Section 2.6.2.4 and it follows by applying Eqns (1.95) and (1.71) that

$$\forall \beta \in \mathscr{R}^n \qquad \mu_\eta(\psi_i I_\beta) = \mu_{\xi\eta}(x_i I_\beta)$$

This is written, the integral in y being n-fold,

$$\int_\beta p_\eta(y) \psi_i(y) \, dy = \int_\beta dy \int_\mathbb{R} x_i p_{\xi\eta}(x, y) \, dx_i$$

The marginal probability density $p_{\xi\eta}$ is obtained from the integral (cf. Eqn (2.7))

$$p_{\xi\eta}(x_i, y) = \int_{\mathbb{R}^{m-1}} p_{\xi\eta}(x, y) \, dx_1 \ldots dx_{i-1} \, dx_{i+1} \ldots dx_m$$

It follows then from Fubini's theorem that we have almost everywhere on the space $(\mathbb{R}^n, \mathscr{R}^n, \mu_0)$ (cf. Chapter 6, Section 4.2.2)

$$p_\eta(y) \psi_i(y) = \int_{\mathbb{R}^m} x_i p_{\xi\eta}(x, y) \, dx_1 \ldots dx_m$$

Let β_0, the Borel set of $\mathbb{R}^n$, be defined by $\beta_0 = \{y | p_\eta(y) = 0\}$. The probability of the event $\{\eta \in \beta_0\}$ is zero since

$$P\{\eta \in \beta_0\} = \mu_\eta(\beta_0) = \int_{\beta_0} p_\eta(y)\,dy = 0$$

Since the r.v. $E[\xi_i|\eta]$ is only a.s. defined it can take any arbitrary value on the set $\{\eta \in \beta_0\}$. The function ψ_i can therefore take arbitrary values on the set β_0 and, to define it, it is sufficient to write

$$\forall y \notin \beta_0 \qquad \psi_i(y) = \int_{\mathbb{R}^m} x_i \frac{p_{\xi\eta}(x,y)}{p_\eta(y)}\,dx_1 \ldots dx_m$$

1.2.4.2. Definition
The **conditional probability density** $p_{\xi|\eta}$ of the $\vec{\text{r.v.}}$ ξ conditioned by the r.v. η is the function

$$p_{\xi|\eta}(x|y) = \frac{p_{\xi\eta}(x,y)}{p_\eta(y)} \qquad (2.12)$$
$$= 0 \qquad y \notin \beta_0$$

1.2.4.3. Properties
The function of x, $p_{\xi|\eta}(x|y)$, is defined over the space $(\mathbb{R}^m, \mathcal{R}^m, \mu_0)$ for all values $y \notin \beta_0$. It is immediate that it is positive, borelian, and integrable with respect to the Lebesgue measure μ_0, and that its integral is equal to unity. It is thus legitimate to regard it as a probability density.

It follows from Eqn (2.12) that

$$\forall y \notin \beta_0 \qquad \psi_i(y) = \int_{\mathbb{R}^m} x_i p_{\xi|\eta}(x|y)\,dx_1 \ldots dx_m \qquad (2.13)$$

This enables us to say, by comparison with Eqn (1.74), that when the $\vec{\text{r.v.}}$ η takes a value $y \notin \beta_0$ the r.v. $E[\xi_i|\eta]$ takes as value the conditional mean value (the conditional expectation) of the r.v. ξ_i.

Repeated application of Eqn (2.12) enables us to factorize the probability density of the $\vec{\text{r.v.}}$ ξ in the following form:

$$p_{\xi_1 \ldots \xi_m}(x_1, \ldots, x_m) = p_{\xi_1}(x_1) p_{\xi_2|\xi_1}(x_2|x_1) \ldots p_{\xi_m|\xi_1 \ldots \xi_{m-1}}(x_m|x_1, \ldots, x_{m-1})$$
$$(2.14)$$

This fundamental formula should be compared with formula (1.14). The probability that the $\vec{\text{r.v.}}$ ξ takes a value belonging to a Borel set $\beta = \beta_1 \times \ldots \times \beta_m$ may then be written (cf. Eqn (2.6))

$$P\{\xi \in \beta\} = \int_{\beta_1} p_{\xi_1}(x_1)\,dx_1 \int_{\beta_2} \ldots$$
$$\ldots \int_{\beta_m} p_{\xi_m|\xi_1 \ldots \xi_{m-1}}(x_m|x_1, \ldots, x_{m-1})\,dx_m \qquad (2.15)$$

72 Random vectors

Example 1.2 Conditional expectation of a gaussian random variable

Let $\xi = (\xi_1, \xi_2)$ be a two-dimensional gaussian r.v. Its probability density $p_{\xi_1 \xi_2}$ may be obtained from Eqn (2.41). The marginal probability density of the r.v. ξ_1 is given by Eqn (1.46). It can also be obtained from Eqn (2.41) by application of Eqn (2.7). By putting

$$m_1 = E[\xi_1] \quad \sigma_1{}^2 = V[\xi_1] \quad m_2 = E[\xi_2]$$

$$\sigma_2{}^2 = V[\xi_2] \quad r_{12} = r_{\xi_1 \xi_2}$$

we get

$$p_{\xi_1 \xi_2}(x_1, x_2) = \frac{1}{2\pi \sigma_1 \sigma_2 (1 - r_{12}{}^2)^{1/2}} \exp\left[-\frac{1}{2(1 - r_{12}{}^2)} \right.$$

$$\left. \times \left\{ \frac{(x_1 - m_1)^2}{\sigma_1{}^2} - 2 r_{12} \frac{(x_1 - m_1)(x_2 - m_2)}{\sigma_1 \sigma_2} + \frac{(x_2 - m_2)^2}{\sigma_2{}^2} \right\} \right]$$

(2.16)

$$p_{\xi_1}(x_1) = \frac{1}{(2\pi \sigma_1{}^2)^{1/2}} \exp\left\{ -\frac{(x_1 - m_1)^2}{2\sigma_1{}^2} \right\}$$

The conditional probability density $p_{\xi_2 | \xi_1}$ of the r.v. ξ_2 conditioned by the value x_1 and taken by the r.v. ξ_1 is obtained from Eqn (2.12). We get

$$p_{\xi_2 | \xi_1}(x_2 | x_1) = \frac{1}{\{2\pi \sigma_2{}^2 (1 - r_{12}{}^2)\}^{1/2}}$$

$$\times \exp\left[-\frac{1}{2\sigma_2{}^2 (1 - r_{12}{}^2)} \left\{ x_2 - m_2 - r_{12} \frac{\sigma_2}{\sigma_1}(x_1 - m_1) \right\}^2 \right] \quad (2.17)$$

It follows from this, by comparison of Eqns (2.17) and (1.46), that the function $p_{\xi_2 | \xi_1}$ is the probability density of a gaussian law with mean value $m_2 + (\sigma_2 / \sigma_1) r_{12}(x_1 - m_1)$. It then follows immediately from Eqn (2.13) that

$$\psi(x_1) = m_2 + r_{12} \frac{\sigma_2}{\sigma_1}(x_1 - m_1)$$

Hence it follows that the conditional expectation $E[\xi_2 | \xi_1]$ is the r.v. (cf. Eqn (1.94))

$$E[\xi_2 | \xi_1] = m_2 + r_{12} \frac{\sigma_2}{\sigma_1}(\xi_1 - m_1) \quad (2.18)$$

The r.v. ξ_2 is clearly square integrable. The r.v. $E[\xi_2 | \xi_1]$ is thus the best approximation, in quadratic mean, to the r.v. ξ_2 by a function of the r.v. ξ_1 (cf. Chapter 1, Section 2.6.3.5). It is interesting to note that this function is linear.

These results can be generalized to the case where ξ_1 and ξ_2 are two jointly gaussian r.v.s (cf. Exercise 1.4).

Random vectors of finite dimension 73

Example 1.3 Hypothesis test and likelihood ratio

Introduction

The general principle of binary hypothesis tests is presented in Chapter 1, Example 1.3. The case in which the test concerns the choice of one or the other of two r.v.s is the subject of Chapter 1, Example 2.10.

The test studied here concerns the choice of one or the other of two real m-dimensional $\vec{\text{r.v.}}$s ξ_0 and ξ_1. The $\vec{\text{r.v.}}$ in question, when ω is the trial realized, takes a value $a \in \mathbb{R}^m$. When hypothesis h_0 applies, $a = \xi_0(\omega)$. When hypothesis h_1 applies, $a = \xi_1(\omega)$. The space $\mathbb{R}^m$ is divided into two regions R_0 and R_1, constituting a partition. When $a \in R_0$, we decide that hypothesis h_0 applies. When $a \in R_1$, we decide that h_1 applies.

The regions R_0 and R_1 are such that the probability of error $P(\varepsilon)$ is minimal. The choice criterion involves the probability densities of the two $\vec{\text{r.v.}}$s, by means of the likelihood ratio. It is obtained by arguing as in the examples quoted.

Let us suppose, in addition, that the $\vec{\text{r.v.}}$s ξ_0 and ξ_1 are conditioned respectively by real n-dimensional $\vec{\text{r.v.}}$s η_0 and η_1. We are led to distinguish two cases, according as we do or do not know the values taken in $\mathbb{R}^n$ by the $\vec{\text{r.v.}}$s η_0 and η_1 when the trial ω is realized.

First case

The $\vec{\text{r.v.}}$s η_0 and η_1 take respectively, with non-zero probabilities, the values b_0 and b_1, and these values are supposed known. Then

$$\xi_0(\omega) = a \text{ or } \xi_1(\omega) = a \qquad \eta_0(\omega) = b_0 \text{ and } \eta_1(\omega) = b_1$$

The distributions of the $\vec{\text{r.v.}}$s ξ_0 and ξ_1 in $\mathbb{R}^m$ are determined by the conditional probability densities:

$$p_{\xi_0|\eta_0}(x|b_0) \quad \text{and} \quad p_{\xi_1|\eta_1}(x|b_1)$$

The choice criterion can then be written (cf. Eqn (1.49))

$$\frac{p_{\xi_0|\eta_0}(a|b_0)}{p_{\xi_1|\eta_1}(a|b_1)} \underset{H_1}{\overset{H_0}{\gtrless}} \frac{p_1}{p_0} \tag{2.19}$$

It defines a partition of the space $\mathbb{R}^m$ into two regions R_0 and R_1. The probability of error may be written (cf. Eqn (1.50))

$$P(\varepsilon) = p_0 \int_{R_1} p_{\xi_0|\eta_0}(x|b_0) \, dx + p_1 \int_{R_0} p_{\xi_1|\eta_1}(x|b_1) \, dx \tag{2.20}$$

Second case

The values taken by the $\vec{\text{r.v.}}$s η_0 and η_1 are unknown and cannot be taken into account in the choice criterion. The distributions of the $\vec{\text{r.v.}}$s ξ_0 and ξ_1 are determined by the probability densities p_{ξ_0} and p_{ξ_1}. The latter may be written

74 Random vectors

(cf. Eqns (2.7) and (2.12))

$$p_{\xi_0}(x) = \int_{\mathbb{R}^n} p_{\eta_0}(y) p_{\xi_0|\eta_0}(x|y) \, dy$$

$$p_{\xi_1}(x) = \int_{\mathbb{R}^n} p_{\eta_1}(y) p_{\xi_1|\eta_1}(x|y) \, dy$$

The choice criterion can now be written

$$\frac{p_{\xi_0}(a)}{p_{\xi_1}(a)} \underset{H_1}{\overset{H_0}{\gtrless}} \frac{p_1}{p_0} \tag{2.21}$$

It defines a partition of the space $\mathbb{R}^m$ into two regions R_0' and R_1'. The probability of error becomes

$$P(\varepsilon) = p_0 \int_{R_1'} p_{\xi_0}(x) \, dx + p_1 \int_{R_0'} p_{\xi_1}(x) \, dx \tag{2.22}$$

The probability of error obtained from Eqn (2.22) is, according to circumstances, greater or less than that obtained from Eqn (2.20).

1.3. COVARIANCE AND CORRELATION (Pfeiffer and Schum, 1973)

1.3.1. *Covariance of two random variables*

1.3.1.1. Definitions

Let ξ_1 and ξ_2 be two complex, square integrable, r.v.s and let $\overset{\circ}{\xi}_1$ and $\overset{\circ}{\xi}_2$ be the centred r.v.s (cf. Eqn (1.68) and Chapter 1, Section 2.4.2).

The **covariance** $C[\xi_1 \xi_2]$ of the pair (ξ_1, ξ_2) is the mean value of the r.v. $\overset{\circ}{\xi}_1 \overset{\circ}{\xi}_2^*$:

$$C[\xi_1 \xi_2] = E[\overset{\circ}{\xi}_1 \overset{\circ}{\xi}_2^*] = E[(\xi_1 - E[\xi_1])(\xi_2 - E[\xi_2])^*] \tag{2.23}$$

The **correlation coefficient** $r_{\xi_1 \xi_2}$ of the pair (ξ_1, ξ_2) is the ratio

$$r_{\xi_1 \xi_2} = \frac{C[\xi_1 \xi_2]}{(V[\xi_1] V[\xi_2])^{1/2}} = \frac{C[\xi_1 \xi_2]}{\sigma_{\xi_1} \sigma_{\xi_2}} \tag{2.24}$$

The r.v.s ξ_1 and ξ_2 are **uncorrelated** when their correlation coefficient, and consequently their covariance, is zero.

1.3.1.2. Properties

Since the r.v.s ξ_1 and ξ_2 are square integrable, the r.v. $\overset{\circ}{\xi}_1 \overset{\circ}{\xi}_2^*$ is integrable and consequently the covariance $C[\xi_1 \xi_2]$ is well defined. It is the scalar product in the space $L^2(\Omega, \mathscr{A}, P)$ of the r.v.s $\overset{\circ}{\xi}_1$ and $\overset{\circ}{\xi}_2$ (cf. Eqn (6.69)). The linearity of the operator E enables us to write an equivalent expression for the covariance:

$$C[\xi_1 \xi_2] = E[\xi_1 \xi_2^*] - E[\xi_1] E[\xi_2^*] \tag{2.25}$$

It follows immediately that

$$C[\xi_2\xi_1] = (C[\xi_1\xi_2])^* \qquad r_{\xi_2\xi_1} = r_{\xi_1\xi_2}{}^* \qquad (2.26)$$

Moreover Schwarz's inequality enables us to place bounds on the covariance and on the correlation coefficient (cf. Eqn (6.70)). We obtain

$$|C[\xi_1\xi_2]|^2 \leq V[\xi_1]V[\xi_2] \qquad |r_{\xi_1\xi_2}| \leq 1 \qquad (2.27)$$

The equalities apply if and only if there exists a constant $c \in \mathbb{C}$ such that the r.v.s $\overset{\circ}{\xi}_1$ and $c\overset{\circ}{\xi}_2$ are a.s. equal (cf. Chapter 6, Section 4.1.7.2). Hence it follows that, if the modulus of the correlation coefficient of two r.v.s is equal to unity, these two r.v.s are a.s. connected by a linear relation (cf. Example 1.5).

It must be noted finally that when $\xi = \xi_1 = \xi_2$ (cf. Eqn (1.70)) we have

$$C[\xi\xi] = V[\xi] \qquad r_{\xi\xi} = 1 \qquad (2.28)$$

Example 1.4   Variance of a sum of random variables

Let $\{\xi_1, \ldots, \xi_n\}$ be a finite sequence of square integrable r.v.s and let η be the square integrable r.v.

$$\eta = \sum_{i=1}^{n} \xi_i$$

The variance of the r.v. η is written (cf. Eqn (1.70))

$$V[\eta] = E\left[\left(\sum_{i=1}^{n} \overset{\circ}{\xi}_i\right)\left(\sum_{j=1}^{n} \overset{\circ}{\xi}_j\right)^*\right]$$

It follows, from linearity of the operator E, that (cf. Eqn (2.26))

$$V[\eta] = \sum_{i=1}^{n} V[\xi_i] + \sum_{i=1}^{n}\sum_{j=1}^{n} 2\,\text{Re}(C[\xi_i\xi_j]) \qquad (i < j) \qquad (2.29)$$

From this it follows that the variance of a sum of pairwise uncorrelated r.v.s is the sum of their variances. When the r.v.s are correlated, the variance of the sum can be, according to circumstances, greater or smaller than the sum of the variances.

Example 1.5   The line of regression

Let ξ and η be two square integrable r.v.s belonging to the space $L^2(\Omega, \mathcal{A}, P)$. In what follows, a constant r.v. cI_Ω is denoted simply by c (cf. Chapter 1, Example 2.1). We seek to define an r.v.

$$\hat{\xi} = a\eta + b \qquad a, b \in \mathbb{C}$$

belonging to the same space, which is the best approximation, in quadratic mean, to the r.v. ξ by a linear function of the r.v. η.

The orthogonal projection theorem gives us the solution (cf. Chapter 6, Section 4.1.8(c)). The r.v. $\hat{\xi}$ is the projection of the r.v. ξ on the closed vector subspace of $L^2(\Omega, \mathcal{A}, P)$ generated by the r.v.s η and 1. We write that the r.v.

$\xi - \hat{\xi}$ is orthogonal to the r.v.s η and 1:

$$E[(\xi - a\eta - b)1] = 0 \qquad E[(\xi - a\eta - b)\eta^*] = 0$$

The constants a and b are the solution of the system

$$\left.\begin{array}{l} aE[|\eta|^2] + bm_\eta^* = E[\xi\eta^*] \\ am_\eta + b = m_\xi \end{array}\right\}$$

By applying Eqns (2.24) and (1.70) we obtain

$$\hat{\xi} = m_\xi + \frac{\sigma_\xi}{\sigma_\eta} r_{\xi\eta}(\eta - m_\eta) \qquad (2.30)$$

This equation defines a straight line in the plane of the r.v.s η and 1. This is the **line of regression** of the r.v. ξ on the r.v. η.

Let us consider the r.v. $\xi - \hat{\xi}$. This is the error committed by the substitution of the r.v. $\hat{\xi}$ for the r.v. ξ. It is immediate that its mean value is zero. Its mean square value may be written (cf. Eqn (2.24))

$$E[|\xi - \hat{\xi}|^2] = \sigma_\xi^2(1 - |r_{\xi\eta}|^2) \qquad (2.31)$$

It is zero when $|r_{\xi\eta}| = 1$. The r.v.s ξ and $\hat{\xi}$ are then a.s. equal (cf. Eqn (1.39)). This implies the equality a.s. of the r.v.s $\overset{\circ}{\xi}$ and $\overset{\circ}{\hat{\xi}}$, as well as that of the r.v.s $\overset{\circ}{\xi}$ and $(\sigma_\xi/\sigma_\eta)\overset{\circ}{\eta}$.

To sum up, we can write

$$|r_{\xi\eta}| = 1 \Leftrightarrow \overset{\circ}{\xi} \underset{\text{a.s.}}{=} \frac{\sigma_\xi}{\sigma_\eta}\overset{\circ}{\eta} \qquad (2.32)$$

Two complex r.v.s are a.s. connected by a linear relation if and only if the modulus of their correlation coefficient is equal to 1.

1.3.2. Covariance and correlation matrices of a random vector

1.3.2.1. Definitions

A **random matrix** A is a matrix whose elements are r.v.s defined over one and the same probability space. The **mean value** $E[A]$ of the matrix A is the matrix which is obtained from it by replacing the r.v.s by their mean values.

Let $\vec{\xi} = (\xi_1, \ldots, \xi_n)$ be a complex n-dimensional r.v. and let X^T be the random matrix $X^T = (\xi_1, \ldots, \xi_n)$ associated with it (cf. Eqn (2.2)). The **covariance matrix** Γ_{XX} of the r.v. $\vec{\xi}$ is the complex $n \times n$ square matrix

$$\Gamma_{XX} = E[(X - E[X])(X - E[X])^{*T}] \qquad (2.33)$$

The **correlation matrix** K_{XX} of the r.v. $\vec{\xi}$ is the complex matrix obtained from Γ_{XX} by replacing the covariances by correlation coefficients.

1.3.2.2. Properties

It follows from Eqn (2.33) that the (i,j)th element of the matrix Γ_{XX} is the covariance $C[\xi_i, \xi_j]$ of the r.v.s ξ_i and ξ_j. The (i,j)th element of the matrix K_{XX} is the correlation coefficient $r_{\xi_i\xi_j}$ of the r.v.s ξ_i and ξ_j. It follows from Eqn (2.24) that

$$C[\xi_i, \xi_j] = r_{\xi_i\xi_j}\sigma_{\xi_i}\sigma_{\xi_j} \tag{2.34}$$

The diagonal elements of the matrix Γ_{XX} are the variances of the r.v.s ξ_i and the diagonal elements of the matrix K_{XX} are equal to unity (cf. Eqn (2.28)). When the r.v.s ξ_i are pairwise uncorrelated, the matrices Γ_{XX} and K_{XX} are diagonal.

It follows immediately from Eqn (2.26) that the matrix Γ_{XX} is hermitian, which means that $\Gamma_{XX}^{T} = \Gamma_{XX}^{*}$. Its eigenvalues are therefore real. Moreover, the matrix Γ_{XX} is non-negative. Indeed, if B is a complex $n \times 1$ matrix, it follows from Eqn (2.33), by putting $E[X] = M_X$, that

$$B^{*T}\Gamma_{XX}B = E[\{B^{*T}(X - M_X)\}\{(X - M_X)^{*T}B\}]$$

We note that the matrix product $(X - M_X)^{*T}B$ is a scalar, which implies that

$$(X - M_X)^{*T}B = B^{T}(X - M_X)^{*}$$

Hence we have

$$B^{*T}\Gamma_{XX}B = E[|B^{*T}(X - M_X)|^2] \geqslant 0$$

The eigenvalues of the matrix Γ_{XX} are therefore positive or zero.

Since the matrix Γ_{XX} is hermitian, there exists a unitary matrix U such that the matrix $U\Gamma_{XX}U^{-1}$ is diagonal, the elements in the diagonal being the eigenvalues of the matrix Γ_{XX}. We recall that a matrix U is unitary when it satisfies $U^{-1} = U^{*T}$. We know also that the linear operator over the vector space $\mathbb{C}^n$ of which U is the matrix with respect to the canonical base is a continuous operator. These results belong to the classical theory of vector spaces of finite dimensions.

1.3.2.3. Diagonalization of a covariance matrix

Consider the mapping $g: \mathbb{C}^n \to \mathbb{C}^n$ whose matrix with respect to the canonical base is the matrix U defined above. This is a continuous mapping and therefore borelian. It enables us to define the complex n-dimensional r.v. $\eta = g \circ \xi$ for which the associated random matrix is written $Y = UX$ (cf. Eqn (2.2)).

It follows immediately that $E[Y] = UE[X]$. The covariance matrix Γ_{XX} is obtained from Eqn (2.33). We obtain

$$\Gamma_{YY} = E[\{U(X - M_X)\}\{U(X - M_X)\}^{*T}]$$
$$= UE[(X - M_X)(X - M_X)^{*T}]U^{*T}$$

From this it follows that, since $U^{-1} = U^{*T}$,

$$\Gamma_{YY} = U\Gamma_{XX}U^{-1}$$

78 Random vectors

The matrix Γ_{XX} is therefore diagonal. The components $\eta_1, \ldots, \eta_n$ of the r.v. $\vec{\eta}$ are therefore pairwise uncorrelated r.v.s and their variances are the eigenvalues of the matrix Γ_{XX}.

To sum up, given a complex n-dimensional r.v. $\vec{\xi}$ it is always possible to transform it by a linear mapping $g:\mathbb{C}^n \to \mathbb{C}^n$ into a complex n-dimensional r.v. $\vec{\eta}$ whose components are pairwise uncorrelated. The matrix of the mapping g with respect to the canonical base is a unitary matrix U which diagonalizes the matrix Γ_{XX}. The random matrices X and Y associated with the r.v.s $\vec{\xi}$ and $\vec{\eta}$ satisfy (cf. Eqn (2.2))

$$Y = UX \qquad E[Y] = UE[X] \qquad \Gamma_{YY} = U\Gamma_{XX}U^{-1} \qquad (2.35)$$

The variances of the components of the r.v. $\vec{\eta}$ are the eigenvalues of the matrix Γ_{XX}.

1.3.3. Covariance and correlation matrices of two random vectors

1.3.3.1. Definitions

Let $\vec{\xi}$ be a complex m-dimensional r.v. and $\vec{\eta}$ a complex n-dimensional r.v. and let X and Y be their associated random matrices (cf. Eqn (2.2)). The covariance matrix Γ_{XY} of the pair $(\vec{\xi}, \vec{\eta})$ is the complex rectangular $m \times n$ matrix.

$$\Gamma_{XY} = E[(X - E[X])(Y - E[Y])^{*T}] \qquad (2.36)$$

The correlation matrix K_{XY} of the pair $(\vec{\xi}, \vec{\eta})$ is the matrix obtained from Γ_{XY} by replacing the covariances by correlation coefficients.

The r.v.s $\vec{\xi}$ and $\vec{\eta}$ are **uncorrelated** when the correlation matrix, and consequently the covariance matrix, is zero.

1.3.3.2. Properties

It follows immediately from Eqn (2.26) that the covariance and correlation matrices satisfy

$$\Gamma_{YX}^* = \Gamma_{XY}^T \qquad K_{YX}^* = K_{XY}^T \qquad (2.37)$$

They are therefore hermitian matrices.

1.4. CHARACTERISTIC FUNCTION OF A REAL RANDOM VECTOR

1.4.1. Definition

Let $\vec{\xi} = (\xi_1, \ldots, \xi_n)$ be a real n-dimensional r.v. with probability law μ_X and let X be its associated random matrix. The characteristic function Θ_X of the r.v. $\vec{\xi}$ is the mean value of the r.v. $\exp\{iU^T X\}$, U being the real matrix such that

$U^T = [u_1 \ldots u_n]$. It may be expressed by the integral

$$\Theta_X(u_1, \ldots, u_n) = E[\exp(iU^T X)]$$

$$= \int_{\mathbb{R}^n} \exp\{i(u_1 x_1 + \ldots + u_n x_n)\} \, d\mu_X(x_1, \ldots, x_n) \quad (2.38)$$

1.4.2. Properties

The properties of the characteristic function of a real $\vec{r.v.}$ are formally the same as those of the characteristic function of a real r.v. (cf. Chapter 1, Section 2.5.1).

The function $\Theta_X(2\pi y_1, \ldots, 2\pi y_n)$ is the inverse Fourier transform of the measure μ_X (cf. Eqn (6.81)). It is bounded and uniformly continuous and $\Theta_X(0, \ldots, 0) = 1$ (cf. Chapter 6, Section 4.3.2).

More generally, the function $\Theta_X(2\pi y_1, \ldots, 2\pi y_n)$ and the measure μ_X are tempered distributions, connected by the Fourier transformation. We can write, defining *ipso facto* the function Φ_X

$$\Phi_X(y_1, \ldots, y_n) = \Theta_X(2\pi y_1, \ldots, 2\pi y_n) \qquad \mu_X \underset{\mathscr{F}}{\overset{\mathscr{F}}{\rightleftarrows}} \Phi_X \quad (2.39)$$

The characteristic function Θ_X characterizes completely, as does the measure μ_X, the class of real $\vec{r.v.}$s a.s. equal to the $\vec{r.v.}$ ξ.

Consider the complex r.v. $\xi_1^{k_1} \times \ldots \times \xi_n^{k_n}$, $k_1 \in \mathbb{N}, \ldots, k_n \in \mathbb{N}$. Its mean value is obtained by successive differentiations of Eqn (2.38) under the integral sign. We get

$$\frac{\partial \Theta_X(0, \ldots, 0)}{\partial u_1^{k_1} \ldots \partial u_n^{k_n}} = i^{k_1 + \ldots + k_n} E[\xi_1^{k_1} \times \ldots \times \xi_n^{k_n}] \quad (2.40)$$

$$k_1 \in \mathbb{N}, \ldots, k_n \in \mathbb{N}$$

This formula should be compared with formula (1.85). It enables us, in particular, to calculate the mean values of the r.v.s ξ_i and the covariances of the pairs (ξ_i, ξ_j).

Example 1.6 Gaussian random vector

An n-dimensional **gaussian** $\vec{r.v.}$ $\xi = (\xi_1, \ldots, \xi_n)$ is a real $\vec{r.v.}$ having probability density function p_X defined by

$$p_X(x_1, \ldots, x_n) = (2\pi)^{-n/2} (\det \Gamma_{XX})^{-1/2} \exp\{-\tfrac{1}{2}(A - M_X)^T \Gamma_{XX}^{-1}(A - M_X)\} \quad (2.41)$$

X being the random matrix associated with the $\vec{r.v.}$ ξ (cf. Eqn (2.2)) and Γ_{XX} its covariance matrix (cf. Eqn (2.33)); here we put

$$M_X = E[X] \qquad A^T = [x_1 \ldots x_n]$$

The probability law of a gaussian r.v. $\vec{\xi}$ is thus defined by its covariance matrix Γ_{XX} and its matrix of means M_X. It is thus defined by the moments of order 1 and 2 associated with its components.

The characteristic function Θ_X of the r.v. $\vec{\xi}$ is obtained from applying Eqn (2.38). We get

$$\Theta_X(u_1, \ldots, u_n) = \exp(iU^T M_X - \tfrac{1}{2} U^T \Gamma_{XX} U) \qquad (2.42)$$

The characteristic function has a simpler expression than the probability density. It is often preferred as a characterization of a gaussian r.v., as in the following example.

Example 1.7 Linear transformation of a gaussian random vector

Let $\xi = (\xi_1, \ldots, \xi_n)$ be a real n-dimensional r.v. and $\eta = (\eta_1, \ldots, \eta_m)$ a real m-dimensional r.v. obtained by linear transformation from the r.v. $\vec{\xi}$ and let X and Y be the associated random matrices (cf. Eqn (2.2)). Let us write

$$Y = BX + C$$

B being a real $m \times n$ matrix and C a real $m \times 1$ matrix. It is easy to verify that the matrix of means M_Y and the covariance matrix Γ_{YY} may be written

$$M_Y = BM_X + C \qquad \Gamma_{YY} = B\Gamma_{XX} B^T \qquad (2.43)$$

We shall show that, if the r.v. $\vec{\xi}$ is gaussian, the r.v. $\vec{\eta}$ is also gaussian. For this it suffices to verify that the characteristic function Θ_Y is of the form of Eqn (2.42). We get from Eqn (2.38)

$$\Theta_Y(U) = E[\exp\{iU^T(BX + C)\}]$$
$$= \exp(iU^T C)\, E[\exp\{i(B^T U)^T X\}]$$

It follows then from Eqn (2.42) that, since the r.v. $\vec{\xi}$ is gaussian,

$$E[\exp\{i(B^T U)^T X\}] = \Theta_X(B^T U)$$
$$= \exp\{iU^T BM_X - \tfrac{1}{2} U^T B\Gamma_{XX} B^T U\}$$

Hence it follows that

$$\Theta_Y(U) = \exp\{iU^T(BM_X + C) - \tfrac{1}{2} U^T(B\Gamma_{XX} B^T) U\}$$

The stated result is thus demonstrated, having taken account of Eqn (2.43). It has considerable importance. It is always possible, in particular, to transform an arbitrary gaussian r.v. into a gaussian r.v. whose components are pairwise uncorrelated. For this it suffices to take $B = U$, U being a unitary matrix which diagonalizes the matrix Γ_{XX} (cf. Eqn (2.35)).

Random vectors of finite dimension 81

1.5. NOTION OF INDEPENDENCE

1.5.1. Independence of a family of random variables

1.5.1.1. Definition
Let $(\xi_i, i \in \mathbb{I})$ be an arbitrary family, finite or infinite, of r.v.s defined over one and the same probability space and let $(\mathcal{B}_{\xi_i}, i \in \mathbb{I})$ be the family of sub-Borel fields generated by the r.v.s ξ_i.

The family (ξ_i) is independent when the family $(\mathcal{B}_{\xi_i})$ is independent (cf. Chapter 1, Section 1.4.3).

Example 1.8 Independent sequences of indicator random variables

Let $(A_n, n \in \mathbb{N})$ be a sequence of events and $(I_{A_n}, n \in \mathbb{N})$ the sequence of indicator r.v.s. The sub-Borel field generated by the r.v. I_{A_n} is the sub-Borel field generated by the event A_n (cf. Chapter 1, Example 2.1). The sequence (I_{A_n}) is therefore independent if and only if the sequence (A_n) is independent (cf. Chapter 1, Section 1.4.3.2). This is expressed by Eqn (1.22). This formula can also be written, by application of Eqns (1.36) and (1.37), as

$$\forall n \in \mathbb{N} \qquad E[I_{A_0} \ldots I_{A_n}] = E[I_{A_0}] \ldots E[I_{A_n}] \qquad (2.44)$$

To sum up, the sequence (A_n) and the sequence (I_{A_n}) are independent if and only if Eqn (2.44) is satisfied.

1.5.1.2. Property: Family of random variables function of an independent family

Let (ξ_i) be an independent family of r.v.s and $(g_i, i \in \mathbb{I})$ a family of Borel functions. The family (η_i) of r.v.s $\eta_i = g_i \circ \xi_i$ is independent.

In fact let $\mathcal{B}_{\eta_i}$ be the sub-Borel field generated by the r.v. η_i. It is contained within the sub-Borel field $\mathcal{B}_{\xi_i}$ generated by the r.v. ξ_i (cf. Chapter 1, Section 2.1.2). It then follows immediately that the family $(\mathcal{B}_{\eta_i})$ is independent and, consequently, that the family (η_i) is independent.

1.5.1.3. Property: Independence of a sequence of random variables

Let $(\xi_n, n \in \mathbb{N}^*)$ be a sequence of real r.v.s and let $\mu_{\xi_1 \ldots \xi_n}$ be the probability law of the real $\vec{\text{r.v.}}$ $(\xi_1, \ldots, \xi_n)$. The sequence (ξ_n) is independent if and only if the probability laws of the $\vec{\text{r.v.s}}$ $(\xi_1, \ldots, \xi_n)$ are the (tensor) products of the probability laws of their components:

$$\forall n \in \mathbb{N}^* \qquad \mu_{\xi_1 \ldots \xi_n} = \mu_{\xi_1} \otimes \ldots \otimes \mu_{\xi_n} \qquad (2.45)$$

When the probability laws $\mu_{\xi_1 \ldots \xi_n}$ are continuous, the preceding relation may be written

$$\forall n \in \mathbb{N}^* \qquad p_{\xi_1 \ldots \xi_n}(x_1, \ldots, x_n) = p_{\xi_1}(x_1) \ldots p_{\xi_n}(x_n) \qquad (2.46)$$

Suppose, to start with, that the sequence (ξ_n) is independent. It follows then

82 Random vectors

from Eqns (2.4) and (1.22) that

$$\mu_{\xi_1 \ldots \xi_n}(\beta_1 \times \ldots \times \beta_n) = \mu_{\xi_1}(\beta_1) \ldots \mu_{\xi_n}(\beta_n)$$

This equality is satisfied for every Borel set of $\mathbb{R}^n$ of the form $\beta_1 \times \ldots \times \beta_n$. The formula (2.45) follows from this by the uniqueness of the product measure (cf. Chapter 6, Section 3.2.1). It can be shown, conversely, that if Eqn (2.45) is satisfied the sequence (ξ_n) is independent.

1.5.1.4. Property: Moments associated with an independent sequence of random variables

Let $\{\xi_1, \ldots, \xi_n\}$ be a finite independent sequence of real r.v.s and let $\mu_{\xi_1 \ldots \xi_n}$ be the probability law of the $\vec{\text{r.v.}}$ $(\xi_1, \ldots, \xi_n)$. Consider the real r.v. $\xi_1^{k_1} \times \ldots \times \xi_n^{k_n}$, $k_1 \in \mathbb{N}, \ldots, k_n \in \mathbb{N}$. Its mean value is

$$E[\xi_1^{k_1} \times \ldots \times \xi_n^{k_n}] = E[\xi_1^{k_1}] \ldots E[\xi_n^{k_n}] \tag{2.47}$$

The mean value of the product is therefore the product of the mean values.

To demonstrate this important property it is sufficient to generalize Eqns (1.71) and (6.55) to the case of a random function of an n-dimensional $\vec{\text{r.v.}}$ by writing

$$E[\xi_1^{k_1} \times \ldots \times \xi_n^{k_n}] = \int_{\mathbb{R}^n} x_1^{k_1} \ldots x_n^{k_n} \, d\mu_{\xi_1 \ldots \xi_n}(x_1, \ldots, x_n)$$

Fubini's theorem (cf. Chapter 6, Section 4.2.2) and Eqn (2.45) then enable us to write

$$\int_{\mathbb{R}^n} x_1^{k_1} \ldots x_n^{k_n} \, d\mu_{\xi_1 \ldots \xi_n}(x_1, \ldots, x_n) = \int_{\mathbb{R}} x_1^{k_1} \, d\mu_{\xi_1}(x_1) \times \ldots \times \int_{\mathbb{R}} x_n^{k_n} \, d\mu_{\xi_n}(x_n)$$

The equality (2.47) follows from this.

Example 1.9 The multinomial law

Let $\{\xi_1, \ldots, \xi_n\}$ be a finite independent sequence of real discrete r.v.s taking their values in the set $\{a_1, \ldots, a_m\}$ and such that

$$\forall i \in [1, n] \quad \forall j \in [1, m] \quad P\{\xi_i = a_j\} = p_j \quad \sum_{j=1}^{m} p_j = 1$$

The $\vec{\text{r.v.}}$ $\xi = (\xi_1, \ldots, \xi_n)$ is discrete and takes m^n values in $\mathbb{R}^n$. Each value can be considered as the result of n independent and identical experiments, at the end of which we observe the values taken by the r.v.s ξ_i.

The independence of the sequence $\{\xi_1, \ldots, \xi_n\}$ implies that

$$P(\{\xi_1 = a_{j_1}\} \cap \ldots \cap \{\xi_n = a_{j_n}\}) = p_{j_1} \ldots p_{j_n}$$

The event $\{\xi_i = a_{j_i}\}$ belongs in fact to the Borel field $\mathscr{B}_{\xi_i}$ generated by the r.v. ξ_i. Let $\eta = (\eta_1, \ldots, \eta_m)$ be the real $\vec{\text{r.v.}}$ such that the r.v. η_j is the number of r.v.s in the sequence $\{\xi_1, \ldots, \xi_n\}$ taking the same value a_j. The r.v.s η_j are not independent since they are connected by the relation $\eta_1 + \ldots + \eta_m = n$. We seek the probability

$$q_{k_1 \ldots k_m} = P\left(\bigcap_{j=1}^{m} \{\eta_j = k_j\}\right) \quad \sum_{j=1}^{m} k_j = n$$

This is the probability that k_1 r.v.s of the sequence $\{\xi_1, \ldots, \xi_n\}$ take the value $a_1, \ldots,$ and that k_m r.v.s of the same sequence take the value a_m. It is easy to see that the event considered is a union of $n!/k_1! \ldots k_m!$ pairwise incompatible events, each having the same probability $p_1^{k_1} \ldots p_m^{k_m}$ in view of the independence of the sequence $\{\xi_1, \ldots, \xi_n\}$. It follows then that

$$q_{k_1 \ldots k_m} = \frac{n!}{k_1! \ldots k_m!} p_1^{k_1} \ldots p_m^{k_m} \quad \sum_{j=1}^{m} k_j = n \qquad (2.48)$$

This formula expresses the **multinomial law**. It obviously reduces to the binomial law when $m = 2$ (cf. Chapter 1, Example 2.6).

1.5.1.5. Property: Independence and non-correlation

Let $(\xi_n, n \in \mathbb{N})$ be an independent sequence of real r.v.s. The r.v.s ξ_i are pairwise uncorrelated. We say also that the sequence is uncorrelated.

In fact, let ξ_i and ξ_j, $i \neq j$, be two of the r.v.s. They are independent and it follows from Eqns (2.25) and (2.47) that

$$C[\xi_i \xi_j] = E[\xi_i]E[\xi_j^*] - E[\xi_i]E[\xi_j^*] = 0$$

The r.v.s ξ_i and ξ_j are therefore uncorrelated.

The converse is not true in the general case. It is true, however, in the case of a gaussian $\vec{r.v.}$ (see Example 1.10).

Example 1.10 Gaussian random sequence

Let $(\xi_n, n \in \mathbb{N})$ be an uncorrelated sequence. It is gaussian when the finite sequences $\{\xi_0, \ldots, \xi_n\}$, $n \in \mathbb{N}$, define gaussian $\vec{r.v.}$s.

Let $\xi = (\xi_0, \ldots, \xi_n)$ be a gaussian $\vec{r.v.}$ and X be its associated random matrix. The covariance matrix Γ_{XX} is diagonal since the r.v.s ξ_i are pairwise uncorrelated. Hence we have

$$\Gamma_{XX} = \begin{bmatrix} \sigma_{\xi_0}^2 & & 0 \\ & \ddots & \\ 0 & & \sigma_{\xi_n}^2 \end{bmatrix}$$

It is easy to verify that the probability density p_X may be written (cf. Eqns (2.41) and (1.46))

$$p_X(x_0, \ldots, x_n) = p_{\xi_0}(x_0) \ldots p_{\xi_n}(x_n)$$

Since this equality is satisfied for all $n \in \mathbb{N}$, the sequence (ξ_n) is independent (cf. Eqn (2.46)).

84 Random vectors

Independence and non-correlation are equivalent for a gaussian sequence. This property is very useful in view of the fact that non-correlation is in general easier to establish than independence.

1.5.1.6. Property: Sum of two independent random variables

Let ξ_1 and ξ_2 be two real independent r.v.s and η the r.v. $\eta = \xi_1 + \xi_2$. The probability law of the r.v. η is the convolution of the probability laws of the r.v.s ξ_1 and ξ_2:

$$\mu_\eta = \mu_{\xi_1} * \mu_{\xi_2} \qquad (2.49)$$

The characteristic functions and the moment-generating functions are connected by the relations (cf. Chapter 1, Section 2.5)

$$\Theta_\eta = \Theta_{\xi_1} \Theta_{\xi_2} \qquad G_\eta = G_{\xi_1} G_{\xi_2} \qquad (2.50)$$

The first relation results from the definitions. In fact the probability law of the r.v. η is the image of the probability law $\mu_{\xi_1 \xi_2}$ of the r.v. (ξ_1, ξ_2) through the borelian mapping from $(\mathbb{R}^2, \mathcal{R}^2)$ into $(\mathbb{R}, \mathcal{R})$ such that $(x_1, x_2) \mapsto x_1 + x_2$ (cf. Section 1.2.3). Since the measure $\mu_{\xi_1 \xi_2}$ is the product measure $\mu_{\xi_1} \otimes \mu_{\xi_2}$ the image measure, by definition, is the convolution of the measures μ_{ξ_1} and μ_{ξ_2} (cf. Chapter 6, Section 3.2.4).

The relations (2.50) are obtained through Fourier (Θ_η) and Laplace (G_η) transformation from Eqn (2.49) (cf. Chapter 6, Section 4.3.3).

Example 1.11 Sum of two independent gaussian random variables

Let ξ_1 and ξ_2 be two independent gaussian r.v.s and let η be the r.v. $\eta = \xi_1 + \xi_2$. The characteristic function Θ_η of the r.v. η may be obtained from Eqns (2.50) and (1.87). We get

$$\Theta_\eta(u) = \exp\left\{i(m_1 + m_2)u - \frac{\sigma_1^2 + \sigma_2^2}{2} u^2\right\} \qquad (2.51)$$

It follows from this that the r.v. η is gaussian. Its mean value is the sum of the mean values and its variance is the sum of the variances, but this is always true for any independent r.v.s (cf. Example 1.4).

1.5.2. Conditional independence of a family of real random variables

1.5.2.1. Definition

Let $(\Omega, \mathcal{A}, P)$ be a probability space, $\mathcal{B} \subset \mathcal{A}$ a sub-Borel field and $(A_n, n \in \mathbb{N})$ a sequence of events. The sequence (A_n) is conditionally independent with respect to the sub-Borel field $\mathcal{B}$ when

$$\forall n \in \mathbb{N} \qquad E[I_{A_0} \times \ldots \times I_{A_n} | \mathcal{B}] = E[I_{A_0} | \mathcal{B}] \ldots E[I_{A_n} | \mathcal{B}] \qquad (2.52)$$

This definition generalizes Eqn (2.44) by involving the conditional expectations. It should be noted that Eqn (2.52) is an a.s. equality between two r.v.s whereas Eqn (2.44) is an equality between two numbers.

Example 1.12 Conditioning on a discrete random variable

Suppose that the conditioning Borel field is generated by the discrete r.v. η considered in Chapter 1, Example 2.19. It follows from Eqn (1.97), by application of Eqns (1.36) and (1.12), that

$$\forall A \in \mathcal{A} \quad E[I_A | \mathcal{B}] = \sum_{k \in \mathbb{N}} P(A | B_k) I_{B_k}$$

This enables us to write Eqn (2.52) in the following equivalent form, taking account of the fact that the sequence of events (B_k) is a partition of the set Ω:

$$\forall n \in \mathbb{N}$$

$$\sum_{k \in \mathbb{N}} P(A_0 \cap \ldots \cap A_n | B_k) I_{B_k} = \sum_{n \in \mathbb{N}} P(A_0 | B_k) \ldots P(A_n | B_k) I_{B_k}$$

It follows from this that the sequence (A_n) is conditionally independent with respect to the sub-Borel field $\mathcal{B}$ when it is conditionally independent with respect to each of the events of the sequence (B_k) (cf. Chapter 1, Section 1.4.1.1).

1.5.2.2. Definitions

Let $(\xi_i, i \in \mathbb{I})$ be an arbitrary family, finite or infinite, of real r.v.s defined over a probability space $(\Omega, \mathcal{A}, P)$ and let $(\mathcal{B}_{\xi_i}, i \in \mathbb{I})$ be the family of sub-Borel fields generated by the r.v.s ξ_i respectively. Let $\mathcal{B} \subset \mathcal{A}$ be a sub-Borel field.

The family $(\mathcal{B}_{\xi_i})$ is conditionally independent with respect to the sub-Borel field $\mathcal{B}$ when every finite sequence of events extracted from every family $(B_i, B_i \in \mathcal{B}_{\xi_i}, i \in \mathbb{I})$ is conditionally independent with respect to the sub-Borel field $\mathcal{B}$ (cf. Chapter 1, Section 1.4.3).

The family (ξ_i) is conditionally independent with respect to the sub-Borel field $\mathcal{B}$ when the family $(\mathcal{B}_{\xi_i})$ is conditionally independent with respect to the sub-Borel field $\mathcal{B}$.

1.5.3. Independence of a family of random vectors

1.5.3.1. Definitions

Let $(\vec{\xi}_i, i \in \mathbb{I})$ be an arbitrary family, finite or infinite, of r.v.s of arbitrary dimensions defined over one and the same probability space and let $(\mathcal{B}_{\vec{\xi}_i}, i \in \mathbb{I})$ be the family of sub-Borel fields generated by the r.v.s $\vec{\xi}_i$ respectively.

The family $(\vec{\xi}_i)$ is independent when the family $(\mathcal{B}_{\vec{\xi}_i})$ is independent (cf. Chapter 1, Section 1.4.3).

1.5.3.2. Properties

The definition of independence of a family of $\vec{r.v.}$s is formally the same as that of independence of a family of r.v.s. The same is true for the definition of conditional independence with respect to a sub-Borel field.

In the same way, the properties stated in Sections 1.5.1.2 and 1.5.1.3 are formally applicable to $\vec{r.v.}$s. We shall therefore not repeat them.

1.5.3.3. Property: Independence of two random vectors

Let $\xi = (\xi_1, \ldots, \xi_m)$ and $\eta = (\eta_1, \ldots, \eta_n)$ be two $\vec{r.v.}$s such that the finite sequence $\{\xi_1, \ldots, \xi_m, \eta_1, \ldots, \eta_n\}$ is independent. The $\vec{r.v.}$s ξ and η are independent.

To prove this it suffices to apply Eqn (2.45) and to verify that the measure $\mu_{\xi\eta}$ over the space $(\mathbb{R}^{m+n}, \mathscr{R}^{m+n})$ is the product measure $\mu_\xi \otimes \mu_\eta$, μ_ξ and μ_η being respectively the probability laws of the $\vec{r.v.}$s ξ and η. It follows from Eqn (2.45) that

$$\mu_{\xi\eta} = \mu_{\xi_1 \ldots \xi_m \eta_1 \ldots \eta_n} = \mu_{\xi_1} \otimes \ldots \otimes \mu_{\xi_m} \otimes \mu_{\eta_1} \otimes \ldots \otimes \mu_{\eta_n}$$

Consider the Borel set

$$\beta = \beta_1 \times \ldots \times \beta_m \times \gamma_1 \times \ldots \times \gamma_n$$

of $\mathbb{R}^{m+n}$, the β_i and γ_j each being Borel sets of $\mathbb{R}$. It follows from the definition of the product measure that

$$\mu_\xi \otimes \mu_\eta(\beta) = \mu_\xi(\beta_1 \times \ldots \times \beta_m)\mu_\eta(\gamma_1 \times \ldots \times \gamma_n)$$

It follows then, since the sub-sequences $\{\xi_1, \ldots, \xi_m\}$ and $\{\eta_1, \ldots, \eta_n\}$ are independent, that

$$\mu_\xi \otimes \mu_\eta(\beta) = \mu_{\xi_1}(\beta_1) \ldots \mu_{\xi_m}(\beta_m)\mu_{\eta_1}(\gamma_1) \ldots \mu_{\eta_n}(\gamma_n)$$

This equality holds for all Borel sets of the type considered. The uniqueness of the product measure therefore enables us to conclude that the measures $\mu_{\xi\eta}$ and $\mu_\xi \otimes \mu_\eta$ are identical. The $\vec{r.v.}$s ξ and η are thus indeed independent.

1.5.3.4. Property: Conditioning of a random variable by an independent random vector

Let ξ be a real r.v. with finite mean value and let η be a real n-dimensional $\vec{r.v.}$, independent of the r.v. ξ and defined over the same probability space as ξ. Then

$$E[\xi | \eta] \underset{\text{a.s.}}{=} E[\xi] \qquad (2.53)$$

Consider, in fact, an event $B \in \mathscr{B}_\eta$. There exists a Borel set $\beta \in \mathscr{R}^n$ such that

$$B = \eta^{-1}(\beta) \Leftrightarrow I_B = I_\beta \circ \eta$$

The r.v. I_B is thus a function of the $\vec{r.v.}$ η, and consequently the r.v.s I_B and ξ are independent (cf. Section 1.5.1.2). This enables us to write

$$\forall B \in \mathscr{B}_\eta \quad E[\xi I_B] = E[\xi]E[I_B] = E[E[\xi]I_B]$$

The property stated follows from this, since the constant r.v. $E[\xi]$ satisfies the general definition (1.93).

Exercises

Exercise 1.1 Linear transformation of a random vector
Let $\vec{\xi} = (\xi_1, \ldots, \xi_n)$ be an n-dimensional r.v., η the r.v. $\eta = \xi_1 + \ldots + \xi_n$, and X and Y the associated random matrices (cf. Eqn (2.2)).
(1) Write the matrix Y in the form $Y = AX$. Hence deduce that the covariance of the r.v. η is the sum of the rows of the covariance matrix Γ_{XX} and obtain Eqn (2.29).
(2) Deduce from the above that the sub-Borel field $\mathscr{B}_\eta$ generated by the r.v. η is contained in the sub-Borel field $\mathscr{B}_\xi$ generated by the $\vec{\text{r.v.}}\ \xi$ (cf. Chapter 6, Exercise 2.4).

Exercise 1.2 Diagonalization of a covariance matrix
Let $\vec{\xi} = (\xi_1, \xi_2)$ be a complex two-dimensional r.v. and let X be the associated random matrix. Put

$$V[\xi_1] = \sigma_1^2 \quad V[\xi_2] = \sigma_2^2 \quad r_{\xi_1 \xi_2} = r_{12} \quad k = \frac{|r_{12}|^2 \sigma_1 \sigma_2}{\lambda_1 - \sigma_1^2}$$

λ_1 being an eigenvalue of the covariance matrix Γ_{XX}.
Let U be the matrix

$$U = \frac{1}{(1+k^2)^{1/2}} \begin{bmatrix} \dfrac{r_{12}^* \sigma_1 \sigma_2}{\lambda_1 - \sigma_2^2} & 1 \\ -1 & \dfrac{r_{12} \sigma_1 \sigma_2}{\lambda_1 - \sigma_1^2} \end{bmatrix}$$

(1) Verify that the matrix U is unitary (cf. Section 1.3.2.2).
(2) Let η be the complex two-dimensional r.v. whose associated random matrix is $Y = UX$. Verify that the covariance matrix of the r.v. η is the diagonal matrix

$$\Gamma_{YY} = \begin{bmatrix} \lambda_1 & 0 \\ 0 & \lambda_2 \end{bmatrix}$$

λ_2 being the other eigenvalue of the matrix Γ_{XX}. (Apply Eqn (2.35).)

Exercise 1.3 Diagonalization of a covariance matrix

Let $\vec{\xi} = (\xi_1, \xi_2)$ be a two-dimensional r.v. and X the associated random matrix. Let $\vec{\eta} = (\eta_1, \eta_2)$ and $\vec{\zeta} = (\zeta_1, \zeta_2)$ be two-dimensional r.v.s whose associated random matrices are $Y = AX$ and $Z = BZ$ respectively, with

$$A = \begin{bmatrix} 1 & 0 \\ -r_{12}\dfrac{\sigma_2}{\sigma_1} & 1 \end{bmatrix} \qquad B = \begin{bmatrix} \dfrac{1}{\sigma_1} & \dfrac{1}{\sigma_2} \\ \dfrac{1}{\sigma_1} & -\dfrac{1}{\sigma_2} \end{bmatrix}$$

(1) Verify that the matrices A and B are not unitary (cf. Section 1.3.2.2).

(2) Suppose that the r.v. $\vec{\xi}$ is real. Verify that the covariance matrices Γ_{YY} and Γ_{ZZ} are diagonal and may be written

$$\Gamma_{YY} = \begin{bmatrix} \sigma_1^2 & 0 \\ 0 & \sigma_2^2(1-r_{12}) \end{bmatrix} \qquad \Gamma_{ZZ} = \begin{bmatrix} 2(1+r_{12}) & 0 \\ 0 & 2(1-r_{12}) \end{bmatrix}$$

Verify that the eigenvalues of the matrices Γ_{YY} and Γ_{ZZ} are not the eigenvalues of the matrix Γ_{XX}.

(3) Suppose now that the r.v. $\vec{\xi}$ is complex. Verify that the matrices Γ_{YY} and Γ_{ZZ} are not diagonal.

Exercise 1.4 Gaussian random vector

Let $\vec{\zeta}$ be an $(m+n)$-dimensional gaussian r.v. $\vec{\zeta} = (\xi_1, \ldots, \xi_m, \eta_1, \ldots, \eta_n)$, let $\vec{\xi}$ and $\vec{\eta}$ be the r.v.s. $\vec{\xi} = (\xi_1, \ldots, \xi_m)$ and $\vec{\eta} = (\eta_1, \ldots, \eta_n)$ and let Z, X and Y be the respective random matrices associated with these r.v.s. Then

$$Z^T = (X^T, Y^T) \qquad X^T = (\xi_1 \ldots \xi_m) \qquad Y^T = (\eta_1 \ldots \eta_n)$$

(1) The probability density $p_{\xi\eta}$ of the r.v. $\vec{\zeta}$ is given by Eqn (2.41). Verify that

$$M_Z = \begin{pmatrix} M_X \\ M_Y \end{pmatrix} \qquad \Gamma_{ZZ} = \begin{pmatrix} \Gamma_{XX} & \Gamma_{XY} \\ \Gamma_{YX} & \Gamma_{YY} \end{pmatrix} \qquad (2.54)$$

(2) Show that the r.v.s $\vec{\xi}$ and $\vec{\eta}$ are gaussian. (Use suitable linear operators to transform the r.v. $\vec{\zeta}$ (cf. Example 1.7).)

(3) The conditional probability density $p_{\xi|\eta}$ of the r.v. $\vec{\xi}$ conditioned by the r.v. $\vec{\eta}$ is written (cf. Eqn (2.12))

$$p_{\xi|\eta}(x_1, \ldots, x_m | y_1, \ldots, y_n) = \frac{p_{\xi\eta}(x_1, \ldots, x_m, y_1, \ldots, y_n)}{p_\eta(y_1, \ldots, y_n)} \qquad (2.55)$$

Suppose that the matrix Γ_{YY} is invertible and show that the density $p_{\xi|\eta}$ is of the gaussian form (2.41) with mean matrix M and covariance matrix Γ such that

$$M = M_X - \Gamma_{XY}\Gamma_{YY}^{-1}(Y - M_Y)$$
$$\Gamma = \Gamma_{XX} - \Gamma_{XY}\Gamma_{YY}^{-1}(Y - M_Y) \qquad (2.56)$$

(4) The probability density $p_{\xi|\eta}$ enables us to determine the marginal conditional probability densities $p_{\xi_i|\eta}$ of the r.v.s ξ_i conditioned by the r.v. η and consequently the values taken by the conditional expectation r.v. $E[\xi_i|\eta]$ (cf. Eqn (2.13)). The conditional expectation $\vec{\text{r.v.}}$ $E[\vec{\xi}|\eta]$ is by definition the $\vec{\text{r.v.}}$ whose components are the r.v.s $E[\xi_i|\eta]$.

Deduce from this that the $\vec{\text{r.v.}}$ $E[\vec{\xi}|\eta]$ is the m-dimensional gaussian $\vec{\text{r.v.}}$ whose associated random matrix is written

$$E[X|Y] = M_X - \Gamma_{XY}\Gamma_{YY}^{-1}(Y - M_Y) \tag{2.57}$$

This generalizes the result expressed by Eqn (2.18) to the case of an $\vec{\text{r.v.}}$

Exercise 1.5 Independent Poisson laws

Let there be two independent Poisson r.v.s with parameters λ_1 and λ_2. Show that their sum is a Poisson r.v. with parameter $\lambda_1 + \lambda_2$. (Use Eqns (2.50) and (1.115).)

Exercise 1.6 Independent Bernoulli laws

Let there be two independent r.v.s having binomial probability laws with parameters (p, n_1) and (p, n_2) (cf. Chapter 1, Example 2.6). Show that their sum has a binomial probability law with parameter $(p, n_1 + n_2)$. (Use Eqns (2.50) and (1.113).)

Exercise 1.7 Gaussian random vector and polar coordinates

Let $\vec{\xi}$ be a gaussian $\vec{\text{r.v.}}$ $\xi = (\xi_1, \xi_2)$, the r.v.s ξ_1 and ξ_2 being independent, centred and of the same variance σ^2. Let $g \circ \xi = (\rho, \vartheta)$ be the r.v. such that

$$\xi_1 = \rho \cos \vartheta \qquad \xi_2 = \rho \sin \vartheta \tag{2.58}$$

The r.v.s ρ and ϑ are the polar coordinates that correspond to the cartesian coordinates ξ_1 and ξ_2 (cf. Example 1.1).

(1) Show that the r.v.s ρ and ϑ are independent and that their marginal probability densities are respectively the functions

$$p_\rho(r) = Y(r)\frac{r}{\sigma^2}\exp\left(-\frac{r^2}{2\sigma^2}\right) \tag{2.59}$$

$$p_\vartheta(\vartheta) = \frac{1}{2\pi} \qquad \vartheta \in [a, a + 2\pi] \tag{2.60}$$

The r.v. ρ has a Rayleigh probability law (cf. Chapter 1, Example 2.9) and the r.v. ϑ is uniformly distributed over the interval $[a, a + 2\pi]$, a being an arbitrary number. (Apply Eqn (2.11).)

(2) Show that the probability density of the r.v. $\eta = \xi_1^2 + \xi_2^2$ is the function

$$p_\eta(y) = Y(y)\frac{1}{2\sigma^2}\exp\left(-\frac{y}{2\sigma^2}\right) \tag{2.61}$$

(Use Eqn (1.91).)

Exercise 1.8 Gaussian random vector and polar coordinates

This exercise continues the previous one. We suppose now that the r.v.s ξ_1 and ξ_2 are not centred and have respective mean values of

$$E[\xi_1] = m_1 \qquad E[\xi_2] = m_2$$

(1) Show that the marginal probability density of the r.v. ρ is the function

$$p_\rho(r) = Y(r) \frac{r}{\sigma^2} \exp\left(-\frac{r^2 + A^2}{2\sigma^2}\right) I_0\left(\frac{rA}{\sigma^2}\right) \tag{2.62}$$

where $A = (m_1^2 + m_2^2)^{1/2}$. (Apply Eqn (2.11) and use the definition

$$I_0(x) = \frac{1}{2\pi} \int_a^{a+2\pi} \exp(x \sin \vartheta)\, d\vartheta \tag{2.63}$$

of the modified Bessel function I_0.)

(2) Show that the marginal probability density of the r.v. ϑ is the function

$$p_\vartheta(\vartheta) = \frac{1}{2\pi} \exp\left(-\frac{A^2}{2\sigma^2}\right) + \frac{A \cos(\vartheta - \vartheta_0)}{(2\pi\sigma^2)^{1/2}} \exp\left\{-\frac{A^2 \sin^2(\vartheta - \vartheta_0)}{2\sigma^2}\right\}$$

$$\times \left[\frac{1}{2} + \Phi\left\{\frac{A \cos(\vartheta - \vartheta_0)}{\sigma}\right\}\right] \qquad -\pi \leqslant \vartheta - \vartheta_0 \leqslant \pi \tag{2.64}$$

putting $A = (m_1^2 + m_2^2)^{1/2}$, Φ being the function defined by Eqn (1.64) (use Eqn (2.11)).

The graph of the function p_ϑ is represented in Fig. 2.3 for different values of the parameter $\alpha = A^2/2\sigma^2$.

(3) Verify that the r.v.s ρ and ϑ are independent if and only if $A = 0$. (Use Eqns (2.10) and (2.11).)

Exercise 1.9 Distribution functions

Let $\{\xi_1, \ldots, \xi_n\}$ be an independent sequence of real r.v.s having the same probability law and the same distribution function F_ξ and the real r.v.s

$$\eta = \sup\{\xi_1, \ldots, \xi_n\} \qquad \zeta = \inf\{\xi_1, \ldots, \xi_n\}$$

(1) Let F_η be the distribution function of the r.v. η. Show that

$$F_\eta = (F_\xi)^n \tag{2.65}$$

(Start from the equality

$$\{\eta < y\} = \bigcap_{i=1}^n \{\xi_i < y\}$$

and apply Eqn (1.22).)

(2) Let F_ζ be the distribution function of the r.v. ζ. Show that

$$F_\zeta = 1 - (1 - F_\xi)^n \tag{2.66}$$

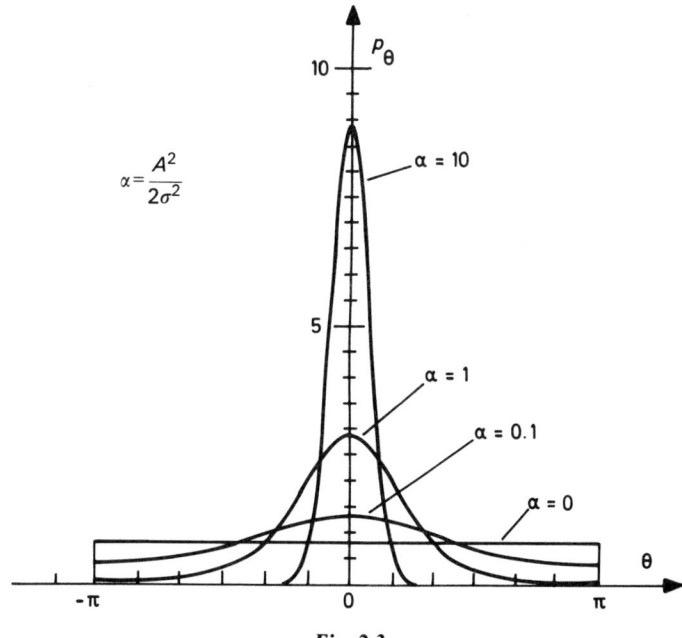

Fig. 2.3

(Start from the equality

$$\{\zeta < y\} = \bigcup_{i=1}^{n} \{\xi_i < y\}$$

and apply Eqn (1.30).)

Exercise 1.10 Exponential law

This exercise complements Exercise 1.9. We suppose now that the r.v.s ξ_i have the same exponential probability law with density

$$p_\xi(x) = Y(x)\, a \exp(-ax) \qquad a \in \mathbb{R}_+^*$$

Show that the probability densities of the r.v.s η and ζ are respectively the functions

$$\begin{aligned} p_\eta(y) &= Y(y)\, na \exp(-ay)\{1 - \exp(-ay)\}^{n-1} \\ p_\zeta(z) &= Y(z)\, na \exp(-naz) \end{aligned} \qquad (2.67)$$

It should be noted that the probability law of r.v. ζ is exponential and has parameter na.

Comment: when the r.v.s ξ_i are lifetimes of identical components put into service at the instant 0, the r.v. ζ is the time to the first breakdown. The probability law of this time is also exponential.

Exercise 1.11 Two-dimensional random vector

Let ξ_1 and ξ_2 be two positive independent r.v.s, having as probability densities the functions p_{ξ_1} and p_{ξ_2}. Let η_1 and η_2 be the r.v.s $\eta_1 = \xi_1$ and $\eta_2 = \xi_1 + \xi_2$.

(1) Show that the probability density of the r.v. $\vec{\eta} = (\eta_1, \eta_2)$ is the function

$$p_{\eta_1\eta_2}(y_1, y_2) = p_{\xi_1}(y_1) p_{\xi_2}(y_2 - y_1) Y(y_1) Y(y_2 - y_1) \quad (2.68)$$

Y denoting the Heaviside function (cf. (6.22)).

(2) Hence deduce

$$\forall y > 0 \quad P\{\eta_1 \leq y \leq \eta_2\} = \int_0^y p_{\xi_1}(y_1) \, dy_1 \int_{y-y_1}^\infty p_{\xi_2}(y_2) \, dy_2$$

$$= Y(y) * p_{\xi_1}(y) * \{\delta(y) - p_{\xi_2}(y)\} \quad (2.69)$$

The symbol $*$ denotes the convolution product in $\mathscr{D}'_+$.

(3) Deduce from Eqn (2.68) that

$$\forall y > 0 \quad P\{\eta_2 < y\} = \int_0^y p_{\xi_1}(y - y_1) F_{\xi_2}(y_1) \, dy_1$$

$$= Y(y) * p_{\xi_1}(y) * p_{\xi_2}(y) \quad (2.70)$$

F_{ξ_2} being the distribution function of the r.v. ξ_2.

(4) Deduce from Eqn (2.68) that the conditional probability density of the r.v. η_2 with respect to the r.v. η_1 may be written

$$p_{\eta_2|\eta_1}(y_2|y_1) = Y(y_2 - y_1) p_{\xi_2}(y_2 - y_1) \quad (2.71)$$

Comment: the expressions (2.69) and (2.70) are particularly interesting when the Laplace transform is used.

Exercise 1.12 Exponential law

This exercise is a sequel to Exercise 1.11. We suppose now that the r.v.s ξ_1 and ξ_2 have the same exponential probability law, with density

$$p_\xi(x) = Y(x) a \exp(-ax) \qquad a \in \mathbb{R}^*_+$$

Show that

$$P\{\eta_1 < y < \eta_2\} = Y(y) a y \exp(-ay) \quad (2.72)$$

(Apply the Laplace transform to (2.69) and use Appendix 3.)

Exercise 1.13 Gaussian random variables

Let ξ and η be two gaussian r.v.s, independent, centred, and with the same variance σ^2. Let φ_0 be an r.v., independent of the r.v.s ξ and η, uniformly distributed over the interval $[-\pi, \pi]$.

(1) Show that the characteristic function Θ_ζ of the r.v. $\zeta = \xi \cos \varphi_0$ may be

written

$$\Theta_\zeta(u) = I_0\left(\frac{u^2\sigma^2}{4}\right)\exp\left(\frac{u^2\sigma^2}{2}\right)$$

Hence deduce that the r.v. ζ is not gaussian. It is only conditionally gaussian with respect to the r.v. φ_0. (Use the expansion

$$\exp(ia\cos\varphi) = \sum_{n\in\mathbb{Z}} i^n J_n(a)\exp(in\varphi)$$

and the integral

$$\int_{\mathbb{R}} p_\xi(x) J_0(ux)\,dx = \exp\left(\frac{u^2\sigma^2}{2}\right) I_0\left(\frac{u^2\sigma^2}{2}\right)$$

for this purpose.)

(2) Show that the r.v. $\zeta = \xi\cos\varphi_0 + \eta\sin\varphi_0$ is centred and gaussian and has variance σ^2. (Calculate its characteristic function and verify that it is of the form of Eqn (1.87).)

Exercise 1.14 Probability law of the product of two random variables

Let $\vec{\xi} = (\xi_1, \xi_2)$ be an r.v. having as probability density the function $p_{\xi_1\xi_2}$ and let η be the r.v. $\vec{\eta} = g\circ\vec{\xi} = (\xi_1, \xi_1\xi_2)$.

(1) Find an expression for the probability density $p_{\eta_1\eta_2}$ of the r.v. $\vec{\eta}$. (Apply the results of Section 1.2.3.)

(2) Hence deduce that the marginal probability density of the r.v. $\zeta = \xi_1\xi_2$ is the function

$$p_\zeta(z) = \int_{\mathbb{R}} p_{\xi_1\xi_2}\left(u, \frac{z}{u}\right)\frac{1}{|u|}\,du \tag{2.73}$$

Exercise 1.15 Ordering of a finite sequence of random variables

Let $\vec{\xi} = (\xi_1, \ldots, \xi_n)$ be a real r.v., the r.v.s ξ_i being independent and uniformly distributed over the interval $[0, a]$. Consider the real r.v. $\vec{\eta} = (\eta_1, \ldots, \eta_n)$ obtained from the r.v. $\vec{\xi}$ by ordering its components. This means that, for every trial $\omega \in \Omega$, the sequence $(\xi_1(\omega), \ldots, \xi_n(\omega))$ is put in order in the form $\xi_{i_1}(\omega) \leq \ldots \leq \xi_{i_n}(\omega)$ and we put $\eta_1(\omega) = \xi_{i_1}(\omega), \ldots, \eta_n(\omega) = \xi_{i_n}(\omega)$.

Show that the probability density of the r.v. $\vec{\eta}$ is the function

$$p_\eta(y_1, \ldots, y_n) = \frac{n!}{a^n} I_A(y_1, \ldots, y_n) \tag{2.74}$$

denoting by I_A the indicator function of the set in $\mathbb{R}^n$

$$A = \{(y_1, \ldots, y_n)\,|\,0 \leq y_1 \leq \ldots \leq y_n\}$$

(Let C be the hypercube in $\mathbb{R}^n$ having as edges the segments $[0, a]$. Verify that

94 Random vectors

putting in order is a transformation from $\mathbb{R}^n$ into $\mathbb{R}^n$ such that every point of the set A is the image of $n!$ points of the set C, and use the expression for the probability density of the r.v. $\vec{\xi}$.)

Exercise 1.16 Statistical estimators of the mean and of the variance
Let $(\xi_i, i \in \mathbb{N}^*)$ be a sequence of pairwise uncorrelated r.v.s, having the same mean m and the same variance σ^2.
(1) Consider the r.v.

$$X_n = \frac{1}{n} \sum_{i=1}^{n} \xi_i$$

Prove that

$$E[X_n] = m \qquad V[X_n] = \frac{\sigma^2}{n} \qquad (2.75)$$

(Apply Eqn (2.29).)
(2) Consider the r.v.s

$$S_n = \frac{1}{n-1} \sum_{i=1}^{n} (\xi_i - X_n)^2 \qquad T_n = \frac{1}{n-1} \left(\sum_{i=1}^{n} \xi_i^2 - n X_n^2 \right)$$

Prove that

$$E[S_n] = E[T_n] = \sigma^2 \qquad (2.76)$$

(Write

$$\xi_i - X_n = \frac{n-1}{n} \left(\xi_i - \frac{1}{n-1} \sum_{\substack{j=1 \\ j \neq i}}^{n} \xi_j \right)$$

and apply Eqn (2.29).)

Comment: these results are used in statistics when the values taken by the r.v.s ξ_i are the results of independent measurements, made under identical circumstances, of one and the same physical magnitude. The r.v. X_n is an unbiased estimator of the mean value m. It is more precise the greater the number n of measurements, since its variance tends to zero when n tends to infinity. The r.v.s S_n and T_n are unbiased estimators of the variance σ^2 (Schwartz, 1963).

Exercise 1.17 The χ^2 law
Let $(\xi_1, \ldots, \xi_n)$ be an independent sequence of centred gaussian r.v.s with the same variance σ^2 and let η be the r.v.

$$\eta = \sum_{i=1}^{n} \xi_i^2$$

(1) Show that the probability density of the r.v. η is the function

$$p_\eta(y) = \frac{1}{(2\sigma^2)^{n/2}} \frac{1}{\Gamma(n/2)} Y(y) y^{n/2-1} \exp\left(-\frac{y}{2\sigma^2}\right) \quad (2.77)$$

This is the chi-squared law (law of χ^2) with n degrees of freedom. It is very widely used in statistics (Bass, 1962). (Calculate the generating function G_η by applying Eqns (2.50) and (1.92), and then take the inverse transform, using Appendix 3.)

(2) Hence deduce that

$$E[\eta] = n\sigma^2 \qquad V[\eta] = 2n\sigma^4 \quad (2.78)$$

(Apply Eqn (1.90).)

Exercise 1.18 Sum of independent uniform random variables

Let $\{\xi_1, \ldots, \xi_n\}$ be an independent sequence of real r.v.s uniformly distributed over the interval $[0, a]$ and let η be the r.v.

$$\eta = \sum_{k=1}^{n} \xi_k$$

(1) Show that the probability density of the r.v. η is the function

$$p_\eta(y) = \frac{1}{a^n} \sum_{k=0}^{n} \binom{n}{k} (-1)^k Y(t-ka) \frac{(t-ka)^{n-1}}{(n-1)!} \quad (2.79)$$

(Calculate the Laplace transform $\mathscr{L}p_\eta$ and then obtain p_η by inversion using the table in Appendix 3. Apply Eqns (2.50) and (3.6).)

(2) Verify directly that $\forall y > na \; p_\eta(y) = 0$. (Expand $(t - ka)^{n-1}$ and use

$$\forall i \leq n-1 \qquad \sum_{k=1}^{n} \binom{n}{k} (-1)^k k^i = 0$$

to achieve the solution.)

(3) Show that the distribution function F_η may be written

$$F_\eta(y) = \frac{1}{a^n} \sum_{k=0}^{n} \binom{n}{k} (-1)^k Y(t-ka) \frac{(t-ka)^n}{n!} \quad (2.80)$$

(4) Verify directly that $\forall y > na \; F_\eta(y) = 1$. Hint: expand $(t - ka)^n$, use the property stated in (2) and the equation

$$\sum_{k=0}^{n} \binom{n}{k} (-1)^k k^n = (-1)^n n!$$

Exercise 1.19 Gaussian random vector
Let $\xi = (\xi_1, \xi_2, \xi_3, \xi_4)$ be a gaussian $\vec{r.v.}$, the r.v.s ξ_i being centred. Show that

$$E[\xi_1 \xi_2 \xi_3] = 0$$

$$E[\xi_1 \xi_2 \xi_3 \xi_4] = E[\xi_1 \xi_2]E[\xi_3 \xi_4] + E[\xi_1 \xi_3]E[\xi_2 \xi_4] \quad (2.81)$$
$$+ E[\xi_1 \xi_4]E[\xi_2 \xi_3]$$

(Apply Eqns (2.40) and (2.42).)

2. Infinite sequences of random variables

An infinite sequence $(\xi_n, n \in \mathbb{N})$ of r.v.s defined over a probability space $(\Omega, \mathcal{A}, P)$ can be considered as a random signal in discrete time. Random signals are studied as such in Chapters 4 and 5. This section has as its subject the study of the principal modes of convergence of infinite sequences of r.v.s.

2.1. MODES OF CONVERGENCE

2.1.1. *Almost sure convergence*

2.1.1.1. Definition
A sequence (ξ_n) of complex r.v.s **converges a.s.** to a complex r.v. ξ when the event

$$\{\xi_n \to \xi\} = \{\omega \,|\, \xi_n(\omega) \to \xi(\omega)\}$$

is almost certain, which may be written

$$P\{\xi_n \to \xi\} = 1 \qquad P\{\xi_n \not\to \xi\} = 0 \quad (2.82)$$

2.1.1.2. Events $\{\xi_n \to \xi\}$ and $\{\xi_n \not\to \xi\}$
The event $\{\xi_n \to \xi\}$ is the set of trials $\omega \in \Omega$ for which the numerical sequences $(\xi_n(\omega))$ converge respectively to the numbers $\xi(\omega)$. In order to calculate the probability of this event, it is necessary to define it in a more precise manner.

Consider the events

$$A_{nk} = \{\omega \,|\, |\xi_n(\omega) - \xi(\omega)| < 1/k\} \qquad B_{nk} = \bar{A}_{nk} \qquad k \in \mathbb{N}^* \quad (2.83)$$

It is then possible to write, taking account of the definition of the convergence of a numeric sequence (cf. Chapter 6, Exercise 1.1),

$$\{\xi_n \to \xi\} = \{\omega \,|\, \forall k \in \mathbb{N}^* \quad \exists N \in \mathbb{N} \quad \forall n \geq N \quad |\xi_n(\omega) - \xi(\omega)| < 1/k\}$$

$$= \bigcap_{n \in \mathbb{N}^*} \bigcup_{N \in \mathbb{N}} \bigcap_{n \geq N} A_{nk} = \bigcap_{k \in \mathbb{N}^*} \liminf A_{nk} \quad (2.84)$$

In the same way

$$\{\xi_n \nrightarrow \xi\} = \bigcup_{k \in \mathbb{N}^*} \bigcap_{N \in \mathbb{N}} \bigcup_{n \geq N} B_{nk} = \bigcup_{k \in \mathbb{N}^*} \limsup B_{nk} \qquad (2.85)$$

2.1.2. Convergence in mean square

2.1.2.1. Definition
A sequence (ξ_n) of square integrable r.v.s **converges in mean square** (m.s.) to the square integrable r.v. ξ when (cf. Chapter 1, Section 2.4.2)

$$\lim_{n \to \infty} E[|\xi_n - \xi|^2] = 0 \qquad (2.86)$$

2.1.2.2. Property
Let (ξ_n) be a sequence converging in m.s. to the r.v. ξ. Then

$$\lim_{n \to \infty} E[\xi_n] = E[\xi] \qquad \lim_{n \to \infty} V[\xi_n] = V[\xi] \qquad (2.87)$$

The proof is left to the reader (cf. Exercise 2.4).

2.1.3. Convergence in probability

2.1.3.1. Definition
A sequence (ξ_n) of complex r.v.s **converges in probability** to the complex r.v. ξ when

$$\forall \varepsilon > 0 \qquad \lim_{n \to \infty} P\{|\xi_n - \xi| \geq \varepsilon\} = 0 \qquad (2.88)$$

2.1.3.2. Property
The sequence (ξ_n) converges in probability to the r.v. ξ if and only if

$$\forall k \in \mathbb{N}^* \qquad \lim_{n \to \infty} P(B_{nk}) = 0 \qquad (2.89)$$

the event B_{nk} being defined by Eqn (2.83). The proof is immediate.

2.1.4. Convergence in law

A sequence (ξ_n) of real r.v.s **converges in law** to a real r.v. ξ when we have, for every point x at which the distribution function F_ξ is continuous,

$$\lim_{n \to \infty} F_{\xi_n}(x) = F_\xi(x) \qquad (2.90)$$

F_{ξ_n} being the distribution function of the r.v. ξ_n (cf. Exercise 2.3).

2.2. COMPARISON OF THE MODES OF CONVERGENCE

2.2.1. *Almost sure convergence and mean square convergence*

The a.s. and m.s. modes of convergence are completely distinct, as is shown by the two following examples.

Example 2.1 **Mean square convergence and almost sure non-convergence**
Let $(\xi_n, n \in \mathbb{N}^*)$ be an independent sequence of binary r.v.s defined by

$$P\{\xi_n = 1\} = \frac{1}{n} \qquad P\{\xi_n = 0\} = 1 - \frac{1}{n}$$

We shall show that the sequence (ξ_n) does not converge a.s. to the constant r.v. $\xi = 0$. The event A_{nk} defined by Eqn (2.83) can be written

$$A_{nk} = \{\xi_n = 0\}$$

For every fixed k, the sequence (A_{nk}) is therefore independent. Hence

$$P\left(\bigcap_{n \geq N} A_{nk}\right) = \prod_{n \geq N} P(A_{nk}) = \prod_{p=0}^{\infty} \left(1 - \frac{1}{N+p}\right)$$

The divergence of the series with general term $1/(N+p)$ implies that the infinite product is zero, whence

$$\forall N \in \mathbb{N}^* \qquad P\left(\bigcap_{n \geq N} A_{nk}\right) = 0$$

It follows from Eqn (2.84) that

$$P(\liminf A_{nk}) \leq \sum_{N \in \mathbb{N}} P\left(\bigcap_{n \geq N} A_{nk}\right) = 0$$

and, since events of zero probability are independent (cf. Chapter 1, Section 1.4.1),

$$P\{\xi_n \to \xi\} = \prod_{k \in \mathbb{N}} P(\liminf A_{nk}) = 0$$

The result stated is thus proved.

In contrast, the sequence (ξ_n) converges in m.s. to the r.v. $\xi = 0$. It is in fact immediate that

$$\lim_{n \to \infty} E[|\xi_n|^2] = \lim_{n \to \infty} \frac{1}{n} = 0$$

Example 2.2 **Almost sure convergence and mean square non-convergence**
Let $(\xi_n, n \in \mathbb{N}^*)$ be an independent sequence of binary r.v.s defined by

$$P\{\xi_n = n\} = \frac{1}{n^2} \qquad P\{\xi_n = 0\} = 1 - \frac{1}{n^2}$$

The sequence (ξ_n) converges a.s. to the constant r.v. $\xi = 0$. The event B_{nk} defined by Eqn (2.83) may be written $B_{nk} = \{\xi_n = n\}$. It follows then for every $N \in \mathbb{N}^*$ and for every $k \in \mathbb{N}^*$ that

$$P(\limsup B_{nk}) \leqslant P\left(\bigcup_{n \geqslant N} B_{nk}\right) \leqslant \sum_{n \geqslant N} P(B_{nk}) = \sum_{n \geqslant N} \frac{1}{n^2}$$

It suffices to make N tend to infinity in order to find that

$$\forall k \in \mathbb{N}^* \qquad P(\limsup B_{nk}) = 0$$

It then follows from Eqn (2.85) that

$$P\{\xi_n \nrightarrow 0\} \leqslant \sum_{k=1}^{\infty} P(\limsup B_{nk}) = 0$$

The a.s. convergence of the sequence (ξ_n) to zero is thus established.

It is immediately obvious, moreover, that $E[|\xi_n|^2] = 1$. The sequence (ξ_n) therefore cannot converge in m.s. to the r.v. $\xi = 0$.

2.2.2. Almost sure convergence and convergence in probability

2.2.2.1. Theorem

If a sequence $(\xi_n, n \in \mathbb{N})$ of r.v.s converges a.s. to an r.v. ξ, it converges in probability to the r.v. ξ.

2.2.2.2. Proof

We write (cf. Eqn (2.83))

$$\forall k \in \mathbb{N}^* \qquad \limsup B_{nk} \subset \left(\bigcup_{k \in \mathbb{N}^*} \limsup B_{nk}\right)$$

Hence (cf. Eqn (2.85))

$$\forall k \in \mathbb{N}^* \qquad P(\limsup B_{nk}) \leqslant P\{\xi_n \nrightarrow \xi\} = 0$$

It then follows from Eqn (6.37) that

$$\forall k \in \mathbb{N}^*$$

$$0 \leqslant \liminf P(B_{nk}) \leqslant \limsup P(B_{nk}) \leqslant P(\limsup B_{nk}) = 0$$

which implies that

$$\forall k \in \mathbb{N}^* \qquad \lim_{n \to \infty} P(B_{nk}) = 0$$

The theorem is thus demonstrated (cf. Eqn (2.89)).

2.2.3. Convergence in mean square and convergence in probability

2.2.3.1. Theorem
If a sequence $(\xi_n, n \in \mathbb{N}^*)$ of r.v.s converges in m.s. to an r.v. ξ, it converges in probability to the r.v. ξ.

2.2.3.2. Proof
We apply Markov's inequality to the r.v. $\xi_n - \xi$ taking $k = 2$ (cf. Eqn (1.75)). It follows that

$$\forall \varepsilon > 0 \quad P\{|\xi_n - \xi| \geq \varepsilon\} \leq \frac{1}{\varepsilon^2} E[|\xi_n - \xi|^2]$$

The right-hand side tends by definition to zero when n tends to infinity, which implies that the left-hand side also tends to zero. The theorem is thus demonstrated.

2.2.4. Convergence in probability and convergence in law

2.2.4.1. Theorem
If a sequence $(\xi_n, n \in \mathbb{N})$ of r.v.s converges in probability to an r.v. ξ, it converges in law to the r.v. ξ.

2.2.4.2. Proof
The proof is left to the reader (cf. Exercise 2.3).

2.2.5. Comparative table

The comparison of the various modes of convergence of sequences of r.v.s is summed up by the following diagram:

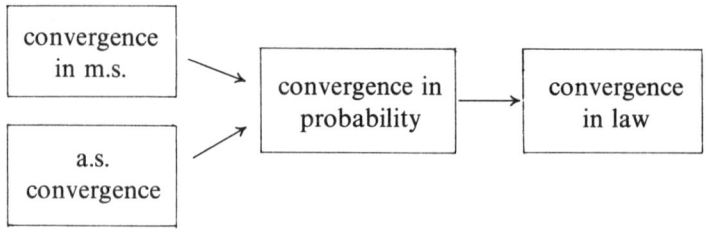

2.3. THE WEAK LAW OF LARGE NUMBERS

The following theorem expresses a form of the weak law of large numbers.

2.3.1. Theorem

Let $(\xi_n, n \in \mathbb{N}^*)$ be a sequence of square integrable pairwise uncorrelated r.v.s such that

$$\forall n \in \mathbb{N}^* \qquad E[\xi_n] = m \qquad V[\xi_n] \leq k$$

Let

$$X_n = \frac{1}{n} \sum_{i=1}^{n} \xi_i$$

be the r.v. X_n. The sequence $(X_n, n \in \mathbb{N}^*)$ converges in m.s. to the constant r.v. $\xi = m$.

2.3.2. Proof

We write, since the r.v.s ξ_n are pairwise uncorrelated (cf. Eqn (2.29)),

$$\forall n \in \mathbb{N}^* \qquad E[X_n] = m \qquad V[X_n] = \frac{1}{n^2} \sum_{i=1}^{n} V[\xi_i] \leq \frac{k}{n}$$

From this it follows that

$$\lim_{n \to \infty} V[X_n] = \lim_{n \to \infty} E[|X_n - m|^2] = 0$$

This establishes the convergence in m.s., and consequently the convergence in probability, of the sequence (X_n) to the constant r.v. $\xi = m$.

Example 2.3 Binary random sequences
Let $(\Omega^{\mathbb{N}^*}, \mathcal{P}(\Omega^{\mathbb{N}^*}), P)$ be the probability space defined in Chapter 1, Example 1.6, and let B_n be the event consisting of the set of binary sequences whose nth symbol is 1. Since the sequence $(B_n, n \in \mathbb{N}^*)$ is independent, so is the sequence (I_{B_n}) of indicator r.v.s (cf. Example 1.8). Furthermore (cf. Eqns (1.26) and (1.37))

$$\forall n \in \mathbb{N}^* \qquad E[I_{B_n}] = p \qquad V[I_{B_n}] = 1 - p$$

The above theorem allows us to state that the sequence of r.v.s

$$\frac{1}{n} \sum_{i=1}^{n} I_{B_i}$$

converges in m.s., and therefore in probability, to the constant r.v. p; this is written

$$\forall \varepsilon > 0 \qquad \lim_{n \to \infty} P\left(\left| \frac{1}{n} \sum_{i=1}^{n} I_{B_i} - p \right| \geq \varepsilon \right) = 0 \qquad (2.91)$$

102 *Random vectors*

It is easy to see that the r.v.

$$\frac{1}{n}\sum_{i=1}^{n} I_{B_i}$$

represents the relative number of symbols 1 or again the relative frequency of the symbol 1 among the first n symbols of an infinite binary sequence. We recall that such a sequence is a trial in the space $\Omega^{\mathbb{N}^*}$. Formula (2.91) shows that the probability that the relative frequency of the symbol 1 differs from its probability p by more than ε tends to zero when n tends to infinity.

This allows us to relate, in practice, the relative frequency and the probability of a symbol (cf. Chapter 1, Section 1.1.4).

2.4. THE STRONG LAW OF LARGE NUMBERS

The following theorem expresses the strong law of large numbers.

2.4.1. *Theorem*

Let $(\xi_n, n \in \mathbb{N}^*)$ be an independent sequence of r.v.s all with the same probability law and let $(X_n, n \in \mathbb{N}^*)$ be the sequence of r.v.s defined by

$$\forall n \in \mathbb{N}^* \qquad X_n = \frac{1}{n}\sum_{i=1}^{n} \xi_i \qquad (2.92)$$

When the mean value $E[\xi]$ of the r.v.s ξ_n is finite, the sequence (X_n) converges a.s. to the constant $E[\xi]$.

2.4.2. *Proof*

We shall accept the difficult proof of this theorem (Neveu, 1970).

Example 2.4 Binary random sequences

The strong law of large numbers can obviously be applied to the sequence of indicator r.v.s I_{B_n} defined in Example 2.3. Hence, it follows that

$$\lim_{n \to \infty} \frac{1}{n}\sum_{i=1}^{n} I_{B_i} \underset{\text{a.s.}}{=} p \qquad (2.93)$$

The relative frequency of the symbol 1 in an infinite binary sequence is a.s. equal to its probability p.

To understand the difference between the strong and weak laws, consider an infinite binary sequence. This is a trial in the set $\Omega^{\mathbb{N}^*}$. The weak law states that, if n is sufficiently large, the relative frequency of the symbol 1 over the first n symbols of the sequence is a.s. equal to p. It allows the possibility, however, at least in theory, that this relative frequency differs from p over the remainder of the sequence. This possibility is excluded by the strong law, which states that the relative frequency of the symbol 1 is a.s. equal to p over the entire infinite sequence.

2.5. THE CENTRAL LIMIT THEOREM

2.5.1. Theorem

Let $(\xi_n, n \in \mathbb{N}^*)$ be an independent sequence of centred r.v.s, having the same probability law μ_ξ and the same variance σ^2. Let $(Y_n, n \in \mathbb{N}^*)$ be a sequence of r.v.s defined by

$$\forall n \in \mathbb{N}^* \quad Y_n = \frac{1}{\sigma n^{1/2}} \sum_{i=1}^{n} \xi_i \tag{2.94}$$

The probability law of the r.v. Y_n tends, when n tends to infinity, to a gaussian law with mean zero and variance unity, which is called the **standard normal law** or the law $N(0, 1)$.

2.5.2. Proof

Let us calculate the characteristic function of the r.v. Y_n. It follows by applying Eqn (2.50), Θ_ξ being the characteristic function associated with the probability law μ_ξ, that (cf. Eqn (1.82))

$$\Theta_{Y_n}(u) = \left\{ \Theta_\xi \left(\frac{u}{\sigma n^{1/2}} \right) \right\}^n$$

It follows then from Eqn (1.86) that

$$\Theta_{Y_n}(u) = \left\{ 1 - \frac{u^2}{2n} + o\left(\frac{u^2}{\sigma^2 n} \right) \right\}^n$$

From this we obtain the result

$$\lim_{n \to \infty} \Theta_{Y_n}(u) = \exp\left(-\frac{u^2}{2} \right)$$

The probability law of the r.v. Y_n therefore tends to a gaussian law with mean zero and variance unity (cf. Eqn (1.87)). It must be noted that this holds whatever the law μ_ξ. This is the reason why the central limit theorem plays a very important role in practical work, especially in statistics.

2.5.3. Corollary

Suppose now that the r.v.s ξ_i have mean value m and let us consider the sequence $(X_n, n \in \mathbb{N}^*)$ of r.v.s defined by

$$\forall n \in \mathbb{N}^* \qquad X_n = \frac{1}{n} \sum_{i=1}^{n} \xi_i \qquad (2.95)$$

The r.v. X_n is the arithmetic mean of the r.v.s $\xi_1, \ldots, \xi_n$. It is easy to verify that it has mean value m and variance σ^2/n. It follows from the above theorem that, when n is sufficiently large, X_n is an approximately gaussian r.v., with mean m and variance σ^2/n. More precisely (Dupraz, 1969, p. 14)

$$\lim_{n \to \infty} p_{X_n}(x) = \lim_{n \to \infty} \frac{1}{(2\pi\sigma^2/n)^{1/2}} \exp\left\{-\frac{(x-m)^2}{2\sigma^2/n}\right\} = \delta(x-m)$$

It is then possible to apply (1.63) to write (cf. Chapter 1, Example 2.13)

$$P\{|X_n - m| < \varepsilon\} \approx 2\Phi\left(\frac{\varepsilon n^{1/2}}{\sigma}\right) \qquad (2.96)$$

The approximation improves with increasing n.

The strong law of large numbers enables us to state, in addition, that the sequence (X_n) tends a.s. to the constant r.v. m when n tends to infinity.

Example 2.5 Limits of the binomial law

Binomial law

Consider the probability space $(\Omega^{\mathbb{N}^*}, \mathscr{P}(\Omega^{\mathbb{N}^*}), P)$ defined in Chapter 1, Example 1.6, and studied in Chapter 1, Example 2.6, and in Example 2.3. We recall that this is the mathematical model of an infinite sequence of identical and independent experiments, the result of which is the observation of the occurrence or non-occurrence of an event of probability p. In this space, a trial is an infinite sequence of binary symbols 0 or 1. Let B_i be the event consisting of the set of sequences whose ith symbol is a 1. The discrete r.v.

$$\xi_n = \sum_{i=1}^{n} I_{B_i}$$

is the number of symbols 1 in the first n symbols of a sequence and the r.v. $X_n = \xi_n/n$ is their relative frequency.

The r.v. ξ_n has a binomial probability law, with parameters (n, p) (cf. Chapter 1, Example 2.6). Its mean and variance are respectively (cf. Eqn (1.114))

$$E[\xi_n] = np \qquad V[\xi_n] = npq = np(1-p)$$

Tendency towards a gaussian (normal) law

The sequence of r.v.s I_{B_i} is independent (cf. Example 2.3). The central limit theorem thus applies. It follows from this that the binomial law with parameters (n, p) tends, when n tends to infinity, p remaining fixed, to a gaussian law with mean np and variance npq.

In the same way, the probability law of the relative frequency X_n tends to a gaussian law, with mean p and variance pq/n. It is in this sense that we can state that a probability may be identified with a relative frequency (cf. Chapter 1, Section 1.1.4).

Tendency towards a Poisson law

The characteristic function of the r.v. ξ_n may be written (cf. Chapter 1, Exercise 2.6)

$$\Theta_{\xi_n}(u) = \{p \exp(iu) + q\}^n \tag{2.97}$$

Suppose that n tends to infinity and p tends to 0 according to the law

$$p = \frac{\lambda}{n} + o\left(\frac{1}{n}\right) \tag{2.98}$$

λ being a fixed parameter. This implies that the product np tends to λ. It follows from this that

$$\Theta_{\xi_n}(u) = \left[1 + \frac{\lambda}{n} \{\exp(iu) - 1\} + o\left(\frac{1}{n}\right)\right]^n$$

and then that

$$\lim_{n \to \infty} \Theta_{\xi_n}(u) = \exp[\lambda\{\exp(iu) - 1\}] \tag{2.99}$$

This enables us to state that the binomial law with parameters (n, p) tends, under the stated conditions, to a Poisson law with parameter λ (cf. Chapter 1, Exercise 2.7).

Remark

The preceding results point to the possibility that the Poisson law of parameter λ becomes assimilated to the gaussian law of mean value and of variance both equal to λ when λ is large enough. This can indeed be verified for practical purposes with $\lambda > 10$. It is an extremely useful possibility in some cases.

Exercises

Exercise 2.1 The strong law of large numbers

Let $(\xi_n, n \in \mathbb{N}^*)$ be an independent sequence of real r.v.s having the same probability law and finite, strictly positive, mean value. Let $(S_n, n \in \mathbb{N}^*)$ be the sequence of real r.v.s defined by

$$\forall n \in \mathbb{N}^* \quad S_n = \sum_{i=1}^{n} \xi_i$$

Show that

$$\forall a > 0 \quad P\left(\bigcup_{N \in \mathbb{N}^*} \bigcap_{n \geq N} \{a < S_n\}\right) = 1 \quad (2.100)$$

This means that, from a certain stage onwards, *all* the r.v.s S_n take a.s. a value greater than a. This interesting result is a consequence of the fact that the mean value of the r.v.s ξ_n is strictly positive. (Start by proving using the notation of Eqn (2.83) that there exists $k \in \mathbb{N}^*$ such that

$$\left(\bigcup_{N \in \mathbb{N}^*} \bigcap_{n \geq N} A_{nk}\right) \subset \left(\bigcup_{N \in \mathbb{N}^*} \bigcap_{n \geq N} \{a < S_n\}\right)$$

and then apply the strong law of large numbers to the sequence (ξ_n).)

Exercise 2.2 Almost sure convergence

Prove that a sequence $(\xi_n, n \in \mathbb{N}^*)$ of complex r.v.s tends a.s. to 0 if the following condition is realized:

$$\exists \alpha > 0 \quad \sum_{n \in \mathbb{N}^*} E[|\xi_n|^\alpha] < +\infty$$

(Use Eqn (2.85) and apply the Borel–Cantelli lemma; cf. Chapter 1, Example 1.5.)

Exercise 2.3 Convergence in law

Let $(\xi_n, n \in \mathbb{N}^*)$ be a sequence of real r.v.s converging in probability to an r.v. ξ.

(1) Show that

$$\forall x \in \mathbb{R} \quad \forall \varepsilon > 0 \quad \forall n \in \mathbb{N}$$

$$F_{\xi_n}(x) \leq F_\xi(x + \varepsilon) + P\{|\xi_n - \xi| \geq \varepsilon\}$$

Hint: use Eqns (1.56) and (1.7) and note that

$$\{\xi_n < x\} \cap \{\xi \geq x + \varepsilon\} \subset \{|\xi_n - \xi| \geq \varepsilon\}$$

(2) Show also that

$$F_{\xi_n} \geq F_\xi(x - \varepsilon) - P\{|\xi_n - \xi| \geq \varepsilon\}$$

(3) Hence conclude that for every point at which the function F_ξ is continuous

$$\lim_{n \to \infty} F_{\xi_n}(x) = F_\xi(x) \quad (2.101)$$

We say then that the sequence (ξ_n) converges in law to the r.v. ξ.

Exercise 2.4   Convergence in mean square

Let $(\xi_n, n \in \mathbb{N})$ be a sequence of r.v.s converging in m.s. to an r.v. ξ.
(1) Show that

$$\lim_{n \to \infty} E[\xi_n] = E[\xi]$$

(2) Hence deduce that the sequence $(\overset{\circ}{\xi}_n, n \in \mathbb{N})$ converges in m.s. to the r.v. $\overset{\circ}{\xi}$.
(3) Show that

$$\lim_{n \to \infty} V[\xi_n] = V[\xi]$$

(Apply Schwarz's inequality (6.70).)

Exercise 2.5   Sequences of gaussian random variables

Let $(\xi_n, n \in \mathbb{N})$ be a sequence of gaussian r.v.s with means m_n and variances σ_n^2, converging in m.s. to an r.v. ξ. Show that the r.v. ξ is gaussian. (Use the results of Exercises 2.3 and 2.4.)

References

BASS J. (1956): *Cours de Mathématiques*, p. 415, Masson, Paris.
BASS J. (1962): *Elements de Calcul des Probabilités*, p. 161, Masson, Paris.
DUPRAZ J. (1969): *La Théorie des Distributions et ses Applications*, Collection ENSTA, p. 14, Cepadues, Toulouse.
NEVEU J. (1970): *Cours de Probabilité de l'Ecole Polytechnique*, Ecole Polytechnique, Paris.
PFEIFFER P. and SCHUM D. (1973): *Introduction to Applied Probability*, p. 296, Academic Press, New York.
SCHWARTZ D. (1963): *Méthodes Statistiques à l'Usage des Medecins et des Biologistes*, p. 117, Editions médicales Flammarion.

Chapter 3
Deterministic signals

A deterministic signal (d.s.) is a tempered distribution over the time space. The advantage of such a definition is that it enables us to operate in the space of tempered distributions, taking advantage of the properties of this space, notably of the Fourier transform.

Linear filtering of d.s.s is a fundamental operation. We can study it in analogue form or in numerical form without stating how the filters may actually be realized. A filter is defined by its impulse response which is a particular deterministic signal and the filtering is a convolution operation. Two very simple examples enable us to compare analogue and numerical filtering: the low-pass RC filter and the recursive filter of order 1. Another example is devoted to the study of conditions under which a filter transmits a d.s. without distorting it.

Sampling of a narrow band d.s. is also an important operation. It is founded on the celebrated sampling theorem and it is an essential tool in modern telecommunications. It is studied in detail and illustrated by examples.

1. Generalities

1.1. DEFINITIONS

(a) A **deterministic signal** is a **tempered distribution** (or again a slowly increasing one) defined over $\mathbb{R}$. The set $\mathbb{R}$ is then considered as the time space in which the variable t takes its values. Generally we shall denote a d.s. by s or more precisely, when it is necessary, by $s(t)$, $t \in \mathbb{R}$, although this last notation, to be completely rigorous, is improper for a distribution.

(b) Tempered distributions constitute a vector space, denoted by $\mathscr{S}'$, of linear forms continuous over a topological vector space of test functions (Dupraz, 1977, p. 14; Guichardet, 1969, p. 128). In what follows we shall not have occasion to use test functions.

The d.s.s with which we shall be concerned will generally be tempered distributions defined either by real or complex functions defined over $\mathbb{R}$ or by positive bounded measures defined over the measurable space $(\mathbb{R}, \mathscr{R})$ (cf. Chapter 6, Section 3) (Guichardet, 1969, p. 130). In what follows, we shall designate in the same way the functions or the measures and the distributions which they define. We recall that a distribution defined by a function does not change when we change the values of the function over an at most denumerable set of points (Dupraz, 1977, p. 10). The corresponding d.s. is then defined, in fact, by a class of functions equal almost everywhere (a.e.) (cf. Chapter 6, Section 4.1.4.3). A tempered distribution belonging to the space of distributions $\mathscr{D}'_+$ whose support is contained in $\mathbb{R}_+$ defines a d.s. which does not exist for negative times. This is a **causal** d.s.

(c) The fundamental advantage in defining a d.s. as a tempered distribution arises from the fact that the Fourier transforms $\mathscr{F}$ and $\bar{\mathscr{F}}$ are continuous, linear, mutually inverse mappings from the space $\mathscr{S}'$ into itself (Guichardet, 1969, p. 134). We can thus associate, in a one-to-one relationship, a d.s. s defined over the time space and a tempered distribution $\mathscr{F}s$ defined over the frequency space. A d.s. has thus a time expression and a frequency expression.

The Fourier transforms $\mathscr{F}$ and $\bar{\mathscr{F}}$ are for this reason fundamental tools in signal theory. In the particular case of tempered distributions belonging to $\mathscr{D}'_+$, we often prefer to use the Laplace transform $\mathscr{L}$ (Dupraz, 1977, p. 115).

(d) D.s.s defined by tempered distributions have in general only an abstract existence. Thus they cannot be actually observed. This is the case notably with the Dirac d.s. which is the limit of an impulse whose duration tends to zero while its amplitude tends to infinity (Dupraz, 1977, p. 14).

Ideal d.s.s represented by tempered distributions must therefore be considered as mathematical models of d.s.s which have a concrete existence.

(e) In the absence of indications to the contrary, a d.s. is either an electromotive force or a potential difference at the terminals of a resistance of $1\,\Omega$ or a current supply or electric current in a resistance of $1\,\Omega$. A d.s. has thus in general a physical dimension.

1.2. FUNDAMENTAL RULES OF CALCULATION IN THE SPACE $\mathscr{S}'$

It is useful to recall some fundamental rules of calculation in the space $\mathscr{S}'$. They are used continually in practice to manipulate tempered distributions which are either d.s.s or Fourier transforms of d.s.s. This is why we use the general variable x to signify either the time t or the frequency f.

Throughout, we shall refer to the tempered distribution defined by Dirac measure δ_a as the Dirac distribution, and we shall denote it by $\delta(x - a)$ (cf. Chapter 6, Example 3.1).

110 *Deterministic signals*

1.2.1. Convolution with $\delta(x-a)$

Convolution with the Dirac distribution $\delta(x-a)$ is a translation of amount a:

$$\forall U \in \mathscr{S}' \quad U(x) * \delta(x-a) = U(x-a) \quad a \in \mathbb{R} \quad (3.1)$$

This is illustrated by Fig. 3.1.

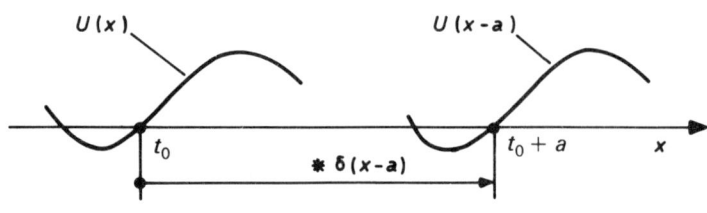

Fig. 3.1

1.2.2. Convolution with $\delta^{(n)}(x)$

Convolution with the derivative of order n of the Dirac distribution $\delta(x)$ is a derivation of order n (in the distribution sense):

$$\forall U \in \mathscr{S}' \quad U(x) * \delta^{(n)} = U^{(n)}(x) \quad n \in \mathbb{N}^* \quad (3.2)$$

1.2.3. Multiplication by $\delta(x-a)$

Consider a tempered distribution defined by an infinitely differentiable slowly increasing function g (Dupraz, 1977, p. 2). Then

$$\forall a \in \mathbb{R} \quad g(x)\delta(x-a) = g(a)\delta(x-a) \quad (3.3)$$

Let U be a tempered distribution equal, in a neighbourhood of the point a, to an infinitely differentiable function g. Then

$$U(x)\delta(x-a) = g(a)\delta(x-a) \quad (3.4)$$

Multiplication by $\delta(x-a)$ is sampling at the point a. This is illustrated by Fig. 3.2.

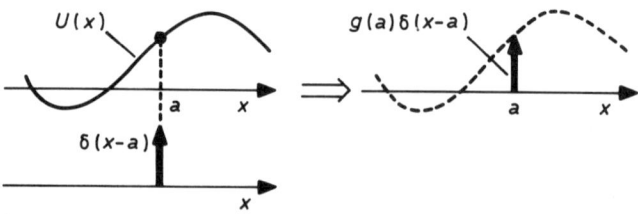

Fig. 3.2

1.2.4. Interchange of convolution and multiplication

Let U_1 and U_2 be two tempered distributions. Then

$$U_1 * U_2 \underset{\mathscr{F}}{\overset{\mathscr{F}}{\rightleftarrows}} \mathscr{F} U_1 \, \mathscr{F} U_2 \tag{3.5}$$

The Fourier transform interchanges convolution and multiplication when the two initial members are both well defined.

Let U_1 and U_2 be two tempered distributions belonging to the space $\mathscr{D}'_+$. Then

$$U_1 * U_2 \underset{\mathscr{L}^{-1}}{\overset{\mathscr{L}}{\rightleftarrows}} \mathscr{L} U_1 \, \mathscr{L} U_2 \tag{3.6}$$

The Laplace transform converts the convolution into multiplication of the transforms and vice versa.

1.2.5. Remark

The product (direct) and the convolution of two arbitrary tempered distributions are not defined in general and Eqn (3.5) cannot be applied when one of the two members is not defined. In contrast, the convolution of two distributions belonging bo $\mathscr{D}'_+$ is a distribution belonging to $\mathscr{D}'_+$ and Eqn (3.6) applies, even when the distributions are not tempered.

2. Complex spectrum

2.1. GENERAL DEFINITIONS

(a) Let s be a complex d.s. Its **complex spectrum** is the Fourier transform $\mathscr{F}s$. Conversely, the d.s. s is obtained from its complex spectrum $\mathscr{F}s$ by taking the inverse transform $\bar{\mathscr{F}}$. Putting $\mathscr{F}s = S$, we shall therefore write the fundamental relation

$$s(t) = \bar{\mathscr{F}} S(t) \underset{\mathscr{F}}{\overset{\mathscr{F}}{\rightleftarrows}} \mathscr{F} s(f) = S(f) \tag{3.7}$$

A d.s. is therefore defined either in the time domain (variable t) or in the frequency domain (variable f).

(b) The complex spectrum of a d.s. is a tempered distribution which is in practice (i) a real or complex function, continuous or discontinuous, (ii) a combination of Dirac distributions or (iii) the sum of a function and a combination of Dirac distributions.

A complex spectrum containing only Dirac distributions is a **discrete spectrum** or alternatively a **line spectrum**.

(c) The complex spectrum S can always be decomposed into the form

$$S = R + iX \tag{3.8}$$

R and X being real tempered distributions. When the complex spectrum is a function, it can always be written

$$S(f) = A(f)\exp\{i\Phi(f)\} \tag{3.9}$$

A being its modulus and Φ its phase.

2.2. BAND-LIMITED DETERMINISTIC SIGNALS

2.2.1. Definitions

The duration of a d.s. is defined by the magnitude of its **time support** supp s. A d.s. has finite duration when its time support is bounded.

The **frequency support** of a d.s. s is the support supp S of its complex spectrum S. It defines the **spectral band** occupied by the d.s. s in the frequency space. A d.s. is **band limited** when its frequency support is bounded.

The Fourier transform of a distribution with bounded support is an infinitely differentiable function whose support is $\mathbb{R}$ (Schwartz, 1979, p. 211). From this it follows that, to be completely rigorous, a d.s. with finite duration cannot be band limited and conversely that a band-limited d.s. cannot have finite duration.

2.2.2. Approximations

The complex spectra with which we are concerned do not, in most cases, have bounded support. However, the complex spectra of d.s.s having concrete existence are functions whose modulus tends to zero, more or less quickly, at infinity.

This leads us to define, for such d.s.s a positive frequency f_c such that the modulus of the complex spectrum is practically negligible for frequencies f satisfying $|f| \geqslant f_c$. The frequency f_c is the **upper cut-off frequency** of the d.s. It is then possible to accept that the d.s. considered is band limited by writing

$$\mathrm{supp}\, S = [-f_c, +f_c]$$

The upper cut-off frequency is an essential characteristic of a d.s. We shall see why in Section 5.

We are also led, in practice, to use d.s.s having simultaneously finite duration and limited bandwidth. This is an approximation which enables us to reason heuristically and to simplify certain calculations considerably.

Example 2.1 Complex spectrum of a real deterministic signal

Let s be a complex d.s. The complex spectrum $\mathscr{F}(s^*)$ of the conjugate complex d.s. may be expressed as a function of the complex spectrum $\mathscr{F}s = S$. It follows from Appendix 2 that

$$\mathscr{F}(s^*) = \check{S}^* \tag{3.10}$$

where we denote by $\check{U}$ the distribution $\check{U}(t) = U(-t)$.

When the d.s. s is real, we get from Eqn (3.10)

$$S = \check{S} \Leftrightarrow R = \check{R} \qquad X = -\check{X} \Leftrightarrow A = \check{A} \qquad \Phi = -\check{\Phi} \tag{3.11}$$

The modulus of the complex spectrum of a real d.s. is thus even whereas its phase is odd. This is a property which it is important to remember inasmuch as concrete d.s.s, by definition, are real. It is convenient, however, to discuss complex d.s.s notably in the study of modulations.

Example 2.2 Analytic deterministic signal

(a) Let s be a complex d.s. The **Hilbert conjugate** d.s. $\mathscr{H}s$ of the d.s. s is defined by the convolution

$$\mathscr{H}s(t) = -s(t) * \frac{1}{\pi t} \tag{3.12}$$

For simplicity, we have written $1/\pi t$ for the tempered distribution principal value (p.v.) $1/\pi t$ (Dupraz, 1977, p. 211). We shall do the same in what follows.

Conversely, the d.s. s is defined, starting from the d.s. $\mathscr{H}s$, by the convolution

$$s(t) = \bar{\mathscr{H}}\mathscr{H}s(t) = \mathscr{H}s(t) * \frac{1}{\pi t} \tag{3.13}$$

The reciprocal transforms $\mathscr{H}$ and $\bar{\mathscr{H}}$ are respectively the direct and inverse Hilbert transforms (Dupraz, 1977, p. 45).

The complex spectrum of the d.s. $\mathscr{H}s$ is obtained by applying Eqn (3.5) and using the table in Appendix 2. We obtain

$$\mathscr{F}[\mathscr{H}s](f) = i\mathscr{F}s(f)\operatorname{sgn} f \tag{3.14}$$

The function sign, which is denoted by sgn, is defined by

$$\begin{aligned} \operatorname{sgn} f &= +1 & f > 0 \\ \operatorname{sgn} f &= -1 & f < 0 \end{aligned} \tag{3.15}$$

The d.s. $\mathscr{H}s$ exists only if the right-hand side of Eqn (3.14) has meaning. For this it is necessary and sufficient that the complex spectrum $\mathscr{F}s$ be equal to a function in the neighbourhood of the origin since the function sgn is discontinuous at this point.

(b) Consider the d.s.

$$\sigma(t) = s(t) - i\mathscr{H}s(t) \tag{3.16}$$

Its complex spectrum may be written

$$\mathscr{F}\sigma(f) = \mathscr{F}s(f)(1 + \text{sgn } f) = 2Y(f)\mathscr{F}s(f) \tag{3.17}$$

where Y is the Heaviside function (cf. Eqn (6.22)). We can thus associate with every d.s. s having a Hilbert conjugate d.s., a d.s. σ whose complex spectrum $\mathscr{F}\sigma$ is zero for negative frequencies and equal to $2\mathscr{F}s$ for positive frequencies. The d.s. σ is the **analytic** d.s. associated with the d.s. s.

Example 2.3 Causal deterministic signal

A **causal** d.s. is a tempered distribution belonging to the space $\mathscr{D}'_+$ of distributions whose support is contained in $\mathbb{R}_+$. It is thus a d.s. which does not exist for negative time.

Let s be a causal d.s. equal to a function in the neighbourhood of the origin. It is then permissible to write $s = Ys$. It follows from Eqn (3.5), by putting $\mathscr{F}s = R + iX$ (cf. Appendix 2), that

$$R(f) + iX(f) = \{R(f) + iX(f)\} * \left\{\tfrac{1}{2}\delta(f) + \frac{1}{i2\pi f}\right\}$$

Hence we obtain the **causality relations**

$$R(f) = X(f) * \frac{1}{\pi f} = -\mathscr{H}X(f)$$

$$X(f) = -R(f) * \frac{1}{\pi f} = \mathscr{H}R(f) \tag{3.18}$$

2.3. TIME TRANSLATION

The notion of a d.s. implies that we have at our disposal an absolute time reference. The expression for a d.s. depends therefore on the time origin.

Let s_ϑ be the d.s. obtained from the d.s. s by a translation ϑ of the time origin. We can write, using Eqn (3.1),

$$s_\vartheta(t) = s(t) * \delta(t - \vartheta) = s(t - \vartheta)$$

Its complex spectrum can be obtained by applying Eqn (3.5). We get (cf. Appendix 2)

$$S_\vartheta(f) = S(f)\exp(-i2\pi\vartheta f) \tag{3.19}$$

A translation of the time origin does not change the modulus of the complex spectrum but changes its phase linearly.

3. Types of deterministic signal

This section is devoted to the study of four types of d.s. very commonly used in practice.

3.1. INTEGRABLE DETERMINISTIC SIGNALS

3.1.1. *Definitions*

An integrable d.s. s is a tempered distribution defined by a complex function s, defined over $\mathbb{R}$, which is integrable with respect to Lebesgue measure μ_0 (cf. Chapter 6, Example 3.3).

The d.s. s does not change if we change the values taken by the function s on an at most denumerable set of points. A d.s. is thus an element of the space $L^1(\mathbb{R}, \mathscr{R}, \mu_0)$ (cf. Chapter 6, Section 4.1.6).

3.1.2. *Complex spectrum*

Let s be an integrable d.s. Its complex spectrum S is defined by the integral

$$S(f) = \int_{\mathbb{R}} \exp(-i2\pi ft) s(t) \, dt \qquad (3.20)$$

This is a continuous bounded function tending to zero at infinity but not necessarily integrable (Dupraz, 1977, p. 75).

Suppose that the function s has continuous and integrable derivatives up to order n inclusive. We then get from Eqns (3.2) and (3.5), using Appendix 2,

$$\mathscr{F}[s^{(n)}](f) = (i2\pi f)^n S(f)$$

Hence we have the result, since the left-hand side is a bounded function,

$$\forall f \in \mathbb{R} \quad |S(f)| \leqslant M/f^n \qquad (3.21)$$

From this we must conclude that, the more regular a d.s. is, the faster is the decrease at infinity of its complex spectrum. This is an important property which is conventionally illustrated by Fig. 3.3, which shows a decrease in the modulus of the complex spectrum of $6n$ decibels per octave.

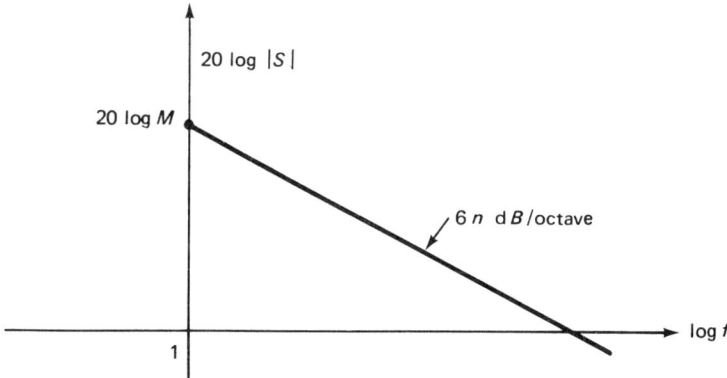

Fig. 3.3

3.1.3. Correlation functions

3.1.3.1. Cross-correlation function

The **cross-correlation function** $R_{s_1 s_2}$ of two integrable d.s.s s_1 and s_2 is the convolution of the functions s_1 and $\check{s}_2{}^*$ (where we denote $s_2(-t)$ by $\check{s}_2(t)$):

$$R_{s_1 s_2}(\tau) = \int_{\mathbb{R}} s_1(t) s_2{}^*(t-\tau)\, dt = s_1(\tau) * \check{s}_2{}^*(\tau) \tag{3.22}$$

The complex function $R_{s_1 s_2}$ is defined a.e. over $\mathbb{R}$ and is integrable with respect to the measure μ_0 (Dupraz, 1977, p. 41). It follows immediately that it satisfies

$$R_{s_1 s_2} = \check{R}_{s_1 s_2}{}^* \tag{3.23}$$

Let $G_{s_1 s_2} = \mathscr{F} R_{s_1 s_2}$ be the Fourier transform of the cross-correlation function. This is a complex function and it follows from Eqns (3.22) and (3.5), using the properties of the Fourier transform given in Appendix 2, that

$$G_{s_1 s_2} = S_1 S_2{}^* \tag{3.24}$$

3.1.3.2. Autocorrelation function

The **autocorrelation function** R_{ss} of an integrable d.s. s is defined by Eqn (3.22), taking $s_1 = s_2 = s$. We obtain

$$R_{ss}(\tau) = \int_{\mathbb{R}} s(t) s^*(t-\tau)\, dt = s(\tau) * \check{s}^*(\tau) \tag{3.25}$$

The function R_{ss} satisfies, from the above,

$$R_{ss} = \check{R}_{ss}{}^* \qquad \mathscr{F} R_{ss} = G_{ss} = |S|^2 \tag{3.26}$$

This is an even function when the d.s. s is real. It is defined at the origin only when the d.s. s has finite energy (cf. Section 3.2).

Example 3.1 Impulse response of a realizable filter

The impulse response of a realizable linear filter having concrete existence is generally a function h which is real, causal and integrable (cf. Example 4.3). This is therefore an important example of an integrable d.s.

The complex spectrum $H = \mathscr{F} h$ of the d.s. h is the transfer function of the filter. It is a continuous function which tends to zero, more or less quickly, at infinity. We write

$$H(f) = R(f) + iX(f) = A(f) \exp\{i\Phi(f)\} \tag{3.27}$$

Since the d.s. h is real, it follows from Eqn (3.11) that the modulus A is an even function and the phase Φ is an odd function. The fact that these are continuous functions allows the possibility of measuring experimentally $A(f)$ and $\Phi(f)$ for every value of the frequency. The function h is causal and consequently the transfer function H satisfies the causality relations (3.18).

Types of deterministic signal

The group propagation time γ of the filter is defined by the derivative

$$\gamma(f) = -\frac{1}{2\pi}\frac{d\Phi(f)}{dt} \qquad (3.28)$$

We shall see in Example 4.4 the role which it plays.

The **autocorrelation function** of the filter is the autocorrelation function R_{hh} of its impulse response. It is defined by the general formula (3.25). It is easy to verify that it may also be written, since the function h is causal and real, in the form

$$R_{hh}(\tau) = \int_0^\infty h(t)h(t+|\tau|)\,dt \qquad (3.29)$$

The function R_{hh} is then even. Its Fourier transform is an even function which may be written (cf. Eqns (3.26) and (3.11))

$$G_{hh} = |H|^2 = H\check{H} \qquad (3.30)$$

3.2. FINITE ENERGY DETERMINISTIC SIGNALS

3.2.1. Definitions

A d.s. s with **finite energy** E_s is a tempered distribution defined by a complex function s, defined over $\mathbb{R}$, and square integrable with respect to Lebesgue measure μ_0 (cf. Chapter 6, Section 4.1.7). The energy E_s is given by the integral

$$E_s = \int_\mathbb{R} |s(t)|^2\,dt \qquad (3.31)$$

When the d.s. s is real and represents a voltage at the terminals of a resistance of $1\,\Omega$, E_s is the energy dissipated by the d.s. s in the resistance.

The d.s. s does not change if we change the values taken by the function s over an at most denumerable set of points. A finite energy d.s. thus belongs to the space $L^2(\mathbb{R}, \mathcal{R}, \mu_0)$ (cf. Chapter 6, Section 4.1.7).

Finite energy d.s.s play a considerable part in practice. In fact, the finite duration d.s.s with which we are concerned have finite energy.

3.2.2. Complex spectrum

Let s be a finite energy d.s. Its complex spectrum S is the Fourier transform of the function s. It is a complex function of integrable square, and Parseval's theorem enables us to write (Dupraz, 1977, p. 95)

$$E_s = \int_\mathbb{R} |S(f)|^2\,df \qquad (3.32)$$

118 Deterministic signals

The function $|S|^2$ is then the density of a bounded positive measure, defined over the space $(\mathbb{R}, \mathcal{R})$, with respect to which the function 1 is integrable and has as integral the energy E_s (cf. Chapter 6, Section 4.1.5). It is the spectral density of the energy E_s or again, more simply, the **energy spectrum** of the d.s. s. It is a measure of the way in which the energy E_s is distributed over the frequency space. It should be noted that the energy spectrum does not change when the time origin is translated. This is an immediate result of Eqn (3.19).

3.2.3. Correlation functions

3.2.3.1. Cross-correlation function

The cross-correlation function $R_{s_1 s_2}$ of two finite energy d.s.s s_1 and s_2 is the convolution of the functions s_1 and $\check{s}_2^*$ ($\check{s}_2(t) = s_2(-t)$):

$$R_{s_1 s_2}(\tau) = \int_{\mathbb{R}} s_1(t) s_2^*(t - \tau) \, dt = s_1(\tau) * \check{s}_2^*(\tau) \tag{3.33}$$

The complex function $R_{s_1 s_2}$ is continuous (Dupraz, 1977, p. 41) and it is bounded, since Schwarz's inequality permits us to write (cf. Eqn (6.70))

$$\forall \tau \in \mathbb{R} \quad |R_{s_1 s_2}(\tau)| \leq (E_{s_1} E_{s_2})^{1/2} \tag{3.34}$$

It follows immediately that it satisfies Eqn (3.23).

Let $G_{s_1 s_2} = \mathscr{F} R_{s_1 s_2}$ be the Fourier transform of the cross-correlation function. This is a complex function and it follows from Eqns (3.33) and (3.5) that

$$G_{s_1 s_2} = S_1 S_2^* \tag{3.35}$$

3.2.3.2. Autocorrelation function

The autocorrelation function of a finite energy d.s. s is defined by Eqn (3.33), taking $s_1 = s_2 = s$. We get then from the above

$$R_{ss} = \check{R}_{ss}^* \qquad \forall \tau \in \mathbb{R} \quad |R_{ss}(\tau)| \leq R_{ss}(0) = E_s \qquad G_{ss} = |S|^2 \tag{3.36}$$

When the d.s. s is real, its autocorrelation function is real. It is thus even, as is its Fourier transform. The energy spectrum of a real d.s. is therefore an even function.

Example 3.2 The 'gate' signal

The **gate** signal is the real d.s. defined by the function $s = A\mathring{I}_T$, denoting by $\mathring{I}_T$ the indicator function of the interval $\lceil -T/2, T/2 \rceil$. It is a rectangular impulse of amplitude A and duration T, centred at the time origin. The complex spectrum S is given by Appendix 2. We obtain

$$s(t) = A\mathring{I}_T(t) \qquad S(f) = AT \frac{\sin(\pi T f)}{\pi T f} \tag{3.37}$$

The gate signal and its complex spectrum are illustrated in Fig. 3.4.

Types of deterministic signal

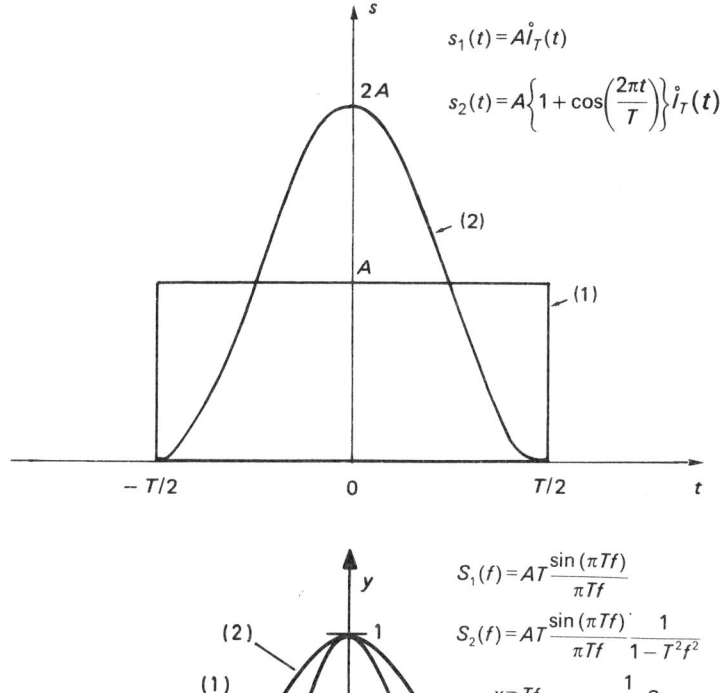

Fig. 3.4

The spectrum vanishes for frequencies n/T, $n \in \mathbb{Z}^*$, and passes through its maximum, equal to AT, at $f = 0$. The total energy is $E_s = A^2 T$, while the energy distributed over the spectral band $[-n/T, n/T]$ is given by the integral

$$E_{s,n} = 2A^2 T^2 \int_0^{n/T} \left\{ \frac{\sin(\pi T f)}{\pi T f} \right\}^2 df = \frac{2A^2 T}{\pi} \text{Si}(2n\pi) \quad (3.38)$$

We get, numerically, $E_{s,1} = E_s \times 0.9028$ and $E_{s,2} = E_s \times 0.9499$.

Thus 90% of the total energy is concentrated on the interval $[-1/T, 1/T]$ in the central lobe of the spectrum. This enables us to say that the spectral band occupied by a rectangular impulse of duration T is of the order of $1/T$ hertz (and not of $2/T$ hertz) and to take $1/T$ as the value of the cut-off frequency f_c of the gate signal. When a better approximation is desired we take a higher value for f_c, between $1/T$ and $2/T$ for example.

120 Deterministic signals

Example 3.3 The cosine-squared signal
The **cosine-squared** signal is the real d.s. defined by the function

$$s(t) = A\left\{1 + \cos\left(\frac{2\pi t}{T}\right)\right\}\mathring{I}_T(t) \qquad (3.39)$$

This is a d.s. of finite duration T. Its complex spectrum S is obtained by using the table in Appendix 2. We obtain

$$S(f) = A\frac{\sin(\pi T f)}{\pi f} * \left\{\delta(f) + \frac{1}{2}\delta\left(f - \frac{1}{T}\right) + \frac{1}{2}\delta\left(f + \frac{1}{T}\right)\right\}$$

$$= AT\frac{\sin(\pi T f)}{\pi T f}\frac{1}{1 - T^2 f^2} \qquad (3.40)$$

The cosine-squared signal and its complex spectrum are represented in Fig. 3.4. The spectrum vanishes for frequencies n/T, $n \in \{2, 3, \ldots\}$ and passes through its maximum, equal to AT, at $f = 0$. The total energy is $E_s = 3A^2T/2$, while the energy distributed over the spectral band $[-2/T, +2/T]$ is about 98% of the total energy. The energy is thus almost entirely concentrated in the central lobe of the spectrum, which enables us to say, to a good approximation, that the cut-off frequency of the cosine-squared signal is $f_c = 2/T$ hertz.

It is interesting to compare the gate signal and the cosine-squared signal. Both signals have as time support the interval $[-T/2, T/2]$. Thus they have duration T. The central lobe of the spectrum of the cosine-squared signal is twice as wide as that of the gate signal, but the energy is more concentrated in it. This is expressed by a decrease at infinity of the spectrum of the cosine-squared signal as $1/f^3$, whereas it is as $1/f$ for the gate signal. This follows from the fact that the cosine-squared signal is more regular than the gate signal, which is not even continuous (cf. Section 3.1.2).

3.3. DETERMINISTIC SIGNAL OF FINITE MEAN POWER

3.3.1. Definitions

A d.s. of **finite mean power** P_s is defined by a complex function s of locally integrable square such that

$$\lim_{T \to \infty} \frac{1}{2T} \int_{-T}^{T} |s(t)|^2 \, dt = P_s < +\infty \qquad (3.41)$$

We can show, by applying Schwarz's inequality (cf. Eqn (6.70)), that

$$\lim_{T \to \infty} \frac{1}{2T} \int_{-T}^{T} |s(t)| \, dt \leqslant P_s^{1/2}$$

This signifies that the mean value of the d.s. is also finite.

The complex spectrum S of the d.s. s is the Fourier transform of the function s. It is a tempered distribution which is not necessarily defined by a function, in contrast with the preceding cases (cf. Example 3.4).

3.3.2. Correlation functions

3.3.2.1. Cross-correlation function

The cross-correlation function $R_{s_1 s_2}$ of two d.s.s s_1 and s_2 of finite mean power is defined by

$$R_{s_1 s_2}(\tau) = \lim_{T \to \infty} \frac{1}{2T} \int_{-T}^{T} s_1(t) s_2^*(t - \tau) \, dt \qquad (3.42)$$

It satisfies $R_{s_2 s_1} = \check{R}_{s_1 s_2}{}^*$, and Schwarz's inequality enables us to write

$$\forall \tau \in \mathbb{R} \qquad |R_{s_1 s_2}(\tau)| \leqslant (P_{s_1} P_{s_2})^{1/2} \qquad (3.43)$$

Being bounded, the cross-correlation function defines a tempered distribution and hence has a Fourier transform. This is not in general the product of the complex spectra of the two d.s.s.

3.3.2.2. Autocorrelation function

The autocorrelation function of a d.s. s of finite mean power is defined by Eqn (3.42), taking $s_1 = s_2 = s$. Hence we get from the above

$$R_{ss} = \check{R}_{ss}{}^* \qquad \forall \tau \in \mathbb{R} \qquad |R_{ss}(\tau)| \leqslant R_{ss}(0) = P_s \qquad (3.44)$$

Let G_{ss} be the Fourier transform of the autocorrelation function. This is a positive bounded measure which is not necessarily defined by a function. We can show, by the same method as in Chapter 5, Section 1.2.1.2, that the function 1 is integrable with respect to the measure G_{ss} and that its integral is the mean power P_s. This can be written

$$P_s = \int_{\mathbb{R}} dG_{ss}(f) \qquad (3.45)$$

The measure G_{ss} is therefore the spectral density of the mean power P_s. This is the **power spectrum**, or simply the spectrum, of the d.s. s. It should be noted that the spectrum G_{ss} is not, in general, the square of the modulus of the complex spectrum S (cf. Example 3.4).

Example 3.4 The 'unit step' signal
The Heaviside function Y is defined by (cf. Eqn (6.22))

$$\begin{aligned} Y(x) &= 1 & x > 0 \\ Y(x) &= 0 & x \leqslant 0 \end{aligned} \qquad (3.46)$$

The **unit step** d.s. Y is the causal d.s. defined by the function Y. It must be noted that the function Y is discontinuous at the origin but that the distribution Y does not depend on the value given to $Y(0)$. The unit step d.s. is a d.s. of finite mean power since

$$\lim_{T \to \infty} \frac{1}{2T} \int_{-T}^{T} Y^2(t) \, dt = \frac{1}{2} \qquad (3.47)$$

Its complex spectrum is the distribution (cf. Appendix 2)

$$\mathscr{F}Y(f) = \frac{1}{2}\delta(f) + \frac{1}{i2\pi f} \qquad (3.48)$$

It is easy to verify that the autocorrelation function and the power spectrum of the d.s. Y are respectively

$$R_{YY}(\tau) = \tfrac{1}{2} \qquad G_{YY}(f) = \tfrac{1}{2}\delta(f) \qquad (3.49)$$

The real part R and the imaginary part X of the complex spectrum $\mathscr{F}Y$ are represented in Fig. 3.5 with the spectrum G_{YY}. It follows immediately that they satisfy the causality relations (3.18).

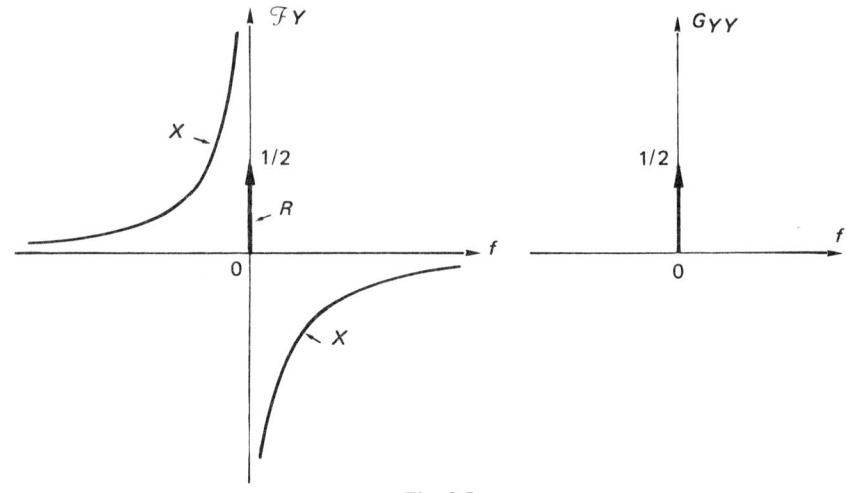

Fig. 3.5

The function Y is infinitely differentiable at every point other than the origin. It follows then from Eqn (3.4) that

$$\forall a \in \mathbb{R}^* \qquad Y(x)\delta(x-a) = \delta(x-a) \qquad (3.50)$$

Finally, the d.s. Y is differentiable and has as derivative the Dirac d.s. $\delta(x)$ (Dupraz, 1977, p. 18):

$$Y' = Y * \delta' = \delta \qquad (3.51)$$

3.4. PERIODIC DETERMINISTIC SIGNALS

3.4.1. Definitions

A **periodic** d.s. is a periodic distribution. We note that a periodic distribution is tempered (Dupraz, 1977, p. 59). A periodic d.s. s of period T can always be written

$$s(t) = s_T(t) * \sum_{n \in \mathbb{Z}} \delta(t - nT) \tag{3.52}$$

s_T being a d.s. of finite duration defined over the interval $[-T/2, T/2[$ by truncation of the d.s. s.

The distribution s can be expanded in the space $\mathscr{S}'$ in a Fourier series of the form (Dupraz, 1977, p. 60)

$$s(t) = \sum_{n \in \mathbb{Z}} c_n \exp(in\omega t) \qquad \omega = 2\pi/T \tag{3.53}$$

The Fourier coefficients c_n are defined as functions of the d.s. s in Appendix 1. Appendix 1 also contains a table giving the coefficients c_n for a certain number of common periodic d.s.s.

3.4.2. Complex spectrum

The complex spectrum S of the d.s. s is obtained by taking the Fourier transform, term by term, of the expansion (3.53). We obtain

$$S(f) = \sum_{n \in \mathbb{Z}} c_n \delta(f - n/T) \tag{3.54}$$

The complex spectrum is discrete, the lines being distributed at the frequencies n/T, $n \in \mathbb{Z}$, with complex mass c_n.

When the d.s. s_T is a function which is m times continuously differentiable the coefficients c_n tend to zero when n tends to infinity according to the law

$$c_n = o\left(\frac{1}{|n|^m}\right) \qquad |n| \to \infty$$

This result should be compared with that expressed in Eqn (3.21).

The complex spectrum S can also be obtained by taking the Fourier transform of both sides of Eqn (3.52). We obtain

$$S(f) = \frac{1}{T} \mathscr{F} s_T(f) \sum_{n \in \mathbb{Z}} \delta(f - n/T) \tag{3.55}$$

It can be seen from this form that the 'envelope' of the complex spectrum S is proportional to the complex spectrum $\mathscr{F} s_T$ of the truncated d.s. s_T and that the

Fourier coefficient c_n can be written

$$c_n = \frac{1}{T} \mathscr{F} s_T\left(\frac{n}{T}\right) \quad (3.56)$$

The envelope is continuous since the d.s. s_T is of finite duration (cf. Section 2.2.1).

3.4.3. Autocorrelation function

Suppose that the periodic d.s. s has finite mean power P_s. Its autocorrelation function R_{ss} can be obtained from Eqn (3.42) by using the expansion (3.53). We obtain, by taking for the T of Eqn (3.42) k times the period T,

$$R_{ss}(\tau) = \lim_{k \to \infty} \frac{1}{2kT} \int_{-kT}^{kT} \left[\sum_n \sum_m c_n c_m^* \exp\{i(n-m)\omega t\} \exp(im\omega\tau) \right] d\tau$$

$$= \sum_{n \in \mathbb{Z}} |c_n|^2 \exp\left(\frac{in2\pi\tau}{T}\right) \qquad \omega = \frac{2\pi}{T} \quad (3.57)$$

The autocorrelation function of a periodic d.s. is therefore periodic, with the same period, and its Fourier coefficient of order n is $|c_n|^2$.

The mean power P_s is written

$$P_s = R_{ss}(0) = \sum_{n \in \mathbb{Z}} |c_n|^2 \quad (3.58)$$

which implies the convergence of the series with general term $|c_n|^2$. The power spectrum G_{ss} is obtained by taking the Fourier transform of both sides of Eqn (3.57). It follows that

$$G_{ss}(f) = \sum_{n \in \mathbb{Z}} |c_n|^2 \delta\left(\frac{f-n}{T}\right) \quad (3.59)$$

The spectrum G_{ss} is therefore a line spectrum. The mean power concentrated in the component with frequenty n/T is equal to $|c_n|^2$.

Example 3.5 Ideal sampling signal

The **ideal sampling** d.s. is the periodic d.s.

$$s(t) = \sum_{n \in \mathbb{Z}} \delta(t - nT - \vartheta) \qquad \vartheta \in [-T/2, T/2[\quad (3.60)$$

This is a **train** of Dirac distributions, of period T and phase ϑ. It plays a fundamental role in the theory of signal sampling. Its expansion in a Fourier series can be written (cf. Appendix 1, case 23)

$$\sum_{n \in \mathbb{Z}} \delta(t - nT - \vartheta) = \frac{1}{T} \sum_{n \in \mathbb{Z}} \exp\left(-in2\pi\frac{\vartheta}{T}\right) \exp\left(in2\pi\frac{t}{T}\right) \quad (3.61)$$

Its complex spectrum can be written

$$S(f) = \frac{1}{T} \sum_{n \in \mathbb{Z}} \exp\left(-in2\pi \frac{\vartheta}{T}\right) \delta\left(f - \frac{n}{T}\right) \qquad (3.62)$$

It should be noted that this is also a periodic train of Dirac distributions of period $1/T$. Note that the components all have modulus equal to $1/T$ and that the spectrum does not tend to zero at infinity, which expresses the fact that the ideal sampling signal does not exist in the real world.

Example 3.6 Rectangular clock signal
The **rectangular clock** d.s. is the periodic function

$$s(t) = A\mathring{I}_a(t) * \sum_{n \in \mathbb{Z}} \delta(t - nT - \vartheta) \qquad \vartheta \in [-T/2, T/2] \qquad (3.63)$$

which is a sequence of rectangular pulses of duration $a < T$, period T and phase ϑ. It is represented in Fig. 3.6 for $\vartheta = 0$.

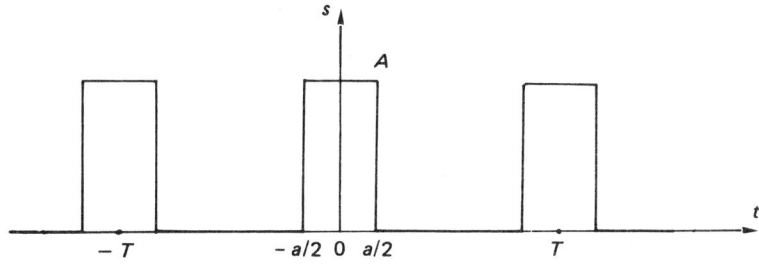

Fig. 3.6

The truncated d.s. s_T is the gate d.s. $A\mathring{I}_a$. The complex spectrum can be obtained from Eqn (3.54). We obtain (cf. Appendix 1, case 1)

$$S(f) = A\frac{a}{T} \sum_{n \in \mathbb{Z}} \frac{\sin(n\pi a/T)}{n\pi a/T} \exp\left(-in2\pi \frac{\vartheta}{T}\right) \delta\left(f - \frac{n}{T}\right) \qquad (3.64)$$

Its modulus is represented in Fig. 3.7. Its decrease at infinity is as $1/f$ and its envelope is the continuous function

$$\frac{A}{T} \left|\frac{\sin(\pi a f)}{\pi f}\right|$$

The autocorrelation function R_{ss} may be obtained from Eqn (3.57). It can be written

$$R_{ss}(\tau) = \frac{A^2 a}{T} \frac{2a}{2T} \sum_{n \in \mathbb{Z}} \frac{\sin^2(n\pi 2a/2T)}{(n\pi 2a/2T)^2} \exp\left(-in2\pi \frac{\tau}{T}\right) \qquad (3.65)$$

In this form, it is seen that it is the Fourier series expansion of the periodic function represented in Fig. 3.8 (cf. Appendix 1, case 3).

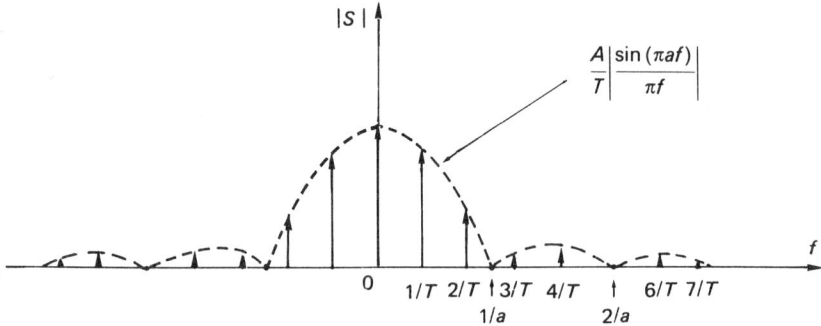

Fig. 3.7

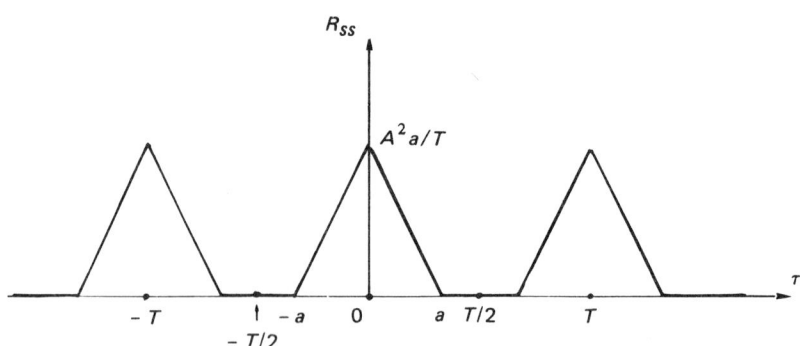

Fig. 3.8

Exercises

Exercise 3.1 Deterministic analytic signal

Consider the d.s. $s(t) = A \cos(2\pi f_0 t + \varphi_0)$. Prove that the associated analytic d.s. is written

$$\begin{aligned} f_0 > 0 & \quad \sigma(t) = A \exp\{i(2\pi f_0 t + \varphi_0)\} \\ f_0 < 0 & \quad \sigma(t) = A \exp\{-i(2\pi f_0 t + \varphi_0)\} \end{aligned} \quad (3.66)$$

(Argue in terms of complex spectra and apply Eqns (3.14) and (3.16).)

Exercise 3.2 Deterministic analytic signal

Consider the d.s.

$$s(t) = x(t) \cos(2\pi f_0 t + \varphi_0) + y(t) \sin(2\pi f_0 t + \varphi_0) \quad (3.67)$$

We suppose that the d.s.s x and y are band limited by a cut-off frequency f_c such that

$$\text{supp } \mathscr{F}x = \text{supp } \mathscr{F}y = [-f_c, f_c] \quad 0 < f_c < f_0$$

(1) Prove that the Hilbert conjugate d.s. $\mathscr{H}s$ may be written

$$\mathscr{H}s(t) = y(t)\cos(2\pi f_0 t + \varphi_0) - x(t)\sin(2\pi f_0 t + \varphi_0) \qquad (3.68)$$

(Argue in terms of complex spectra and apply Eqn (3.14).)

(2) Deduce from the above that the analytic d.s. σ associated with the d.s. s may be written

$$\sigma(t) = \{x(t) - iy(t)\}\exp\{i(2\pi f_0 t + \varphi_0)\} \qquad (3.69)$$

(Apply Eqn (3.16).)

Exercise 3.3 Hilbert conjugate deterministic signal

Consider the gate d.s. $s(t) = A\mathring{I}_T(t)$. Prove that the Hilbert conjugate d.s. $\mathscr{H}s$ may be written

$$\mathscr{H}s(t) = \frac{A}{\pi}\log\left|\frac{t-T/2}{t+T/2}\right| \qquad (3.70)$$

(Apply Eqn (3.12).)

4. Linear filtering

A **linear filter** is a device which establishes a linear relation between a d.s. applied at its input and the d.s. which is obtained at its output.

Linear filtering plays a very important role in signal theory. It is studied in its functional aspect in this section but the manner in which the filters are realized in practice is not touched on. The whole section refers to analogue as opposed to numerical filtering, which is the subject of Section 5.3.

4.1. IMPULSE RESPONSE

(a) The **impulse response** of a linear filter is the d.s. $h(t, t_0)$ resulting from the application of the d.s. $\delta(t - t_0)$ at the input of the filter. The filter is **invariant** in time when its impulse response is a function only of the difference $t - t_0$. We shall confine ourselves in what follows to considering only filters of this type, taking $t_0 = 0$.

(b) The impulse response of an invariant filter is the d.s. h resulting from the application of the Dirac d.s. $\delta(t)$ at the input of the filter (cf. Fig. 3.9). An invariant filter is **causal** when its impulse response is a causal d.s. The d.s. h is then a tempered distribution belonging to the space $\mathscr{D}'_+$ (cf. Example 2.3).

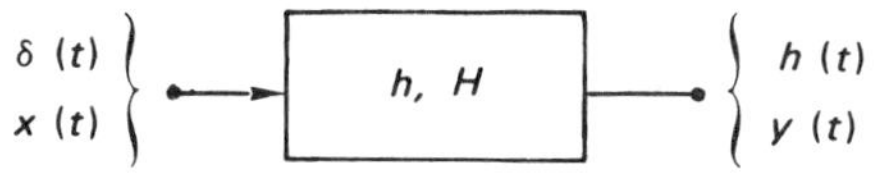

Fig. 3.9

128 Deterministic signals

An invariant filter is **stable** if the output d.s. is a bounded function when the input d.s. is a bounded function. The necessary and sufficient condition for an invariant causal filter to be stable is that its impulse response be of the form (cf. Example 4.3)

$$h(t) = \sum_{i=1}^{n} a_i \delta(t - t_i) + k(t)$$

$$a_i \in \mathbb{R} \qquad t_i \in \mathbb{R}_+ \qquad k \in L^1 \cap \mathscr{D}'_+ \qquad (3.71)$$

where k is an integrable causal function.

An invariant filter which is causal and stable is said to be **realizable**. When the filter has a concrete existence, its impulse response is real. There is no disadvantage, however, from the mathematical point of view, in supposing that the impulse response is complex. Unless the contrary is stated, all the results which follow are valid in this case.

4.2. TRANSFER FUNCTION

(a) The **transfer function** of an invariant filter is the Fourier transform $\mathscr{F}h = H$ of its impulse response. When the filter has a concrete existence, its impulse response is real and its transfer function satisfies (cf. Eqn (3.11))

$$H^* = \check{H} \qquad (3.72)$$

The transfer function of a realizable filter is the Fourier transform of a d.s. of the form of Eqn (3.71). It is therefore a continuous bounded function which tends to zero at infinity only when h is an integrable function.

(b) The Laplace transform of a d.s. of the form of Eqn (3.71) is a function of the complex variable p, holomorphic in the half-plane $\operatorname{Re} p \geq 0$. The transforms $\mathscr{L}h$ and $\mathscr{F}h$ are then connected by

$$\mathscr{L}h(i2\pi f) = \mathscr{F}h(f) \qquad (3.73)$$

The transfer function of a realizable filter is therefore defined either with respect to the real variable f (transfer function in f) or with respect to the complex variable p (transfer function in p).

4.3. RESPONSE TO AN ARBITRARY DETERMINISTIC SIGNAL

Consider a realizable filter, with impulse response h, to whose input a d.s. x is applied. The response of the filter is the output d.s. y. We seek an expression for it, distinguishing successively the case in which we are interested in the transient state and that in which only the permanent state is of interest.

4.3.1. Transient response

(a) Suppose that the filter is initially at rest and that a causal d.s. x is applied at its input. The output d.s. is the causal d.s. y defined by the convolution in $\mathscr{D}'_+$ (Dupraz, 1977, p. 50)

$$y(t) = x(t) * h(t) \tag{3.74}$$

This follows from the facts that the mapping $s \mapsto s * h$ from $\mathscr{D}'_+$ into $\mathscr{D}'_+$ is linear and continuous and that the distribution x can be defined as a limit of linear combinations of Dirac distributions and from the definition of the impulse response h. The output d.s. is therefore the convolution of the input d.s. with the impulse response of the filter.

When the d.s. x is a causal function, the d.s. y is the causal function defined by the integral

$$\forall t \in \mathbb{R}_+ \qquad y(t) = \int_0^t h(u) x(t-u) \, du \tag{3.75}$$

The situation is represented schematically in Fig. 3.9.

(b) The transform $\mathscr{L}y$ is obtained as a function of the transform $\mathscr{L}x$ by taking the Laplace transform of the left-hand and right-hand sides of Eqn (3.74). We obtain (cf. Eqn (3.6))

$$\mathscr{L}y = \mathscr{L}x \mathscr{L}h \tag{3.76}$$

The complex spectrum $\mathscr{F}y$ is obtained as a function of the complex spectrum $\mathscr{F}x$ by taking the Fourier transform of the left-hand and right-hand sides of Eqn (3.74). We obtain (cf. Eqn (3.5))

$$\mathscr{F}y = \mathscr{F}x \mathscr{F}h = \mathscr{F}x H \tag{3.77}$$

These relations are very important. It must be noted that Eqn (3.77) is not always satisfied whereas Eqn (3.76) *is* always satisfied (cf. Section 1.2.5).

(c) The impulse response is a dimensional physical quantity. The dimensional equation follows from Eqn (3.75):

$$[h] = [x]^{-1}[y][T] \tag{3.78}$$

It must be noted that, when $x(t) = c\delta(t)$, the dimensions of the d.s. x and the constant c are connected by the relation

$$[x] = [c][T]^{-1}$$

Example 4.1 Rational transfer function

A case which is extremely important in practice is that in which the relation between the d.s. x and the d.s. y is a linear differential equation with constant coefficients. When the initial conditions are zero and the d.s. x is causal, this is a convolution equation in $\mathscr{D}'_+$ (cf. Example 4.5). Its solution, when $x(t) = \delta(t)$, is the impulse response of the filter.

130 Deterministic signals

The use of the Laplace transform (in preference to the Fourier transform in order to avoid a division in $\mathscr{S}'$) enables us to arrive at an algebraic equation which has as solution a rational fraction $\mathscr{L}h(p) = A(p)/B(p)$. This can always be decomposed into the form

$$\mathscr{L}h(p) = \frac{A(p)}{B(p)} = \sum_{l=0}^{m} b_l + \sum_{i=1}^{n} \sum_{j=1}^{k_i} \frac{a_{ij}}{(p - p_i)^j}$$

The required impulse response h is the unique inverse in $\mathscr{D}'_+$ of the rational fraction $\mathscr{L}h$. It follows (cf. Appendix 3) that

$$h(t) = \sum_{l=0}^{m} b_l \delta^{(l)}(t) + \sum_{i=1}^{n} \sum_{j=1}^{k_i} a_{ij} Y(t) \frac{t^{j-1} \exp(p_i t)}{(p - p_i)^j} \qquad (3.79)$$

We can conclude immediately, by comparing Eqns (3.79) and (3.71), that the filter is stable if and only if

$$m = 0 \qquad \forall i \in [1, n] \qquad \operatorname{Re} p_i < 0$$

A filter whose transfer function $\mathscr{L}h(p)$ is a rational fraction is thus stable if and only if (i) the degree of the numerator is less than or equal to that of the denominator and (ii) the poles of the rational fraction have a strictly negative real part. The zeros of the rational fraction can be *a priori* arbitrary. In practice, in the case of a filter with concrete existence, the zeros cannot be situated on the imaginary axis and have zero real part, as is shown in the following section (Example 4.2). The situation is illustrated in Fig. 3.10.

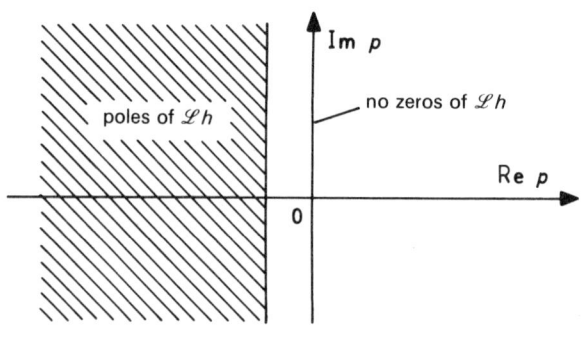

Fig. 3.10

The transfer function $\mathscr{F}h(f)$ is obtained from Eqn (3.79) by applying Eqn (3.73). It is a rational fraction with complex coefficients that are functions of the real variable f. The values taken by the function $\mathscr{F}h$ are therefore the values taken by the function $\mathscr{L}h$ on the imaginary axis.

4.3.2. Permanent response

When the filter is realizable, the relation (3.74) may be applied in the case where the d.s. x is a non-causal but bounded function. The d.s. y is then a bounded function (cf. Example 4.3) which is written

$$y(t) = \int_{-\infty}^{t} x(u)h(t-u)\,du = \int_{0}^{\infty} h(u)x(t-u)\,du \tag{3.80}$$

The response of the filter at time t depends only on the values taken by the input signal before time t. This is a result of the causality of the impulse response.

Example 4.2 Response at frequency f_0
Let $x(t)$ be the sinusoidal complex d.s. $x(t) = \exp\{i(2\pi f_0 t + \varphi_0)\}$. It follows from Eqn (3.80) that

$$y(t) = \exp\{i(2\pi f_0 t + \varphi_0)\} \int_{0}^{\infty} \exp(-i2\pi f_0 u)h(u)\,du$$

$$= x(t)H(f_0) \tag{3.81}$$

The d.s. $\exp\{i(2\pi f_0 t + \varphi_0)\}$ is thus an eigenfunction of the linear convolution operator defined by $s \mapsto s * h$, having $H(f_0)$ as its eigenvalue.

Suppose that there exists a frequency f_0 such that $H(f_0) = 0$. The d.s. y would then be zero, by virtue of Eqn (3.81), which would mean that the filter would have infinite attenuation at the frequency f_0. Now a filter having concrete existence inevitably has losses and thus cannot have infinite attenuation at any frequency. Hence we must conclude that, in practice, the transfer function $\mathscr{F}h$ does not vanish for any frequency or, equivalently, that the transfer function $\mathscr{L}h$ does not vanish on the imaginary axis (cf. Example 4.1).

Example 4.3 Conditions for stability of a filter
Consider a filter having impulse response of the form of Eqn (3.71) and suppose that the d.s. x applied at its input is a bounded function. We get from Eqns (3.75) and (3.1)

$$y(t) = \sum_{i=1}^{n} a_i x(t - t_i) + \int_{0}^{t} k(u)x(t-u)\,du$$

Hence we have the result

$$\forall t \in \mathbb{R} \qquad |y(t)| \leq \left(\sum_{i=1}^{n} a_i + \|k\|_1\right)M$$

The output d.s. y is therefore a bounded function and Eqn (3.71) is a sufficient condition for stability of the filter.

Let us now take as input d.s. the bounded function $x(t) = \text{sgn}\{k(t_0 - t)\}$. It is discontinuous at least at the point t_0. Suppose that the impulse response h contains the term $\delta^{(m)}(t - t_i)$, $t_i > 0$, $m \geq 1$. The corresponding term of the output d.s. is the convolution $x(t) * \delta^{(m)}(t - t_i)$. This is the derivative of order m, translated by t_i, of the d.s. x. It cannot be a bounded function since the function x is discontinuous. Suppose next that the function k is not integrable. The corresponding term of the output d.s. at the instant t_0 may be written

$$\int_0^\infty k(u) x(t_0 - u) \, du = \int_0^\infty k(u) \, \text{sgn}\{k(u)\} \, du$$

$$= \int_0^\infty |k(u)| \, du$$

$$= +\infty$$

It is therefore not bounded. It follows from the above that the condition (3.71) is also necessary.

Example 4.4 Distortion-free filtering

(a) Consider a d.s. x and an invariant linear filter having a transfer function H. We seek the condition that H must satisfy for the output d.s. y to be equal to the input d.s. x, to within a gain of A and a delay of γ. We say therefore that the filter transmits the d.s. x **without distortion.** Let us write

$$y(t) = A x(t - \gamma) = A x(t) * \delta(t - \gamma) \qquad A > 0 \qquad \gamma > 0$$

Then taking the Fourier transforms

$$\mathscr{F} y(f) = A \mathscr{F} x(f) \exp(-i 2\pi \gamma f) = H(f) \mathscr{F} x(f)$$

The condition we seek is thus written

$$\forall f \in \text{supp } \mathscr{F} x \qquad H(f) = A \exp(-i 2\pi \gamma f) \qquad (3.82)$$

The modulus of the transfer function is constant over the support of the complex spectrum of the d.s. to be transmitted. This necessary and sufficient condition is very important in practice. When it is not satisfied, the filtering is accompanied by linear distortion.

It follows from Eqn (3.82) that

$$\forall f \in \text{supp } \mathscr{F} x \qquad \gamma = -\frac{1}{2\pi} \frac{d\Phi}{df} \qquad (3.83)$$

The **group propagation time** of the filter is thus constant and equal to the delay γ over the whole spectral band occupied by the d.s. x (cf. Eqn (3.28)).

(b) Suppose that the d.s. s is band limited and that its complex spectrum satisfies $\text{supp } \mathscr{F} x = [-f_c, +f_c]$. Suppose moreover that the distribution $\mathscr{F} x$ is equal to a function in the neighbourhood of the frequencies $\pm f_c$, which implies the absence of lines at these frequencies. We can then take as a transfer

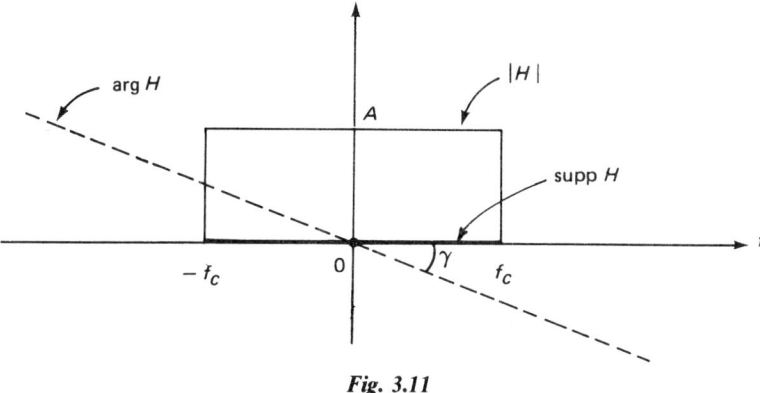

Fig. 3.11

function satisfying Eqn (3.82)

$$H(f) = A\mathring{1}_{2f_c}(f)\exp(-i2\pi\gamma f) \qquad (3.84)$$

The modulus and the argument of the function H are represented in Fig. 3.11. The corresponding impulse response is the translated function

$$h(t) = A\frac{\sin(2\pi f_c t)}{\pi t} * \delta(t-\gamma) \qquad (3.85)$$

The function H is not continuous and moreover the impulse response h is not causal. The filter is therefore not realizable. It is very often used, however, as a model of an ideal low-pass filter. Of all filters which transmit the d.s. x without distortion, this is the one that occupies the smallest spectral band.

Example 4.5 Low-pass RC filter

Consider the electric circuit in Fig. 3.12. It represents a **low-pass RC filter**. When the initial state is zero (capacity discharged), the input current $i(t)$, the input voltage $e(t)$ and the output voltage $s(t)$ are connected, for all $t \geq 0$, by the

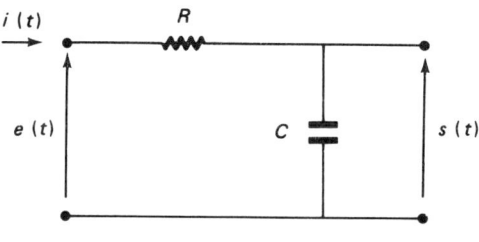

Fig. 3.12

convolution equations in $\mathscr{D}'_+$:

$$e(t) = Ri(t) + \frac{1}{C}\int_0^t i(u)\,\mathrm{d}u = \left\{R\delta(t) + \frac{1}{C}Y(t)\right\} * i(t)$$

$$s(t) = \frac{1}{C}Y(t) * i(t)$$

It follows, by taking Laplace transforms (cf. Appendix 3 and Eqn (3.6)), that

$$\mathscr{L}h(p) = \frac{\mathscr{L}s(p)}{\mathscr{L}e(p)} = \frac{1}{1+RCp}$$

$\mathscr{L}h$ being the transfer function (of p) linking the output and input voltages. It is dimensionless. The transfer function (of f) $H = \mathscr{F}h$ and the impulse response h are therefore (cf. Eqn (3.73) and Appendix 2)

$$h(t) = Y(t)\frac{1}{RC}\exp\left(-\frac{t}{RC}\right)$$

$$H(f) = \frac{1}{1+i2\pi RCf} \tag{3.86}$$

The autocorrelation function R_{hh} is the inverse Fourier transform of the function $|H|^2$ (cf. Eqn (3.30)). It follows (cf. Appendix 2) that

$$R_{hh} = \frac{1}{2RC}\exp\left(-\frac{|\tau|}{RC}\right) \tag{3.87}$$

The corresponding graphs are shown in Fig. 3.13.

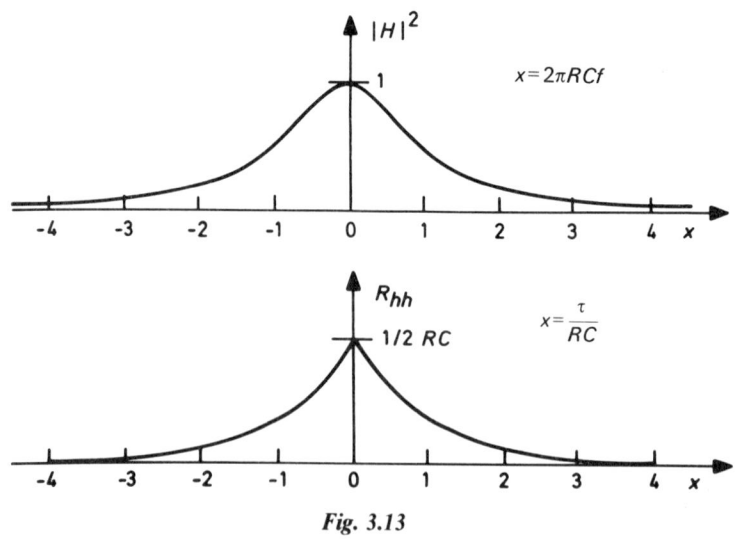

Fig. 3.13

Exercises

Exercise 4.1 Step response of a filter
Consider an invariant linear filter with impulse response h. The unit step d.s. $x = Y$ is applied at the input of the filter (cf. Example 3.4). The **step response** of the filter is the output d.s. y.

(1) Verify that, when the impulse response is an integrable function, the step response is defined by the integral

$$y(t) = \int_0^t h(u)\,du = h(t) * Y(t) \tag{3.88}$$

(2) Verify that the step response of a low-pass RC filter is the function (cf. Example 4.5)

$$y(t) = Y(t)\left\{1 - \exp\left(-\frac{t}{RC}\right)\right\} \tag{3.89}$$

Exercise 4.2 Linear envelope demodulator
The block diagram of a **linear envelope demodulator** is represented in Fig. 3.14. It is constructed by placing in series the non-linear absolute value function and a linear low-pass filtering function.

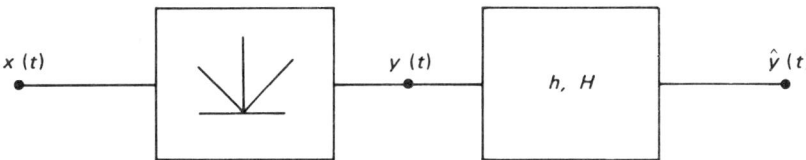

Fig. 3.14

Suppose that the input d.s. is

$$x(t) = a(t)\cos(2\pi f_0 t + \varphi_0) \qquad f_0 > 0 \tag{3.90}$$

φ_0 being a phase constant and a a d.s. such that $a(t) \geq 0$. The d.s. a is the **real envelope** of the d.s. x.

(1) Show that the complex spectrum $\mathscr{F}y$ of the d.s. $y = |x|$, produced by the absolute function, can be written

$$\mathscr{F}y(f) = \frac{2}{\pi}\mathscr{F}a(f) * \sum_{n \in \mathbb{Z}}(-1)^{n+1}\frac{\exp(in2\varphi_0)}{4n^2 - 1}\delta(f - 2nf_0) \tag{3.91}$$

(cf. Appendix 1, case 12).

(2) Suppose that $\operatorname{supp}\mathscr{F}a = [-f_c, +f_c]$ and that the signal y is filtered by an ideal low-pass filter with transfer function $H = \mathring{I}_{2b}$. Suppose also that

$b < f_c < f_0$. Let $\hat{y}$ be the filtered d.s. Prove that

$$\hat{y}(t) = \frac{2}{\pi} a(t) \tag{3.92}$$

The d.s. $\hat{y}$ is therefore proportional to the envelope a of the d.s. x. This explains the name linear envelope demodulator. (Use complex spectra.)

Exercise 4.3 Perfect amplitude limiter

The block diagram of a perfect **amplitude limiter** is represented in Fig. 3.15. It is constructed by placing in series the non-linear function sign and a linear band-pass filtering function. Suppose that the input d.s. is

$$x(t) = a(t) \cos\{2\pi f_0 t + \varphi_0 + \varphi(t)\} \qquad f_0 > 0 \tag{3.93}$$

where φ_0 is a phase constant, φ a real d.s. and a a d.s. such that $a(t) \geq 0$.

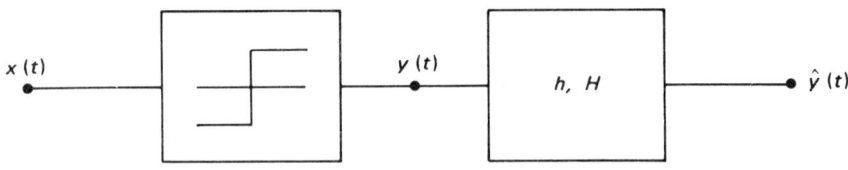

Fig. 3.15

(1) Write the expression for the d.s. $y = \operatorname{sgn} x$, produced by the function sign, in the form of an expansion in series of imaginary exponentials. Use the representation of the function sign (cf. Eqn (3.15)) by an (improper) Fourier integral (cf. Appendix 2) as well as the results

$$\exp(iu \cos \vartheta) = \sum_{n \in \mathbb{Z}} i^n J_n(u) \exp(-in\vartheta)$$

$$\forall n \in \mathbb{Z} \qquad \int_0^\infty \frac{J_{n+1}(t)}{t} dt = \frac{1}{2n+1}$$

(2) Deduce from this expansion the expression for the complex spectrum

$$\mathscr{F} y(f) = \frac{2}{\pi} \sum_{n \in \mathbb{Z}} \frac{(-1)^n}{2n+1} \exp\{-i(2n+1)\varphi_0\} \psi_{2n+1}(f) * \delta\{f + (2n+1)f_0\}$$

$$\psi_{2n+1}(f) = \mathscr{F}[\exp\{-i(2n+1)\varphi(t)\}](f)$$

(3) The d.s. y is filtered by an ideal band-pass filter having transfer function

$$H(f) = \mathring{I}_B(f) * \{\delta(f + f_0) + \delta(f - f_0)\}$$

with $B/2 < f_0$ and $\operatorname{supp} \psi_1 \subset [-B/2, B/2]$. Let $\hat{y}$ be the filtered d.s. Prove that

$$\hat{y}(t) = \frac{4}{\pi} \cos\{2\pi f_0 t + \varphi_0 + \varphi(t)\} \tag{3.94}$$

Draw a diagram representing the supports of the complex spectra and show that the approximation (3.94) becomes better as the support of ψ_1 becomes narrower. The phase of the d.s. $\hat{y}$ is identical with that of the d.s. x but its envelope is constant. The amplitude limiter has thus obliterated the *amplitude modulation* while preserving entirely the *angular modulation*.

Exercise 4.4 Ideal dephaser
Consider the real d.s.

$$s(t) = A \cos(\omega_0 t + \varphi_0) \qquad \hat{s}(t) = A \cos(\omega_0 t + \varphi_0 + \varphi) \qquad f_0 > 0$$

φ_0 and φ being arbitrary phases.

Prove that the transfer function H_φ of the filter such that the d.s. $\hat{s}$ is obtained, for every positive frequency f_0, by filtering of the d.s. s may be written

$$H_\varphi(f) = \exp(i\varphi \operatorname{sgn} f) \qquad (3.95)$$

(Apply Eqn (3.77).)

This filter is an **ideal dephaser** of algebraic angle φ. It is not realizable since its transfer function is discontinuous. The modulus and the argument of H_φ are represented in Fig. 3.16.

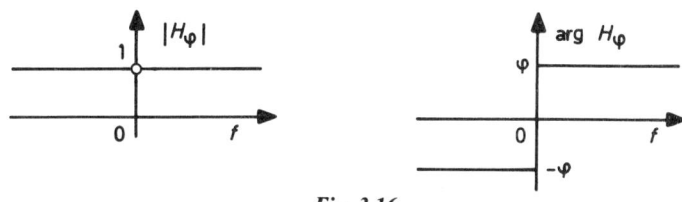

Fig. 3.16

Exercise 4.5 Complex envelope and matched filter
Consider a complex d.s. a and the real d.s.

$$x(t) = \operatorname{Re}[a(t) \exp\{i(\omega_0 t + \varphi_0)\}] \qquad \omega_0 = 2\pi f_0 \qquad \varphi_0 \in [-\pi, +\pi] \qquad (3.96)$$

The d.s. a is the **complex envelope** of the d.s. x. Let $X = \mathscr{F}x$ and $A = \mathscr{F}a$ be the complex spectra of the two d.s.s. We suppose that the energy E_a of the d.s. a is finite and that its frequency band is bounded by a cut-off frequency f_c such that supp $A = [-f_c, +f_c]$.

(1) Prove that

$$|X(f)|^2 = \tfrac{1}{4}|A(f)|^2 * \delta(f - f_0) + \tfrac{1}{4}|\breve{A}(f)|^2 * \delta(f + f_0) \qquad (3.97)$$

Hence deduce that the energy E_x of the d.s. x is finite and equal to $E_a/2$. (Note that the supports of $A(f - f_0)$ and $A(f + f_0)$ are disjoint.)

(2) The d.s. x is filtered by a filter having impulse response h. Let y be the

138 *Deterministic signals*

filtered d.s. and b its complex envelope. Show that

$$b(t) = a(t) * h(t)\exp(-i\omega_0 t)$$
$$B(f) = A(f)H(f+f_0) \tag{3.98}$$

(3) Suppose that the filter used is of band-pass type and has transfer function

$$H(f) = \mathring{I}_{2f_c}(f)\exp(-i2\pi\gamma f) * \{\exp(i\psi_0)\delta(f-f_0) + \exp(-i\psi_0)\delta(f+f_0)\}$$

where $\mathring{I}_{2f_c}$ is the indicator function of the interval $[-f_c, +f_c]$ and ψ_0 is an arbitrary phase angle. Show that the complex envelopes a and b are connected by

$$b(t) = a(t-\gamma)\exp(i\psi_0) \tag{3.99}$$

The complex envelope a is thus transmitted without distortion by the filter H, with delay equal to its group propagation time (cf. Example 4.4).

(4) Suppose now that the impulse response of the filter used is

$$h(t) = \text{Re}[a^*(T-t)\exp\{i(\omega_0 t + \psi_0)\}] \qquad T \in \mathbb{R}$$

(4.1) Show that

$$b(t) = \tfrac{1}{2}R_{aa}^*(t-T)\exp(i\psi_0) \tag{3.100}$$

R_{aa} being the autocorrelation function of the d.s. a (cf. Eqn (3.33)). Hence deduce that

$$|b(T)| = \tfrac{1}{2}R_{aa}(0) = E_x$$

The value of the complex envelope b at the instant T is equal to the energy of the d.s. x. For this reason we say that the filter h is **matched** to the complex envelope of the d.s. x. (Apply Eqn (3.96), using the conditions on the supports.)

(4.2) What condition must be imposed on the phase ψ_0 so as to have also $y(T) = |b(T)| = E_x$?

Show that we then have $h(t) = x(T-t)$. The filter is then matched to the d.s. x itself.

Under what conditions is the filter realizable?
(*Answers*: $\psi_0 = -\omega_0 T - \varphi_0$ and $x(t) = 0$ for all $t > T$.)

Exercise 4.6 Band-pass RLC filter
The electric circuit of a **band-pass RLC filter** is shown in Fig. 3.17.

(1) Suppose that the filter is initially in the zero state (no current in the inductance L, capacitance C discharged).

Show that, under these conditions, the input current i and the output voltage v are connected, for all $t \geq 0$, by the convolution equation in $\mathscr{D}'_+$

$$i(t) = v(t) * \left\{\frac{1}{R}\delta(t) + C\delta'(t) + \frac{1}{L}Y(t)\right\}$$

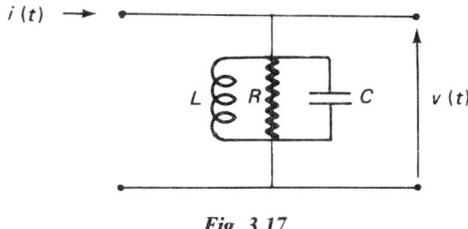

Fig. 3.17

Hence deduce the expression for the transfer function $\mathscr{F}h = H$

$$H(f) = \frac{\mathscr{F}v(f)}{\mathscr{F}i(f)} = \frac{R}{1 + iQ(f/f_0 - f_0/f)} \quad (3.101)$$

$$f_0 = \frac{1}{2\pi(LC)^{1/2}} \quad Q = R\left(\frac{C}{L}\right)^{1/2}$$

The frequency f_0 is the **resonance frequency** of the filter and the number Q is its **power factor**. The three-decibel bandwidth B of the filter is $B = f_0/Q$. (Use Eqn (3.51) and the table in Appendix 2.)

(2) Suppose that the Q factor is large in comparison with 1. This implies that the bandwidth B is small in comparison with the frequency f_0. Hence deduce that, under these conditions, we can give the approximation

$$Q \gg 1 \Rightarrow H(f) \approx \frac{R}{1 + i(2Q/f_0)f} * \{\delta(f + f_0) + \delta(f - f_0)\} \quad (3.102)$$

The transfer function H is then obtained, by two translations, from the transfer function of a low-pass RC filter having cut-off frequency $f_c = f_0/2Q$ (cf. Example 4.5).

5. Sampled deterministic signals

5.1. SAMPLING OF A BAND-LIMITED DETERMINISTIC SIGNAL

(a) Let s be a band-limited d.s. Consider the d.s. $\tilde{s}$ obtained by multiplying the d.s. s by the ideal sampling d.s. defined by Eqn (3.60):

$$\tilde{s}(t) = s(t) \sum_{n \in \mathbb{Z}} \delta(t - nT - \vartheta) \quad (3.103)$$

The multiplication is possible since s is an infinitely differentiable function (cf. Sections 2.2 and 1.2.3). We obtain (cf. Eqn (3.3))

$$\tilde{s}(t) = \sum_{n \in \mathbb{Z}} s_n \delta(t - nT - \vartheta) \quad (3.104)$$

The d.s. $\tilde{s}$ is a train of Dirac distributions with as masses the analogue samples

140 Deterministic signals

$s_n = s(nT + \vartheta)$, drawn periodically from the d.s. s at sampling frequency $f_e = 1/T$ with arbitrary phase ϑ—hence the name of ideal sampled d.s. given to the d.s. $\tilde{s}$.

The complex spectrum $\tilde{S}$ of the d.s. $\tilde{s}$ is obtained from Eqn (3.103) by applying Eqn (3.62). We get, since $f_e = 1/T$,

$$\tilde{S}(f) = S(f) * \frac{1}{T} \sum_{n \in \mathbb{Z}} \exp(-\mathrm{i} n 2\pi \vartheta f_e) \delta(f - n f_e) \quad (3.105)$$

The shape of the complex spectrum $\tilde{S}$ is represented in Fig. 3.18 in the particular case in which the support of the complex spectrum S is centred on the frequency 0. It contains an infinity of terms of the same modulus, obtained from S by similarity and by translations of step $n f_e$, $n \in \mathbb{Z}$. In the general case, the translated spectra overlap partially as shown in Fig. 3.18. It must be noted that the support of the complex spectrum $\tilde{S}$ is not bounded but that $\tilde{S}(f)$ is well defined for all f since it is the sum of a finite number of terms. Sampling is thus an operation which enlarges the spectral band.

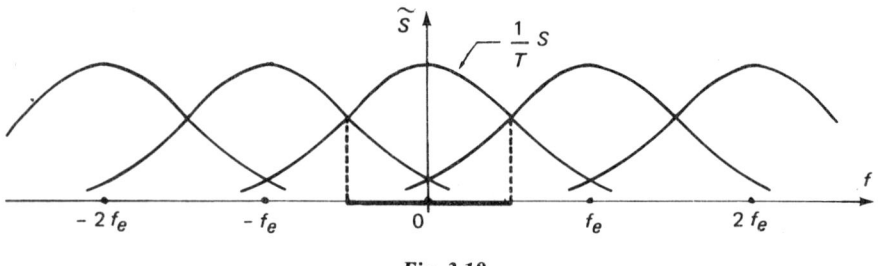

Fig. 3.18

(b) Let f_c be the upper cut-off frequency of the d.s. s. The complex spectrum S then satisfies supp $S \subset [-f_c, +f_c]$.

Suppose that the complex spectrum S contains no lines at the frequencies $\pm f_c$ and let us choose a sampling frequency f_e such that $f_e \geq 2 f_c$. It is obvious that the translated spectra do not overlap and that we can write

$$\tilde{S}(f) \mathring{I}_{2f_c}(f) = \frac{1}{T} S(f)$$

This is illustrated in Fig. 3.19. It follows from this (cf. Eqn (3.77)) that the d.s. s is obtained by filtering of the sampled d.s. $\tilde{s}$ by an ideal low-pass filter having as transfer function

$$H(f) = T \mathring{I}_{2f_c}(f)$$

This effects interpolation between the samples s_n and restores exactly the continuous d.s. s.

When the complex spectrum S contains lines at the frequencies $\pm f_c$, the product $\tilde{S}(f) \mathring{I}_{2f_c}(f)$ no longer has a meaning (cf. Eqn (3.3)). It is easy to verify

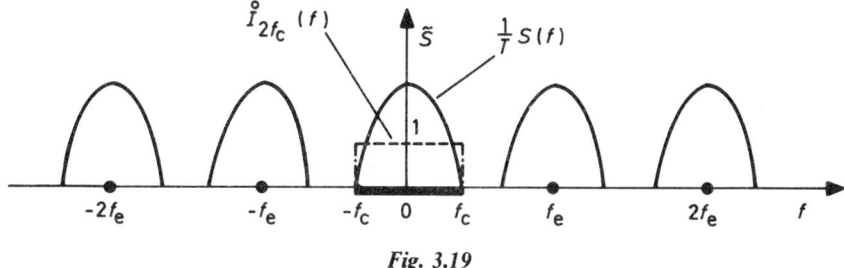

Fig. 3.19

that all these results still apply when we take a sampling frequency strictly greater than $2f_c$ and a filter having as transfer function a gate of width strictly greater than $2f_c$.

These results lead us to state the celebrated sampling theorem, also known as **Shannon's theorem**.

(c) The sampling theorem: A d.s. s with upper cut-off frequency f_c, whose complex spectrum S contains no lines at the frequencies $\pm f_c$, is completely defined by the denumerable sequence of samples $s_n = s(nT + \vartheta), n \in \mathbb{Z}$, drawn at frequency $f_e = 1/T \geqslant 2f_c$ with arbitrary phase ϑ. When the spectrum S contains lines at the frequencies $\pm f_c$ we must take a frequency $f_e > 2f_c$ (cf. Example 5.1).

The sampling theorem has considerable practical importance. Indeed, it is basic in numerical transmissions. In fact, a d.s. is never strictly band limited and the notion of cut-off frequency is an approximation (cf. Section 2.2.2). When we sample a d.s. we always have a situation such as that represented in Fig. 3.18. There is an overlap of translated spectra which prevents us from recovering the d.s. s exactly by filtering. The sampling frequency must therefore be chosen taking account of the form of the complex spectrum in such a way as to ensure that the effect of the overlap can be neglected.

Example 5.1 Sampling of a sinusoidal deterministic signal

Consider a sinusoidal d.s. $s(t) = A \cos(2\pi f_0 t + \varphi_0)$. Its complex spectrum is (cf. Appendix 2)

$$S(f) = \tfrac{1}{2}\{\exp(i\varphi_0)\delta(f - f_0) + \exp(-i\varphi_0)\delta(f + f_0)\} \qquad (3.106)$$

It has as support the set supp $S = \{f_0\} \cup \{-f_0\}$, which is obviously bounded by the cut-off frequency $f_c = f_0$. The presence of lines at the frequencies $\pm f_c$ prevents us from writing $S = S\overset{\circ}{I}_{2f_c}$ and consequently from sampling at frequency $2f_c$. Moreover, it is easy to show graphically that, if we sample at frequency $2f_c$, we obtain alternately positive and negative samples, of the same absolute value, which do not allow us to determine the sinusoid, since an infinity of periodic d.s.s pass through these points. A sinusoidal d.s. must therefore be sampled at a frequency strictly greater than twice its own frequency.

142 Deterministic signals

Example 5.2 Change of frequency by sampling

Let s be a band-limited d.s. whose complex spectrum S is such that

$$\operatorname{supp} S = [-f_2, -f_1] \cup [f_1, f_2] \qquad 0 < f_1 < f_2$$

It is always possible to sample at a frequency f_e greater than or equal to, as the case may be, $2f_2$, but there are other interesting possibilities which use the fact that the interval $[-f_1, f_1]$ is empty. Suppose for simplicity that the d.s. s is real and that its complex spectrum, shown in Fig. 3.20, contains no lines at the frequencies $\pm f_1$ and $\pm f_2$. Put $f_0 = (f_1 + f_2)/2$ and write, defining *ipso facto*

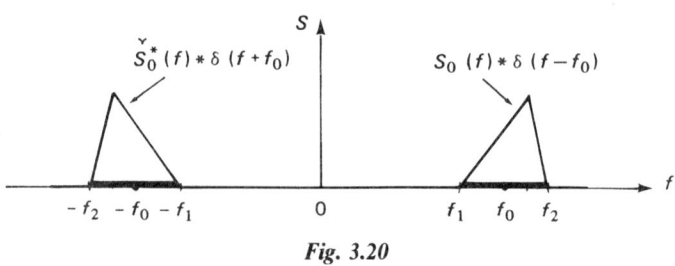

Fig. 3.20

the spectrum S_0 and taking into account the fact that $\check{S} = S^*$, since the d.s. s is real (cf. Eqn (3.11)),

$$S(f) = S_0(f) * \delta(f - f_0) + \check{S}_0^*(f) * \delta(f + f_0) \qquad (3.107)$$

This enables us to write, taking inverse Fourier transforms (cf. Appendix 2),

$$s(t) = s_0(t)\exp(i\omega_0 t) + s_0^*(t)\exp(-i\omega_0 t) \qquad \omega_0 = 2\pi f_0 \qquad (3.108)$$

Let us sample the d.s. at a frequency f_e. The complex spectrum of the sampled signal is obtained by translations of amplitude nf_e, $n \in \mathbb{Z}$, of the spectrum S. The translated spectra do not overlap if there exists an integer such that

$$(-f_1 + kf_e < f_1) \wedge (-f_2 + (k+1)f_e > f_2)$$

For this, the sampling frequency must satisfy

$$f_e \geq 2(f_2 - f_1) \qquad (3.109)$$

Several solutions are then available to choose the integer k.

(a) *Solution 1:* $k = 2p$ Put $\hat{f}_0 = f_0 - pf_e$. The complex spectrum of the sampled d.s. is shown in the band $[-f_e/2, f_e/2]$ in Fig. 3.21.

The two terms of Fig. 3.20 centred on the frequencies $\pm f_0$ are now centred on the frequencies $\pm \hat{f}_0$ without being inverted. The sampling has thus effected a *change of frequency* but has also generated other translated terms. We must therefore filter in order to isolate the two terms situated in the band $[-f_e/2, f_e/2]$.

(b) *Solution 2:* $k = 2p - 1$ Put $\hat{f}_0 = f_0 - pf_e$. The situation is now that of

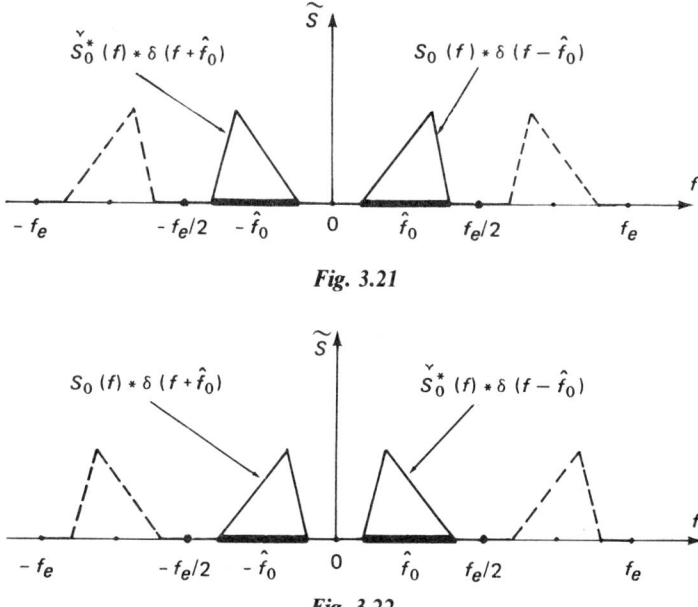

Fig. 3.21

Fig. 3.22

Fig. 3.22. The sampling has thus effected a change of frequency, but this time with inversion of the translated spectra.

Example 5.3   Sampling and hold

The ideal sampled d.s. $\tilde{s}$ defined by Eqn (3.103) can be filtered to generate a sampled d.s. $\hat{s}$ which is a function and for this reason can have a concrete existence. The filter in fact realizes interpolation between the samples. We write

$$\hat{s}(t) = \tilde{s}(t) * h(t) = h(t) * \sum_{n \in \mathbb{Z}} s_n \delta(t - nT - \vartheta) \qquad (3.110)$$

h being the impulse response of the interpolation filter used. The global operation is represented in Fig. 3.23. The complex spectrum $\hat{S}$ of the d.s. $\hat{s}$ may be written (cf. Eqn (3.77))

$$\hat{S} = \tilde{S}H$$

It contains, like the spectrum $\tilde{S}$, an infinity of terms translated by $nf_e, n \in \mathbb{Z}$, but the multiplication of the whole by the function H implies that none of these terms is proportional to the complex spectrum S. The translated spectra are thus distorted.

We have **sampling and hold** (or blocking of order zero) when the impulse response h is a rectangular impulse of the form

$$h(t) = \mathring{I}_a(t) * \delta(t - a/2) \qquad 0 < a < T \qquad (3.111)$$

144 Deterministic signals

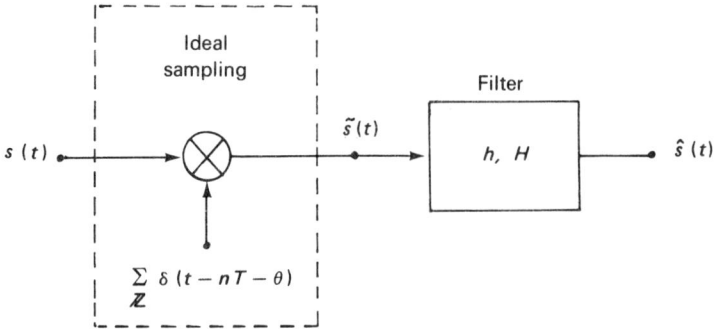

Fig. 3.23

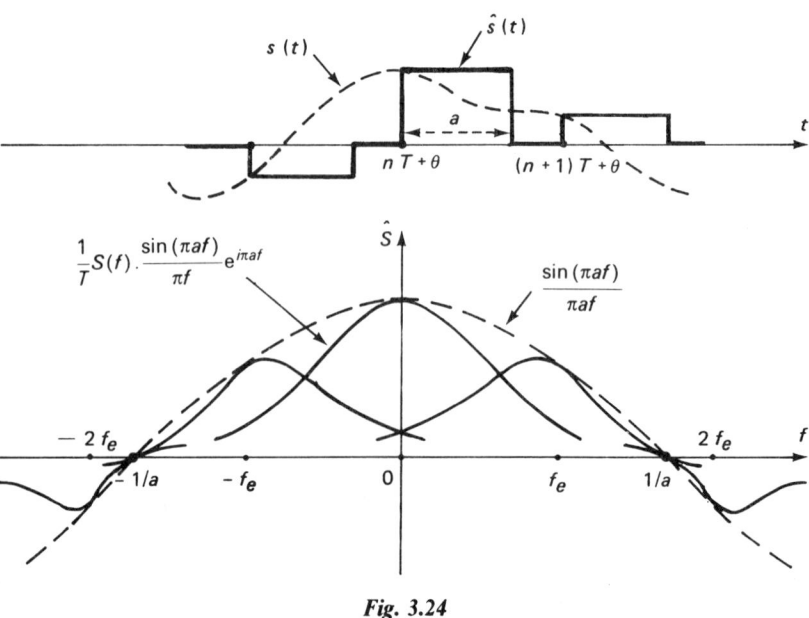

Fig. 3.24

The d.s. and the shape of the complex spectra are shown in Fig. 3.24. This shows the distortion due to the multiplication by the transfer function of the filter.

When the d.s. s is band limited by an upper cut-off frequency f_c, and when the sampling frequency is $f_e = 2f_c$, it suffices to take $h(t) = \sin(\pi f_e t)/\pi f_e t$ to have $\hat{s} = s$. This situation is the subject of the sampling theorem. Interpolation between sample points is then achieved without error by the ideal filter whose transfer function is $H = \mathring{I}_{f_e}/f_e$.

Example 5.4 Analogue sampling

The sampled d.s. $\tilde{s}$ is obtained by multiplication of the d.s. s by the ideal

Sampled deterministic signals 145

sampling d.s. of Eqn (3.60). By multiplying the d.s. s by the clock d.s. of Eqn (3.63) we get the sampled d.s.

$$\hat{s}(t) = s(t)\left\{\mathring{I}_a(t) * \sum_{n \in \mathbb{Z}} \delta(t - nT - \vartheta)\right\} \qquad 0 < a < T \qquad (3.112)$$

Its complex spectrum $\hat{S}$ may be written, since $f_e = 1/T$ (cf. Eqn (3.64)),

$$\hat{S}(f) = \frac{a}{T} S(f) * \sum_{n \in \mathbb{Z}} \frac{\sin(n\pi a f_e)}{n\pi a f_e} \exp(-in2\pi\vartheta f_e) \delta(f - nf_e) \qquad (3.113)$$

It contains, as does $\tilde{S}$, an infinity of terms translated by nf_e, $n \in \mathbb{Z}$, the translation of nf_e being accompanied by multiplication by the constant factor $\sin(n\pi af_e)/n\pi af_e$. The translated terms are thus not distorted, as they are in the case of sampling and hold. The situation is shown in Fig. 3.25. The d.s. $\hat{s}$ is equal to the d.s. s within the pulses cut off by the clock d.s., whence the name **analogue sampling** given to this process.

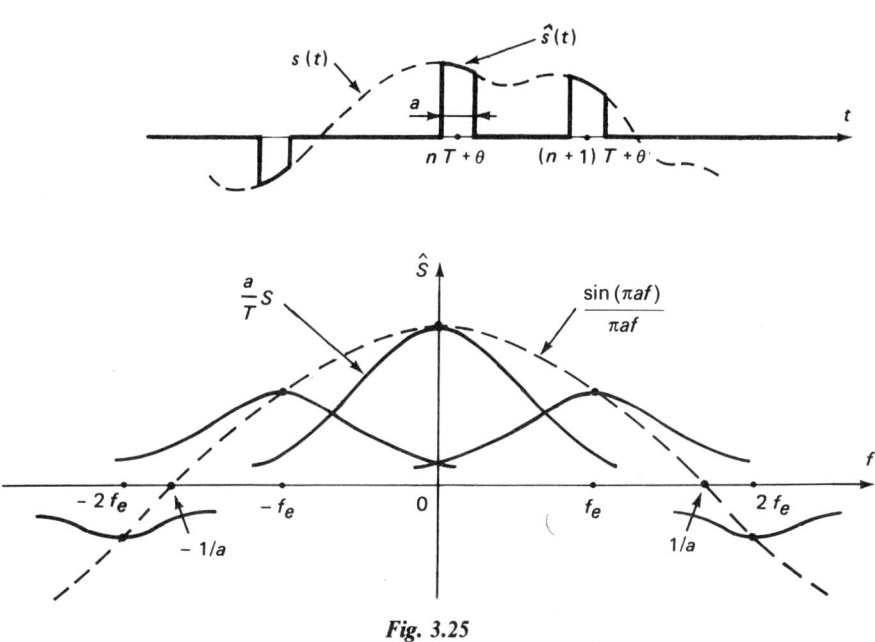

Fig. 3.25

5.2. NUMERICAL SEQUENCES

The sampled d.s.s considered in the preceding section are obtained by sampling of an analogue band-limited d.s. It is possible, however, to define a sampled d.s. without making explicit reference to a band-limited d.s.

5.2.1. Sampled signal associated with a numerical sequence

Consider a real or complex numerical sequence $(s_n, n \in \mathbb{Z})$. When it is bounded, which is always the case in practice, the distribution

$$\sum_{n \in \mathbb{Z}} s_n \delta(t - nT) \qquad T > 0$$

is tempered (Dupraz, 1977, p. 38). We can then associate with it an ideal sampled d.s. $\tilde{s}$ and a complex spectrum $\mathcal{F}\tilde{s}$ defined respectively by

$$\tilde{s}(t) = \sum_{n \in \mathbb{Z}} s_n \delta(t - nT)$$

$$\mathcal{F}\tilde{s}(f) = \sum_{n \in \mathbb{Z}} s_n \exp(-i2\pi nTf)$$
(3.114)

It follows immediately that $\mathcal{F}\tilde{s}$ is a periodic distribution of period $1/T$. It can therefore be expanded in a Fourier series in the space $\mathcal{S}'$ and the uniqueness of the expansion enables us to state that the nth coefficient is the number $c_n = s_{-n}$. Hence the numbers s_n can be obtained from the complex spectrum by the integrals

$$\forall n \in \mathbb{Z} \qquad s_n = T \int_{-1/2T}^{1/2T} \mathcal{F}\tilde{s}(f) \exp(i2\pi nTf) \, df \qquad (3.115)$$

We must note carefully that the integral is to be taken in the distribution sense when the transform $\mathcal{F}\tilde{s}$ is not a function (cf. Appendix 1).

The ideal sampled d.s. defined by Eqn (3.114) has the same properties as the ideal sampled d.s. defined in the preceding section. In particular, it can be filtered to obtain a sampled d.s. $\hat{s}$ which is a function and can have a concrete existence. We can imagine, in practice, that the d.s. $\hat{s}$ is generated by a computer which has in its memory the sequence (s_n) and wishes to transmit it to another computer while it is controlled by a clock of frequency $1/T$.

5.2.2. z transform of a numerical sequence

5.2.2.1. Definition

The z transform of a numerical sequence $(s_n, n \in \mathbb{Z})$ is the Laurent series (Bass, 1956, p. 626)

$$S(z) = \sum_{n \in \mathbb{Z}} s_n z^{-n} \qquad (3.116)$$

5.2.2.2. Remark

Throughout this section $S(z)$ is the z transform of a sequence (s_n) whereas in the preceding sections we have often denoted by $S(f)$ the complex spectrum of a d.s. *s*. This avoids an accumulation of notation, but introduces a risk of confusion. To avoid this inconvenience, the complex spectrum of a d.s. *s* is systematically denoted by $\mathscr{F}s$ throughout this section.

5.2.2.3. Properties

The domain of absolute convergence of the series $S(z)$ is the open annulus defined by $0 < 1/\rho_1 < |z| < \rho_2$ where ρ_1 is the radius of convergence of the power series

$$\sum_{n \in \mathbb{N}} s_n u^n$$

and ρ_2 is the radius of convergence of the power series

$$\sum_{n \in \mathbb{N}} s_{-n} u^n$$

The domain of convergence is not empty when $1/\rho_1 < \rho_2$. The z transform is then a holomorphic function inside the annulus.

The numbers s_n are obtained from the z transform by the contour integrals

$$\forall n \in \mathbb{Z} \quad s_n = \frac{1}{i2\pi} \int_\Gamma S(z) z^{n-1} \, dz \quad (3.117)$$

the closed contour Γ encircling the origin in the positive sense (counterclockwise), while remaining within the annulus of convergence. In most cases these integrals can be evaluated by applying the theorem of residues.

It must be noted that, for a function $S(z)$ to be the z transform of a causal sequence (s_n, $n \in \mathbb{N}$), it is necessary and sufficient that the domain in which it is holomorphic be the exterior of a disc centred at the origin.

When the annulus of convergence contains the circle of radius unity, and only in this case, the complex spectrum $\mathscr{F}\tilde{s}$ is a continuous function connected with the z transform S by

$$\mathscr{F}\tilde{s}(f) = S\{\exp(i2\pi Tf)\} \quad (3.118)$$

5.2.3. *Product of two numerical sequences*

Consider two numerical sequences $(x_n, n \in \mathbb{Z})$ and $(h_n, n \in \mathbb{Z})$ and the numerical sequence $(y_n, n \in \mathbb{Z})$ such that

$$\forall n \in \mathbb{Z} \quad y_n = x_n h_n \quad (3.119)$$

When the sequence (h_n) is such that

$$h_0 = h_1 = \ldots = h_{N-1} = 1$$

the other terms being zero, the sequence (y_n) is the finite sequence obtained by truncation of the sequence (x_n), retaining only N terms.

The complex spectrum $\mathscr{F}\tilde{y}$ of the d.s. $\tilde{y}$ associated with the sequence (y_n) may be written

$$\mathscr{F}\tilde{y}(f) = \sum_{n\in\mathbb{Z}} x_n h_n \exp(-i2\pi nTf)$$

It follows, by introducing the complex spectra $\mathscr{F}\tilde{h}$ and $\mathscr{F}\tilde{x}$ and applying Eqn (3.115), that

$$\mathscr{F}\tilde{y}(f) = \sum_{n\in\mathbb{Z}} x_n \exp(-i2\pi nTf) \left\{ T \int_{-1/2T}^{1/2T} \mathscr{F}\tilde{h}(u) \exp(i2\pi nTu)\, du \right\}$$

$$= \int_{-1/2T}^{1/2T} \mathscr{F}\tilde{h}(f-v) \left\{ \sum_{n\in\mathbb{Z}} x_n \exp(-i2\pi nTv) \right\} dv$$

Hence we have the (improper) convolution integral

$$\mathscr{F}\tilde{y}(f) = T \int_{-1/2T}^{1/2T} \mathscr{F}\tilde{x}(v)\mathscr{F}\tilde{h}(f-v)\, dv \qquad (3.120)$$

When the z transforms of the sequences (x_n), (y_n) and (h_n) exist, the preceding relation becomes (cf. Eqn (3.118))

$$Y\{\exp(i2\pi Tf)\} = T \int_{-1/2T}^{1/2T} X\{\exp(i2\pi Tv)\} H[\exp\{i2\pi T(f-v)\}]\, dv$$

$$(3.121)$$

5.3. NUMERICAL FILTERING

Numerical filtering is an essential tool in signal theory and has been extensively developed. The object of this section is to introduce numerical filtering from the general and functional point of view, without tackling the more technical aspect of realization. The reader interested in this aspect can consult the work cited in Bellanger (1981).

5.3.1. *Convolution of two numerical sequences*

Consider two causal numerical sequences $(x_n, n \in \mathbb{N})$ and $(h_n, n \in \mathbb{N})$ and the associated ideal sampled d.s.

$$\tilde{x}(t) = \sum_{n\in\mathbb{N}} x_n \delta(t-nT) \qquad \tilde{h}(t) = \sum_{n\in\mathbb{N}} h_n \delta(t-nT) \qquad (3.122)$$

Their convolution is always defined. It is the causal sampled d.s.

$$\tilde{y}(t) = \tilde{x}(t) * \tilde{h}(t) = \sum_{m\in\mathbb{N}} \sum_{n\in\mathbb{N}} x_m h_n \delta\{t-(m+n)T\} \qquad (3.123)$$

It is easy to verify that we can also write, defining at the same time the causal numerical sequence $(y_n, n \in \mathbb{N})$,

$$\tilde{y}(t) = \sum_{n \in \mathbb{N}} y_n \delta(t - nT)$$

(3.124)

$$y_n = \sum_{m=0}^{n} x_m h_{n-m} = \sum_{m=0}^{n} h_m x_{n-m}$$

Hence we have the result, by definition of the product of two power series, that the z transform of the sequence (y_n) is the product of the z transforms of the sequences (x_n) and (h_n). It is equivalent to saying that the sequence (y_n) is the convolution of the sequences (x_n) and (h_n). This may be written

$$Y(z) = X(z)H(z) \Leftrightarrow (y_n) = (x_n) * (h_n)$$

(3.125)

5.3.2. *z transform of a numerical filter*

(a) It is natural, in comparing the relations (3.125) and (3.74), to consider that the sequence (y_n) is generated by filtering of the sequence (x_n) by a **linear numerical filter** which has as **impulse response** the sequence (h_n). It then follows immediately that the sequence (h_n) is the response of the filter to the causal sequence (x_n) defined by

$$x_0 = 1 \qquad \forall n \in \mathbb{Z}^* \qquad x_n = 0$$

which thus plays here for the numerical filter a role similar to that played by the Dirac d.s. $\delta(t)$ for analogue filtering. The ideal d.s.s associated with these two sequences are represented in Fig. 3.26 for the case where the sequence (h_n) is real.

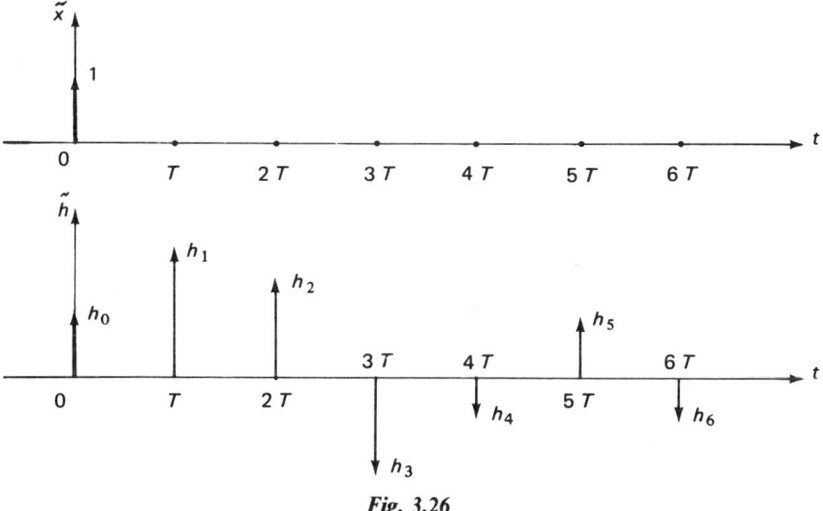

Fig. 3.26

150 Deterministic signals

A causal numerical sequence (h_n) defines an invariant causal numerical filter. The z transform of the filter is the z transform of the sequence. It is the power series

$$H(z) = \sum_{n \in \mathbb{N}} h_n z^{-n} \qquad (3.126)$$

When the sequence (h_n) is finite, the filter is called a **finite impulse response** (FIR) filter or alternatively a **transversal filter**. When the sequence (h_n) is infinite, the filter is called an **infinite impulse response** (IIR) filter or alternatively a **recursive** filter (Bellanger, 1981, pp. 133, 200).

(b) An invariant and causal numerical filter is **stable** when the response to a bounded sequence is a bounded sequence. There is no inconvenience, from the mathematical point of view, in supposing that the sequence (h_n) is complex. In the absence of any statement to the contrary, the results which follow are valid in this case.

Reasoning exactly similar to that which we used with reference to analogue filtering (cf. Example 4.3) shows that an invariant and causal numerical filter is stable if and only if the series with general term h_n is absolutely convergent, which may be written

$$\sum_{n \in \mathbb{N}} |h_n| < +\infty$$

The power series $H(z)$ is then normally convergent for $|z| \geq 1$. Its annulus of convergence has an infinite outer radius and it contains the circumference of radius unity. This is illustrated, in the plane of the complex variable z, by Fig. 3.27.

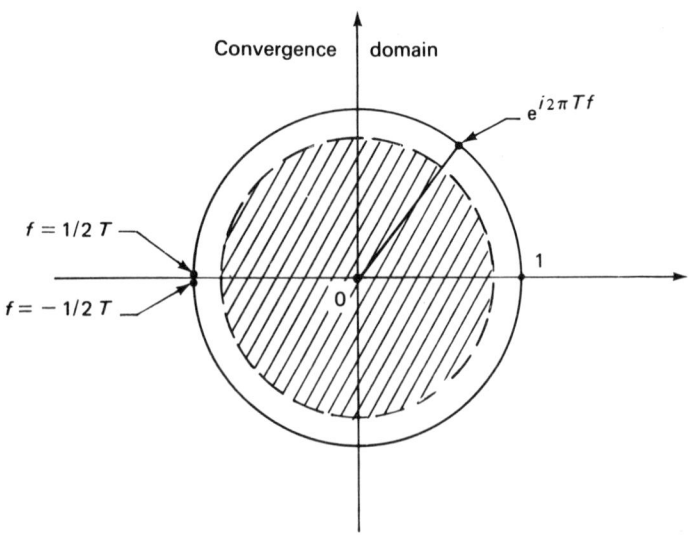

Fig. 3.27

Sampled deterministic signals 151

A numerical transversal filter is always stable because its impulse response is finite. A numerical recursive filter whose z transform is a rational fraction is stable if and only if all its poles are within the circle of radius unity. This results immediately from the previous discussion.

5.3.3. *Transfer function of a numerical filter*

Consider an invariant causal numerical filter defined by its impulse response (h_n). When the filter is stable, the circle of radius unity is contained within the domain of convergence of its z transform. The Fourier transform $\mathscr{F}\tilde{h}$ of the d.s. $\tilde{h}$ associated with the sequence (h_n) is then a continuous function and it follows from Eqns (3.118) and (3.126) that

$$\mathscr{F}\tilde{h}(f) = H\{\exp(i2\pi Tf)\} = \sum_{n \in \mathbb{N}} h_n \exp(-i2\pi n Tf) \quad (3.127)$$

This is the **transfer function** of the numerical filter. It is periodic, of period $1/T$, and it is sufficient to consider it in the interval $[-1/2T, 1/2T[$. When the sequence (h_n) is real it is clear that the transfer function obeys the following relation which is the one corresponding to (3.72):

$$H^*\{\exp(i2\pi Tf)\} = H\{\exp(-i2\pi Tf)\} \quad (3.128)$$

The modulus of the transfer function of a real numerical filter is therefore even and its argument is odd.

From (3.123) and (3.127) we then get the following relation, which is the one corresponding to (3.77):

$$\mathscr{F}\tilde{y}(f) = \mathscr{F}\tilde{x}(f) H\{\exp(i2\pi Tf)\} \quad (3.129)$$

The only difference is that here the three terms are periodic, with period $1/T$.

5.3.4. *Permanent response of a numerical filter*

Consider an invariant causal and stable numerical filter defined by its impulse response (h_n). The relation (3.124) can be extended to the case of non-causal but bounded sequences (x_n). It follows then that

$$\forall n \in \mathbb{Z} \quad y_n = \sum_{m \in \mathbb{N}} h_m x_{n-m} = \sum_{m=-\infty}^{n} x_m h_{n-m} \quad (3.130)$$

Example 5.5 Response at the frequency f_0
Let the sequence be defined by

$$\forall n \in \mathbb{Z} \quad x_n = \exp\{i(2\pi n Tf_0 + \varphi_0)\} \quad (3.131)$$

It is obtained by sampling at frequency $f_e = 1/T$ the sinusoidal d.s. $x(t) =$

exp{i(2πf_0t + $φ_0$)} of frequency f_0 (cf. Eqn (3.104)). It follows that

$$\forall n \in \mathbb{Z} \qquad y_n = x_n H\{\exp(i2\pi T f_0)\} \qquad (3.132)$$

This relation should be compared with relation (3.81) which is the corresponding form for analogue d.s.s. It indicates that the sequence (x_n) is an eigenfunction of the convolution operator defined by $(x_n) \mapsto (x_n) * (h_n)$, the corresponding eigenvalue being $H\{\exp(i2\pi T f_0)\}$.

5.3.5. Inversion of a numerical filter

(a) Consider an invariant causal and stable numerical filter with impulse response the sequence (h_n). The power series $H(z)$ is invertible if and only if the term h_0 is not zero. Then let the inverse series be

$$K(z) = H^{-1}(z) = 1/H(z) \qquad (3.133)$$

When its domain of convergence contains the circle of radius unity, $K(z)$ is the z transform of an invariant causal and stable numerical filter, having as its impulse response an absolutely convergent sequence (k_n). The filter H is then invertible and it has as inverse the filter $K = H^{-1}$.

The transfer function of the filter K is obtained from Eqns (3.127) and (3.133). We get

$$\mathscr{F}\tilde{k}(f) = K\{\exp(i2\pi T f)\} = \frac{1}{H\{\exp(i2\pi T f)\}} \qquad (3.134)$$

(b) Consider an invertible and stable filter H and a bounded sequence (x_n) applied at its input. The response is a bounded sequence (y_n). Conversely, when the inverse filter is stable, the sequence (x_n) can be obtained by filtering the sequence (y_n) by the inverse filter H^{-1}. The z transforms are then connected by

$$X(z) = Y(z)H^{-1}(z) \qquad (3.135)$$

This is illustrated by Fig. 3.28.

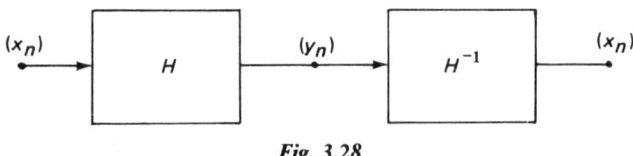

Fig. 3.28

Example 5.6 Recursive filter of order 1

(a) Consider a complex numerical sequence $(x_n, n \in \mathbb{Z})$ and the complex sequence $(y_n, n \in \mathbb{Z})$ defined by the recurrence relation

$$\forall n \in \mathbb{Z} \qquad y_n = \lambda y_{n-1} + \mu x_n \qquad \lambda, \mu \in \mathbb{C} \qquad (3.136)$$

The z transforms of the two sequences can be obtained by multiplying the two sides of Eqn (3.136) by z^{-n} and summing over $\mathbb{Z}$. We obtain

$$Y(z) = H(z)X(z) = \frac{\mu}{1 - \lambda z^{-1}} X(z)$$

The sequence (y_n) is therefore obtained by filtering the sequence (x_n) using the numerical filter whose z transform is the rational fraction

$$H(z) = \frac{\mu}{1 - \lambda z^{-1}} = \frac{\mu z}{z - \lambda} \tag{3.137}$$

This is the **recursive filter of order 1**. Its block diagram is given in Fig. 3.29. The filter is stable if and only if $|\lambda| < 1$ and its impulse response is the infinite

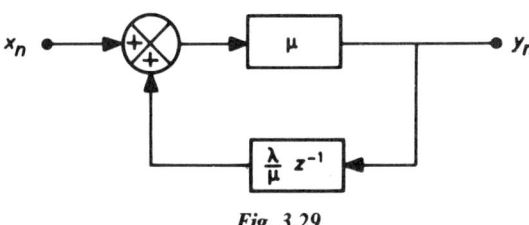

Fig. 3.29

complex causal sequence obtained by series expansion of the rational fraction H. We obtain

$$\forall n \in \mathbb{N} \qquad h_n = \mu \lambda^n \tag{3.138}$$

The transfer function of the filter is the function

$$H\{\exp(i2\pi Tf)\} = \frac{\mu}{1 - \lambda \exp(-i2\pi Tf)} \tag{3.139}$$

(b) Suppose now that the parameters λ and μ are real and that the filter is stable. The modulus of the transfer function may be written

$$|H\{\exp(i2\pi Tf)\}| = |\mu|\{1 + \lambda^2 - 2\lambda \cos(2\pi Tf)\}^{1/2} \tag{3.140}$$

Hence we have

$$|H\{\exp(i2\pi Tf)\}|_{f=0} = \left|\frac{\mu}{1-\lambda}\right| \qquad |H\{\exp(i2\pi Tf)\}|_{f=\pm 1/2T} = \left|\frac{\mu}{1+\lambda}\right|$$

$$\tag{3.141}$$

It follows from this that the recursive filter of order 1 is of low-pass type for $0 < \lambda < 1$.

It is easy to verify that the slope at the origin of the phase of the transfer

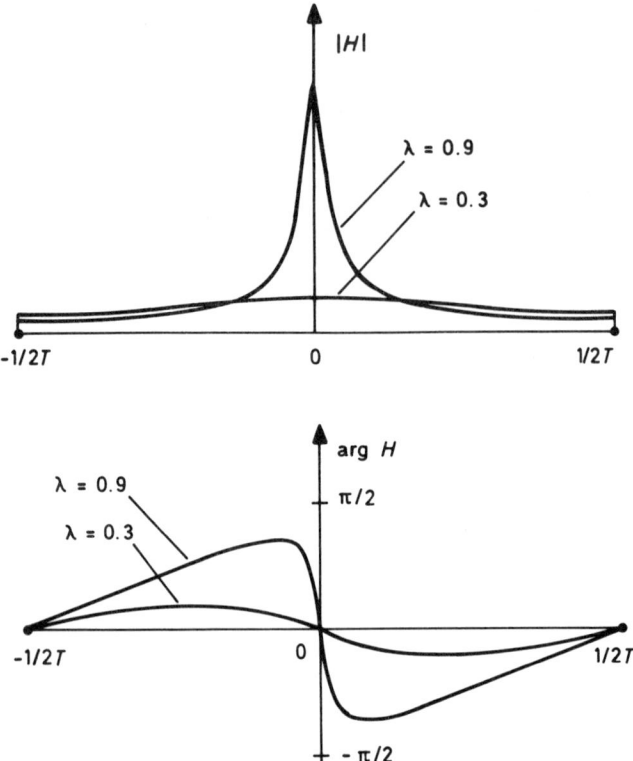

Fig. 3.30

function may be written

$$\frac{d}{df}[\arg H\{\exp(i2\pi Tf)\}]_{f=0} = 2\pi T \frac{\lambda}{1-\lambda} \qquad (3.142)$$

The modulus and phase of the transfer function are represented in Fig. 3.30 for $\mu = 1$ with $\lambda = 0.3$ and $\lambda = 0.9$. It should be noted that the filtering is more effective the closer λ is to unity.

(c) The recursive filter of order 1 is invertible and the inverse filter has as z transform the rational fraction

$$K(z) = \frac{z - \lambda}{\mu z} \qquad (3.143)$$

It has a simple pole at the origin. The inverse filter is thus stable and its impulse response is the finite complex sequence (k_n) defined by

$$k_0 = \frac{1}{\mu} \qquad k_1 = -\frac{\lambda}{\mu} \qquad k_n = 0 \quad n \geq 2 \qquad (3.144)$$

The inverse filter is thus transversal and it follows that (cf. Eqn (3.130))

$$\forall n \in \mathbb{Z} \qquad x_n = \frac{1}{\mu} y_n - \frac{\lambda}{\mu} y_{n-1} \qquad (3.145)$$

The number x_n is a linear combination of the numbers y_n and y_{n-1}. From this fact we say that the filter K works out a **moving average** over the sequence (y_n).

The filter K is of high-pass type when the parameters λ and μ are real and when $0 < \lambda < 1$. This follows immediately from Eqns (3.134) and (3.140).

Exercises

Exercise 5.1 Autocorrelation function of a numerical filter

(1) Consider a complex causal numerical sequence $(h_n, n \in \mathbb{N})$ and the associated d.s. (cf. Eqn (3.114))

$$\tilde{h}(t) = \sum_{n \in \mathbb{N}} h_n \delta(t - nT)$$

We can consider the sequence (h_n) to be the impulse response of a complex numerical filter.

The autocorrelation function $R_{\tilde{h}\tilde{h}}$ of the d.s. $\tilde{h}$ is defined by analogy with Eqn (3.25) by the convolution

$$R_{\tilde{h}\tilde{h}}(\tau) = \tilde{h}(\tau) * \check{\tilde{h}}^*(\tau)$$

We must note that this convolution is not defined in the general case.

(1.1) Prove that

$$R_{\tilde{h}\tilde{h}}(\tau) = \sum_{k \in \mathbb{Z}} R_{hh}(kT) \delta(\tau - kT)$$

by putting

$$\forall k \in \mathbb{Z} \qquad R_{hh}(kT) = \sum_{n \in \mathbb{N}} h_n h_{n-k}^* = \sum_{m \in \mathbb{N}} h_m^* h_{m+k} \qquad (3.146)$$

(1.2) Hence deduce that

$$\forall k \in \mathbb{Z} \qquad R_{hh}(kT) = R_{hh}^*(-kT) \qquad (3.147)$$

(1.3) The sequence $(R_{hh}(kT), k \in \mathbb{Z})$ defines the autocorrelation function of the numerical filter whose impulse response is the sequence (h_n). Show that its z transform may be written

$$R_{hh}(z) = H(z) H_*(z^{-1}) \qquad (3.148)$$

where $H(z)$ is the z transform of the sequence (h_n) and $H_*(z)$ that of the sequence (h_n^*).

156 Deterministic signals

(1.4) Hence deduce that

$$R_{hh}\{\exp(i2\pi Tf)\} = |H\{\exp(i2\pi Tf)\}|^2 \qquad (3.149)$$

(2) Suppose now that the sequence (h_n) is real and that it is the impulse response of a stable filter.

(2.1) Show that under these conditions we have

$$\forall k \in \mathbb{Z} \qquad R_{hh}(kT) = \sum_{n \in \mathbb{N}} h_n h_{n+|k|} \qquad (3.150)$$

(2.2) Prove that the annulus of convergence of the transform $R_{hh}(z)$ contains the circle of radius unity, which implies that the autocorrelation function of a stable filter is well defined and that $R_{hh}\{\exp(i2\pi Tf)\}$ is a continuous function of the frequency.

Exercise 5.2 Step response of a numerical filter

A signal defined by the causal sequence $(x_n = 1 \ \forall n \in \mathbb{N})$ is applied at the input of a stable numerical filter defined by the causal sequence $(h_n, n \in \mathbb{N})$. The output sequence $(y_n, n \in \mathbb{N})$ is the **step response** of the filter.

(1) Show that

$$\forall n \in \mathbb{N} \qquad y_n = \sum_{m=0}^{n} h_m \qquad (3.151)$$

(Apply Eqn (3.124).)

(2) Prove that the step response of the recursive filter of order 1 (cf. Example 5.6) is defined by

$$\forall n \in \mathbb{N} \qquad y_n = \mu \frac{1 - \lambda^{n+1}}{1 - \lambda} \qquad (3.152)$$

Compare the step response of the recursive filter with that of the low-pass RC analogue filter (cf. Exercise 4.1).

Exercise 5.3 Moving average transversal filter

Consider the transversal filter whose impulse response is the real sequence defined by

$$h_n = \frac{1}{N} \qquad n \in \{0, 1, \ldots, N-1\} \qquad N \in \mathbb{N}^*$$
$$= 0 \qquad n \notin \{0, 1, \ldots, N-1\} \qquad (3.153)$$

(1) Show that its z transform is the rational fraction

$$H(z) = \frac{1}{N} \frac{1 - z^{-N}}{1 - z^{-1}} \qquad (3.154)$$

(Apply Eqn (3.116).)

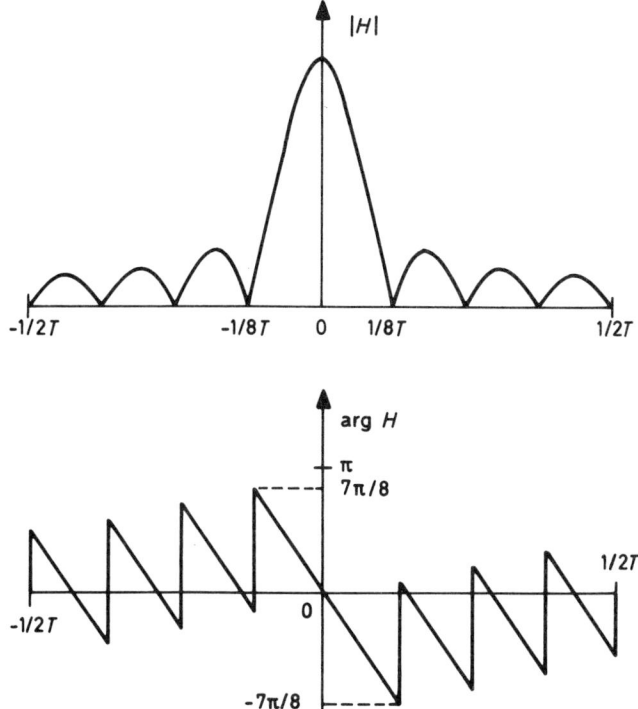

Fig. 3.31

(2) Show that its transfer function may be written

$$H\{\exp(i2\pi Tf)\} = \frac{1}{N} \exp\{-i(N-1)\pi Tf\} \frac{\sin(N\pi Tf)}{\sin(\pi Tf)} \quad (3.155)$$

Hence deduce that the moving average transversal filter is of low-pass type and that its phase varies linearly, section by section, as a function of the frequency.

(3) Draw the graphs of the modulus and of the phase of the transfer function in the case $N = 8$. (The graphs are shown in Fig. 3.31.) Compare these graphs with those of Fig. 3.30.

Exercise 5.4 Recursive filter of order 2

Consider a complex numerical sequence $(x_n, n \in \mathbb{Z})$ and the complex numerical sequence $(y_n, n \in \mathbb{Z})$ defined by the recurrence relation

$$\forall n \in \mathbb{Z} \quad y_n = \lambda_1 y_{n-1} + \lambda_2 y_{n-2} + \mu x_n \quad \lambda_1, \lambda_2, \mu \in \mathbb{C} \quad (3.156)$$

(1) Show that the sequence (y_n) is obtained by filtering the sequence (x_n) by means of a numerical filter having as z transform the rational fraction

158 Deterministic signals

$$H(z) = \frac{\mu z^2}{z^2 - \lambda_1 z - \lambda_2} \tag{3.157}$$

It defines the **recursive filter of order 2**. (Apply Eqn (3.116).)

(2) Show that the impulse response of the filter is the causal sequence defined by

$$\forall n \in \mathbb{N} \qquad h_n = \mu \frac{z_1^{n+1} - z_2^{n+1}}{z_1 - z_2} \tag{3.158}$$

z_1 and z_2 being the poles of $H(z)$. (Use Eqn (3.117) and apply the theorem of residues.)

(3) Verify that the filter is stable only if $|\lambda_2| < 1$.

(4) Show that, inversely, the sequence (x_n) can be generated by numerical filtering of the sequence (y_n) and that

$$\forall n \in \mathbb{N} \qquad x_n = \frac{1}{\mu} y_n - \frac{\lambda_1}{\mu} y_{n-1} - \frac{\lambda_2}{\mu} y_{n-2} \tag{3.159}$$

The inverse filter is thus a **moving average** transversal filter of order 3.

(5) Suppose now that the poles of $H(z)$ are of the form $z_1 = r \exp(i\vartheta)$ and $z_2 = r \exp(-i\vartheta)$, which implies that λ_1 and λ_2 are real and may be written

$$\lambda_1 = 2r \cos \vartheta \qquad \lambda_2 = -r^2$$

(5.1) Verify that the impulse response of the filter is the causal sequence defined by

$$\forall n \in \mathbb{N} \qquad h_n = \mu r^n \frac{\sin\{(n+1)\vartheta\}}{\sin \vartheta} \tag{3.160}$$

(5.2) Suppose that the filter is stable. Verify that the modulus of its transfer function has extrema for the frequencies $f = 0$ and $f = \pm 1/2T$.

Show that, if the condition $|\cos \vartheta| < 2r/(1+r^2)$ is also satisfied, the modulus has a maximum for the frequencies $f = \pm f_0$, where the frequency $f_0 \in [0, 1/2T]$ and the value of the maximum are given respectively by

$$\cos(2\pi T f_0) = \cos \vartheta \frac{1+r^2}{2r}$$

$$|H\{\exp(i2\pi T f_0)\}| = \frac{|\mu|}{1-r^2} \left(\frac{1}{1 - \cos^2 \vartheta} \right)^{1/2} \tag{3.161}$$

The frequency f_0 is the **resonance frequency** of the filter, which is of band-pass type.

(5.3) Draw the graphs of the modulus and of the phase of the transfer function, supposing that $0 < \cos \vartheta < 2r/(1+r^2)$. (The graphs are shown in Fig. 3.32 for the case when $\lambda_1 = -0.8$ and $\lambda_2 = -0.8$.)

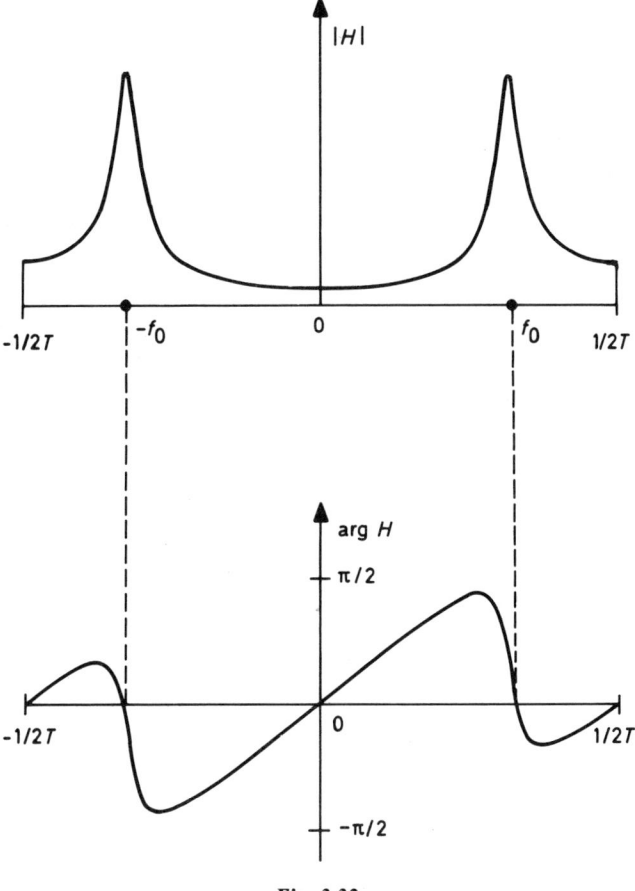

Fig. 3.32

Exercise 5.5 The Nyquist channel

Consider a sequence $(x_n, n \in \mathbb{Z})$ of analogue samples and the associated ideal signal (cf. Eqn (3.114))

$$\tilde{x}(t) = \sum_{n \in \mathbb{Z}} x_n \delta(t - nT)$$

The d.s. $\tilde{x}$ is transmitted by a channel represented by an (analogue) linear filter having impulse response h and transfer function H. The filtered d.s. is sampled at the instants $\gamma + mT, m \in \mathbb{Z}, \gamma > 0$. Let $(y_m, m \in \mathbb{Z})$ be the sequence of sample points obtained and

$$\tilde{y}(t) = \sum_{m \in \mathbb{Z}} y_m \delta(t - mT - \gamma)$$

be the associated ideal d.s. This is illustrated in Fig. 3.33.

160 Deterministic signals

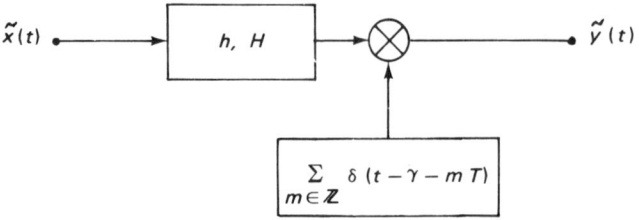

Fig. 3.33

It is required that the sequence (y_m) be identical with the sequence (x_n) to within a delay γ and a gain A. This may be written

$$\tilde{y}(t) = A\tilde{x}(t - \gamma) \qquad \gamma > 0 \qquad A > 0 \tag{3.162}$$

In other words it is required that the sequence (x_n) be transmitted without **interference** between the sample points, since the samples y_m are equal to the samples x_m and are independent of the samples x_k, $k \neq m$.

The object of the exercise is to find conditions which the filter h must satisfy in order that this may be true. These conditions are sometimes called **Nyquist conditions.**

(1) Show that Eqn (3.162) is satisfied, whatever the sequence (x_n) may be, if and only if

$$h(\gamma) = A \qquad \forall k \in \mathbb{Z}^* \quad h(\gamma + kT) = 0 \tag{3.163}$$

(Express Eqn (3.162) in the form

$$\{\tilde{x}(t) * h(t)\} \sum_{m \in \mathbb{Z}} \delta(t - \gamma - mT) = A\tilde{x}(t) * \delta(t - \gamma)$$

Hence deduce that the condition (3.163) is necessary by considering the case of a sequence such that $\tilde{x}(t) = \delta(t)$. Then show that it is sufficient.)

(2) Deduce from Eqn (3.163) the equivalent condition

$$\sum_{n \in \mathbb{Z}} \exp\left(-i2\pi\gamma\frac{n}{T}\right) H\left(f - \frac{n}{T}\right) = AT \exp(-i2\pi\gamma f) \tag{3.164}$$

(Write Eqn (3.163) in the equivalent form

$$h(t) \sum_{n \in \mathbb{Z}} \delta(t - nT - \gamma) = A\delta(t - \gamma)$$

and take the Fourier transform of both sides. This supposes, to be completely rigorous, that h is an infinitely differentiable function, which is the case when the filter is band limited.)

(3) We now impose on the filter used the condition that it has a cut-off frequency f_c such that supp $H = [-f_c, f_c]$.

(3.1) Show that no filter exists satisfying Eqn (3.164) when $1/T > 2f_c$. (Argue in terms of the supports to show that Eqn (3.164) is impossible.)

(3.2) Show that when $1/T = 2f_c$ there exists only one filter satisfying Eqn (3.164), having for transfer function

$$H(f) = AT\mathring{I}_{1/T}(f)\exp(-i2\pi\gamma f) \quad (3.165)$$

This filter defines the **Nyquist channel**.

(3.3) Deduce from the above that a filter meeting the Nyquist conditions necessarily has a cut-off frequency such that $f_c \geq 1/2T$. The Nyquist filter is the one whose spectral bandwidth is minimal. It is unfortunately not realizable since its transfer function is not continuous.

(3.4) Consider the cosine-squared filter having transfer function

$$H(f) = \frac{AT}{2}\mathring{I}_{2/T}(f)\{1 + \cos(\pi Tf)\}\exp(-i2\pi\gamma f) \quad (3.166)$$

Show that it satisfies the Nyquist conditions. It is 'more realizable' than the Nyquist filter since its transfer function is continuous, but its cut-off frequency is twice as large since here $f_c = 1/T$. This is illustrated by Fig. 3.34.

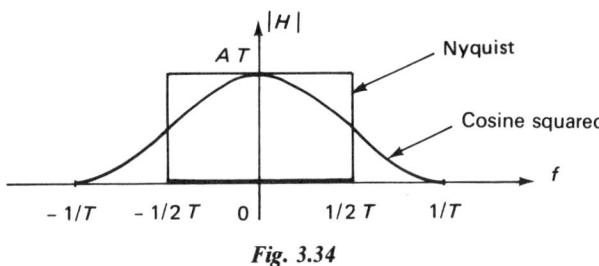

Fig. 3.34

Exercise 5.6 'Partial response' channel of order 2

This exercise is a continuation of the preceding exercise. Consider now a channel, called a **partial response** channel of order 2, defined by a filter whose impulse response h satisfies

$$h(\gamma) = a_0 \quad h(\gamma + T) = a_1 \quad \forall k \notin \{0, 1\} \; h(\gamma + kT) = 0 \quad (3.167)$$

It thus does not satisfy condition (3.163) and consequently a sequence (x_n) is not generally transmitted without interference between successive sample points.

(1) Show that

$$\forall m \in \mathbb{Z} \quad y_m = a_0 x_m + a_1 x_{m-1} \quad (3.168)$$

There is interference between two successive sample points and the ensemble consisting of the filter followed by the sampler (cf. Fig. 3.33) can be represented by a transversal numerical moving average filter of order 2 (cf. Eqn (3.145)).

(2) Show that it is possible to restore the sequence (x_n) by filtering the sequence (y_m) by a recursive filter of order 1 which is stable only if $|a_1/a_0| < 1$ (cf. Example 5.6).

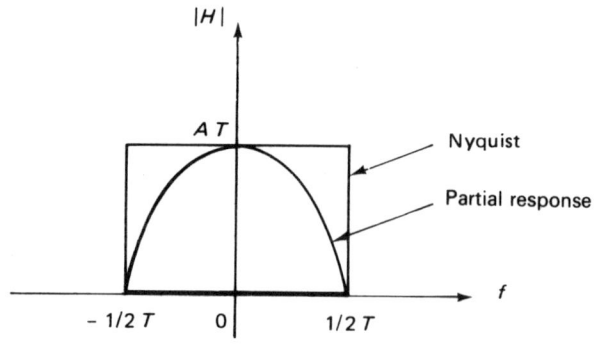

Fig. 3.35

(3) Consider the filter having the transfer function

$$H(f) = \frac{AT}{2} \mathring{1}_{1/T}(f)\{1 + \exp(-i2\pi Tf)\}\exp(-i2\pi\gamma f) \qquad (3.169)$$

It has the same bandwidth as the Nyquist channel (cf. Fig. 3.35). Show that its impulse response satisfies

$$h(\gamma) = \frac{A}{2} \qquad h(\gamma + T) = \frac{A}{2} \qquad \forall k \notin \{0, 1\} \quad h(\gamma + kT) = 0$$

The transfer function (3.169) thus defines a 'partial response' channel of order 2. This is illustrated by Fig. 3.35.

Exercise 5.7 Causality relations

(1) Let $(v_n, n \in \mathbb{Z})$ be the real numerical sequence defined by

$$\forall n < 0 \quad v_n = -\tfrac{1}{2} \qquad \forall n \geqslant 0 \quad v_n = \tfrac{1}{2}$$

Show that the complex spectrum of the sampled d.s. $\tilde{v}$ associated with the sequence (v_n) can be expressed as

$$\mathscr{F}\tilde{v}(f) = \frac{1}{2} - \frac{i}{2}\cot(\pi Tf) \qquad (3.170)$$

(Use Eqn (3.114) and Fourier series expansion 22 in Appendix 1.)

(2) Let $(u_n, n \in \mathbb{Z})$ be the causal numerical sequence defined by

$$\forall n < 0 \quad u_n = 0 \qquad \forall n \geqslant 0 \quad u_n = 1$$

Show that the complex spectrum $\mathscr{F}\tilde{u}$ can be expressed as

$$\mathscr{F}\tilde{u}(f) = \frac{1}{2T}\sum_{n \in \mathbb{Z}}\delta\left(f - \frac{n}{T}\right) + \frac{1}{2} - \frac{i}{2}\cot(\pi Tf) \qquad (3.171)$$

(Write $u_n = v_n + w_n$ with $\forall n \in \mathbb{Z}$, $w_n = \tfrac{1}{2}$, and apply Eqn (3.120).)

(3) Let $(x_n, n \in \mathbb{N})$ be a causal numerical sequence and let $\mathscr{F}\tilde{x} = R + iX$ be

the complex spectrum. Prove the causality relations

$$R(f) = \operatorname{Re} x_0 + T \int_{-1/2T}^{1/2T} x(v) \cot(f - v) \, dv$$

$$X(f) = \operatorname{Im} x_0 - T \int_{-1/2T}^{1/2T} R(v) \cot(f - v) \, dv \qquad (3.172)$$

(Write $x_n = x_n u_n$ and apply Eqn (3.120).)

References

BASS J. (1956): *Cours de Mathématiques*, Masson, Paris.
BELLANGER M. (1981): *Traitement Numérique du Signal*, Masson, Paris.
DUPRAZ J. (1977): *La Théorie des Distributions et ses Applications*, Cepadues, Toulouse.
GUICHARDET A. (1969): *Calcul Intégral*, Colin, Paris.
SCHWARTZ L. (1979): *Méthodes Mathématiques pour les Sciences Physiques*, Hermann, Paris.

Chapter 4
Random signals

The general properties of random signals (r.s.s) form the subject of this chapter. It has a bearing on Chapters 1 and 2 since an r.s. is a family of random variables defined over one and the same probability space. It has a bearing also on Chapter 3 since an r.s. is also a family of deterministic signals.

In addition to notions already introduced, e.g. independence, non-correlation, a supplementary notion is that of equivalent representation. It enables us to represent the same random signal by different families of deterministic signals.

R.s.s of finite mean power, in continuous time, possess particular properties. Under certain conditions, they can be differentiated and integrated. They are important inasmuch as they generally have concrete existence. One section is devoted to them.

Two types of r.s. play an essential role in theory as in practice. The first is the gaussian r.s. This is the model of thermal noise and in general of all noise in nature which has a regular and continuous character. The second is the Poisson r.s. This is the theoretical model for point processes which are regular and devoid of memory. It is widely used in queuing theory. These signals are the subject of numerous examples and exercises.

1. Definitions and general properties

An r.s. is a real or complex **random function** of time. The reader interested in more details about random functions should consult Blanc-Lapierre and Picinbono (1981), Neveu (1964) and Wong (1971, p. 41).

1.1. GENERAL DEFINITIONS

1.1.1. Definition 1

Let $(\Omega, \mathcal{A}, P)$ be a probability space, $I \subset \mathbb{R}$ be a set representing the **time space**, in which the variable t takes its values, and $(\mathbb{R}^I, \mathcal{R}^I)$ be a measurable product space. A real r.s. ξ is a measurable mapping from the measurable space $(\Omega, \mathscr{A})$ into the measurable space $(\mathbb{R}^I, \mathscr{R}^I)$.

1.1.2. Realizations of a random signal

The set $\mathbb{R}^I$ is the set of real functions defined on the set I. The Borel field $\mathscr{R}^I$ is the smallest Borel field of subsets of $\mathbb{R}^I$ containing the Borel fields generated, when t runs through I, by the projections $\varphi_t : \mathbb{R}^I \mapsto (\mathbb{R}_t, \mathscr{R}_t)$, $\mathbb{R}_t$ being the factor space corresponding to the instant $t \in I$ and $\mathscr{R}_t$ being its Borel field (cf. Chapter 6, Section 1.2.2 and Example 1.4).

The mapping ξ produces for every trial $\omega \in \Omega$ a corresponding point of $\mathbb{R}^I$, i.e. a real function defined over the set I, which we shall denote by $\xi(\omega, t)$. This is the **realization** or again the **trajectory** of the r.s. ξ corresponding to the trial ω.

The mapping ξ is measurable if and only if the composite mappings $\xi_t = \varphi_t \circ \xi : (\Omega, \mathscr{A}) \mapsto (\mathbb{R}_t, \mathscr{R}_t)$ are measurable (cf. Chapter 6, Section 2.2.2). These mappings are thus real random variables (r.v.s) defined over the space $(\Omega, \mathscr{A}, P)$.

This leads us to give the following definition. It is equivalent to the preceding definition and is extended to the case of a complex r.s.

1.1.3. Definition 2

A complex r.s. $\xi(t)$, $t \in I$, is a mapping $(\omega, t) \mapsto \xi(\omega, t)$ from the set $\Omega \times I$ into $\mathbb{C}$, such that (i) for every $\omega \in \Omega$, $\xi(\omega, t)$ is a complex function of time defining a d.s. over the set I and (ii) for every $t \in I$, $\xi(\omega, t)$ is a complex r.v., denoted, according to circumstances, by ξ_t or $\xi(t)$, defined over the space $(\Omega, \mathscr{A}, P)$.

1.1.4. Classes of random signal

We use several classes of r.s., according to the nature of the time space I over which the r.s.s are defined:

- r.s. in continuous time, $I = \mathbb{R}$
- causal r.s. in continuous time, $I = \mathbb{R}_+$
- r.s. in discrete time, $I = T\mathbb{Z}, T \in \mathbb{R}_+^*$
- causal r.s. in discrete time, $I = T\mathbb{N}, T \in \mathbb{R}_+^*$

Figures 4.1(a) and 4.1(b) represent respectively the realizations $\xi(\omega_i, t)$ and $\xi(\omega_j, t)$ of an r.s. in continuous time and in discrete time. In the first case, the realizations represented are continuous functions. In the second, we denote by ξ_n the r.v. $\xi(nT)$ and the realizations are denumerable sets of points.

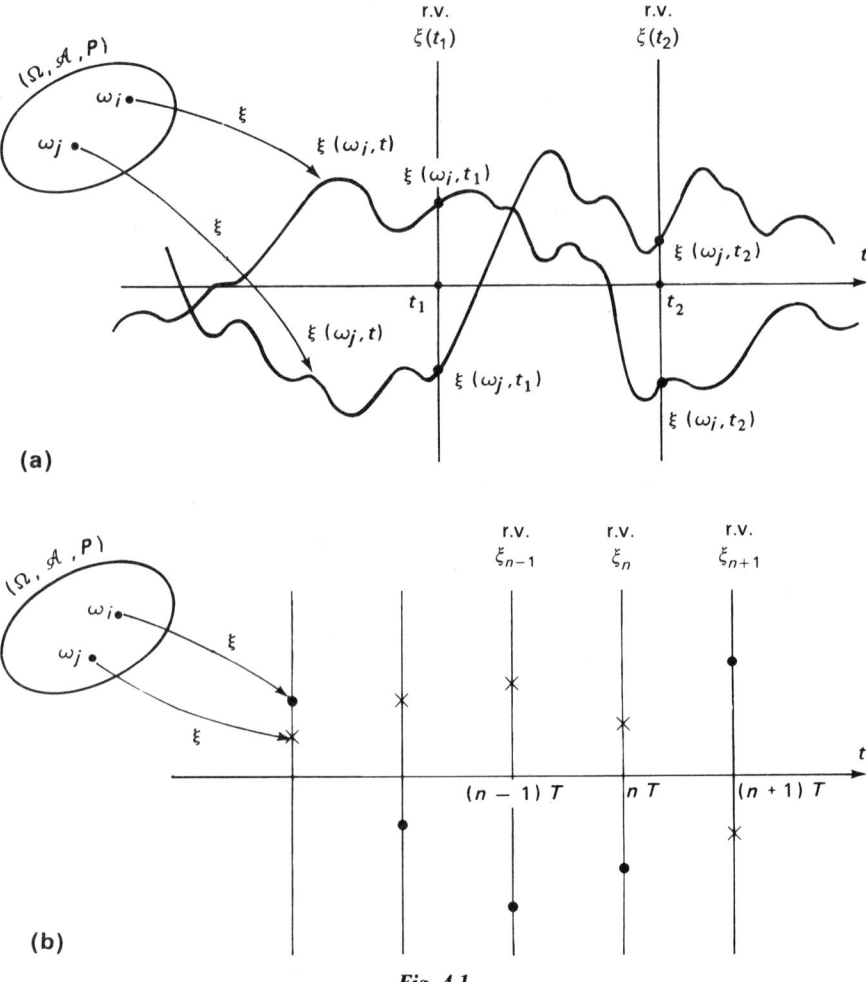

Fig. 4.1

Example 1.1 Random signal of finite mean power

Let $\xi(t)$, $t \in I$, be a real r.s. We interpret the r.v. $\xi(t)$ physically as a voltage measured at the terminals of an ideal resistance of $1\,\Omega$ at time t. The mean square value $E[\xi^2(t)]$ is the mean instantaneous power $P_\xi(t)$ developed in the resistance at time t by the r.s. ξ, the mean being taken over the ensemble of realizations of the r.s.

In the case when the r.s. is complex, we shall say by extension that the mean square value $E[|\xi(t)|^2]$ is the mean instantaneous power of the r.s. ξ. A complex r.s. $\xi(t)$, $t \in I$, is of finite mean power when

$$\forall t \in I \qquad E[|\xi(t)|^2] = P_\xi(t) < +\infty \qquad (4.1)$$

which implies that the r.v.s $\xi(t)$ are square integrable.

It is then possible to define an r.s. of finite mean power as a mapping $t \mapsto \xi(t)$ from the set I into the Hilbert space $L^2(\Omega, \mathscr{A}, P)$ of square integrable r.v.s (cf. Chapter 1, Section 2.4.2). The mean instantaneous power $P_\xi(t)$ is the square of the norm of the r.v. $\xi(t)$.

R.s.s of finite mean power play an important role in practice, since they can have a concrete existence. They form the subject of Section 3. We note that an r.s. of finite mean power is also said to be of order 2.

Example 1.2 Sinusoidal random signal

Let $\xi(t)$, $t \in \mathbb{R}$, be the r.s. whose realizations may be written

$$\xi(\omega, t) = A(\omega) \cos\{2\pi f_0 t + \varphi_0(\omega)\} \qquad f_0 > 0 \qquad (4.2)$$

where A is a real positive r.v. and φ_0 is a real r.v. distributed over the interval $[-\pi, \pi]$. It is completely defined by the probability law of the random vector (r.v.) (A, φ_0). Its realizations are sinusoids of the same frequency f_0 but of different amplitudes and phases. The realizations $\xi(\omega_i, t)$ and $\xi(\omega_j, t)$ are represented in Fig. 4.2.

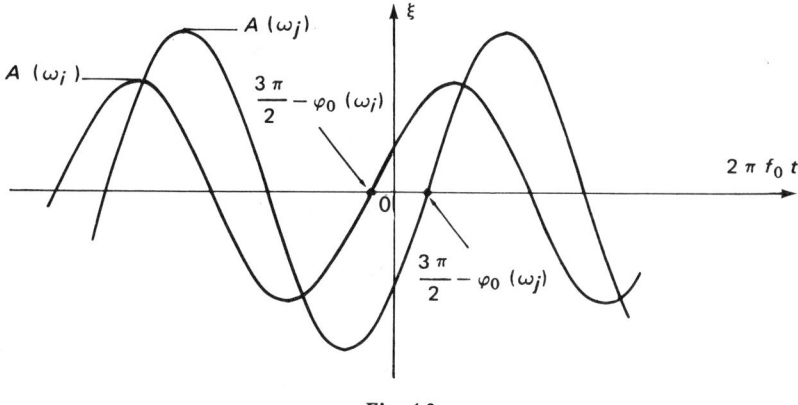

Fig. 4.2

Example 1.3 Poisson random signal (Cox and Miller, 1967)

Renewal process

Let $(I_n, n \in \mathbb{N}^*)$ be a strictly increasing sequence of r.v.s defined over the probability space $(\Omega, \mathscr{A}, P)$, the r.v. T_1 being strictly positive. We then have, for every trial ω, $0 < T_1(\omega) < \ldots < T_n(\omega) \ldots$. The sequence (T_n) defines the instants at which certain **point random events** occur.

Let $N(t), t \in \mathscr{R}_+^*$, be the real causal r.s. defined by the counting function of the sequence (T_n). The r.v. $N(t)$ represents the number of point events realized in the interval $]0, t]$. It can be written

$$N(t) = \sum_{n \in \mathbb{N}_+^*} I_{\{T_n \leq t\}} \qquad (4.3)$$

168 *Random signals*

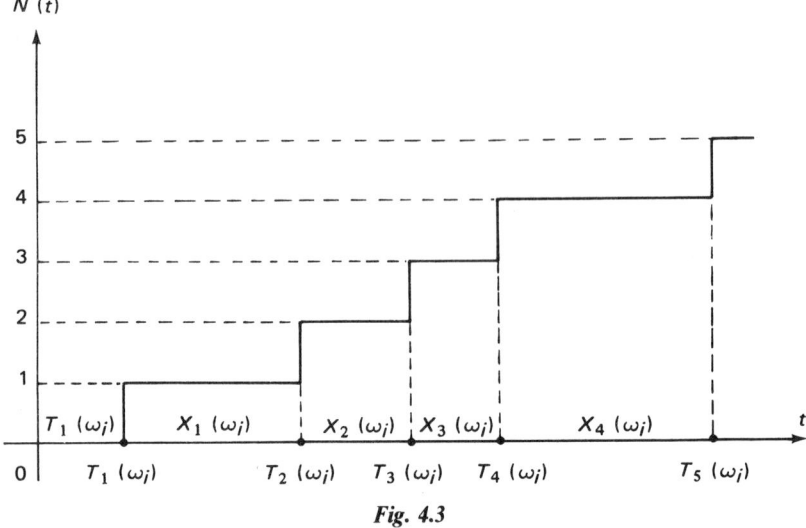

Fig. 4.3

$I_{\{T_n \leq t\}}$ being the indicator r.v. of the event $\{T_n \leq t\}$. The realizations of the r.s. N are step functions with unit step. The realization $N(\omega_i, t)$ is represented in Fig. 4.3.

Consider the sequence $(X_n, n \in \mathbb{N}^*)$ of positive r.v.s defined by

$$\forall n \in \mathbb{N}^* \qquad X_n = T_{n+1} - T_n$$

The sequence (T_n) defines a **renewal process** when the sequence $T_1, X_1, X_2, \ldots$ is independent and when the r.v.s X_n all have the same probability law. The probability density of the n-dimensional $\vec{\text{r.v.}}$ $(T_1, X_1, \ldots, X_{n-1})$ is then the n-fold tensor product of the probability densities of its components. This may be written (cf. Eqn (2.46))

$$p_{T_1, X_1, \ldots, X_{n-1}}(t_1, x_1, \ldots, x_{n-1}) = p_{T_1}(t_1) p_X(x_1) \ldots p_X(x_{n-1}) \qquad (4.4)$$

p_X being the probability density of the r.v.s X_n. Another definition of a renewal process is the subject of Exercise 1.10.

Poisson renewal process

An extremely important case is that in which the r.v. T_1 has the same probability law as the r.v.s X_n and this law is exponential and defined by the density (cf. Chapter 1, Example 2.8)

$$p_X(x) = Y(x) \vartheta \exp(-\vartheta x) \qquad \vartheta \in \mathbb{R}_+^* \qquad (4.5)$$

The sequence (T_n) then defines a **Poisson renewal process.**

The r.v. T_n in this case is the sum of n independent r.v.s all having probability density p_X. The probability density p_{T_n} is thus the n-fold convolution of the densities p_X (cf. Chapter 2, Section 1.5.1.6). It follows, by using the generating

Definitions and general properties

functions (cf. Eqn (2.50) and Appendix 3), that

$$\forall n \in \mathbb{N}^* \qquad P_{T_n}(t) = Y(t)\vartheta^n \frac{t^{n-1}}{(n-1)!} \exp(-\vartheta t) \qquad (4.6)$$

The probability density $p_{T_1,\ldots,T_n}$ of the n-dimensional r.v. $(T_1, \ldots, T_n)$ can be obtained from Eqn (4.4) by change of variables. We obtain (cf. Chapter 2, Exercise 1.15)

$$p_{T_1,\ldots,T_n}(t_1, \ldots, t_n) = I_n(t_1, \ldots, t_n)\vartheta^n \exp(-\vartheta t_n) \qquad (4.7)$$

denoting by I_n the indicator function of the set of $\mathbb{R}_+^n$:

$$\{(t_1, \ldots, t_n) \mid 0 \leqslant t_1 \leqslant \ldots \leqslant t_n\}$$

Poisson random signal

Let $\{N(t) = n\}$ be the event consisting of the realization of n point events T_i in the interval $]0, t]$. It may be written

$$\{N(t) = n\} = \{T_n \leqslant t\} \cap \{T_{n+1} > t\}$$

It defines a region in the space $\mathbb{R}_+^{n+1}$ and we can obtain its probability by integrating the density $p_{T_1,\ldots,T_{n+1}}$ over that region:

$$P\{N(t) = n\} = \int_0^t dt_1 \int_{t_1}^t dt_2 \ldots \int_{t_{n-1}}^t dt_n \int_t^\infty \vartheta^{n+1} \exp(-\vartheta t_{n+1}) \, dt_{n+1}$$

We get, by applying Eqn (4.53),

$$\forall t \in \mathbb{R}_+ \qquad \forall n \in \mathbb{N}^* \qquad P\{N(t) = n\} = \frac{(\vartheta t)^n}{n!} \exp(-\vartheta t) \qquad (4.8)$$

The r.v. $N(t)$ is therefore a Poisson r.v. with parameter ϑt (cf. Chapter 1, Example 2.5). For this reason the causal r.s. $N(t), t \in \mathbb{R}_+^*$, is called a **Poisson r.s.**

Comment

Many physical phenomena characterized by the realization of point random events are represented, to a good approximation, by the Poisson renewal process model. In particular, this is the case with the emission of photons by a monochromatic source and with the arrival of calls at a central telephone exchange. More generally, the theory of queues makes extensive use of the Poisson model to represent random arrivals of customers queuing at a service point (Kleinrock, 1975). Poisson processes are the subject of many of the exercises at the end of this chapter.

1.2. PROBABILITY LAW OF A REAL RANDOM SIGNAL

1.2.1. *Definition*

The probability law of a real r.s. ξ is the measure μ_ξ over the measurable space $(\mathbb{R}^I, \mathcal{R}^I)$, which is the image of the measure P through the measurable mapping ξ (cf. Chapter 6, Section 3.1.4).

1.2.2. *Probability of a set of realizations*

The measure μ_ξ is positive and normed. The space $(\mathbb{R}^I, \mathcal{R}^I, \mu_\xi)$ is therefore a probability space and the mapping ξ brings this space into correspondence with the space $(\Omega, \mathcal{A}, P)$, as shown schematically in Fig. 4.4.

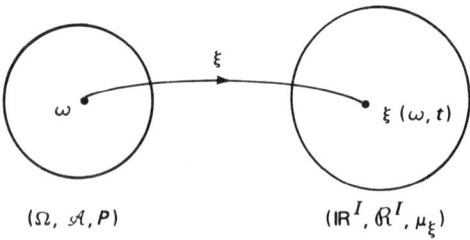

Fig. 4.4

Let $\beta \in \mathcal{R}^I$ be a Borel set. This is a set of real functions defined over the set I. The probability that the realization of the r.s. ξ corresponding to the chosen trial belongs to the set β may be obtained by applying Eqn (6.28). We obtain

$$\mu_\xi(\beta) = P\{\xi^{-1}(\beta)\} \tag{4.9}$$

Let us take as Borel set β a measurable block

$$\prod_{t \in I} \beta_t \qquad \beta_t \in \mathcal{R}_t$$

with $\beta_t = \mathbb{R}_t$ except for a finite set of instants $J = \{t_1, \ldots, t_n\}$. We get from Eqn (4.9), remembering always that $\xi_t = \varphi_t \circ \xi$ and noting that $\xi^{-1}(\mathbb{R}_t) = \Omega$,

$$\mu_\xi(\beta) = P\left\{\bigcap_{t \in J} \xi_t^{-1}(\beta_t)\right\} \tag{4.10}$$

In the following we shall use the notation

$$\forall i \in [1, n] \qquad \xi(t_i) = \xi_i \qquad \beta_{t_i} = \beta_i$$

The Borel set β considered is the set of real functions defined over the set I

Definitions and general properties 171

which pass through the Borel sets β_i at the instants t_i. We have

$$\mu_\xi(\beta) = P\left\{\bigcap_{i=1}^n \xi_i^{-1}(\beta_i)\right\} = P(\{\xi_1 \in \beta_1\} \cap \ldots \cap \{\xi_n \in \beta_n\}) \qquad (4.11)$$

Taking account of Eqn (2.4), we get from this

$$\mu_\xi(\beta) = \mu_{\xi_1 \ldots \xi_n}(\beta_1 \times \ldots \times \beta_n) \qquad (4.12)$$

The measure μ_ξ is therefore projected onto the space $(\mathbb{R}^J, \mathscr{R}^J)$ according to the probability law $\mu_{\xi_1 \ldots \xi_n}$ of the r.v. $\vec{(\xi_1, \ldots, \xi_n)}$ (cf. Chapter 6, Section 3.2.2). The number $\mu_\xi(\beta)$ is the probability that the r.s. ξ takes values in the Borel sets $\beta_1, \ldots, \beta_n$ at instants $t_1, \ldots, t_n$ respectively. All this is illustrated in Fig. 4.5.

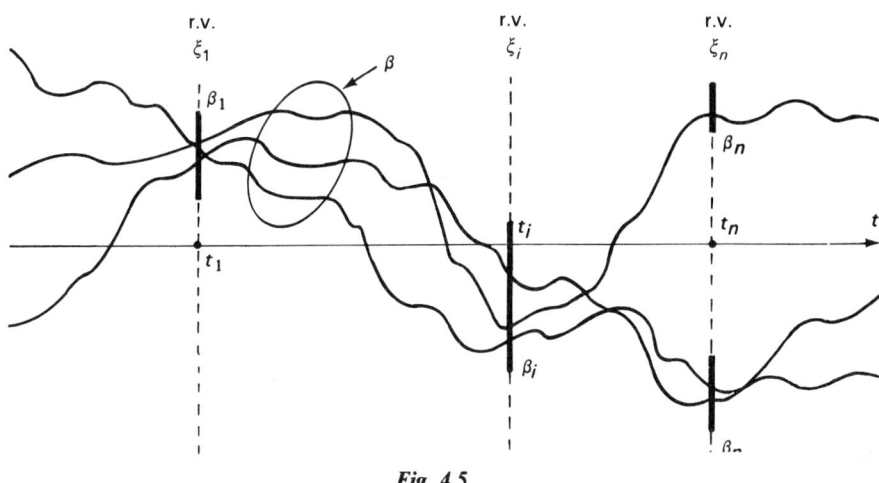

Fig. 4.5

Remark: Consider a function $g: I \to \mathbb{R}$. It is a point of $\mathbb{R}^I$ which we can consider as an intersection of measurable blocks by writing

$$g = \bigcap_{t \in I} (\ldots \times \mathbb{R} \times \{g(t)\} \times \mathbb{R} \times \ldots)$$

When the set I is not denumerable, the intersection is not necessarily measurable. This means that the point considered does not necessarily belong to the Borel field $\mathscr{R}^I$ and that consequently its inverse image through ξ is not necessarily an event belonging to the Borel field $\mathscr{A}$. When this is not an event, it is impossible to determine the probability that the function g is a realization of the r.s. ξ.

1.2.3. Family of projective probability laws

Let $\{J\}$ be the family of finite sets of instants belonging to the set I and $\{\mu_J\}$ be the family of measures obtained by projection of the measure μ_ξ onto the measurable spaces $(\mathbb{R}^J, \mathcal{R}^J)$. Consider a set $K \subset J$. It is easy to verify that the measure μ_K is the projection of the measure μ_J onto the measurable space $(\mathbb{R}^K, \mathcal{R}^K)$. For this reason we say that the family $\{\mu_J\}$ is **projective**.

Conversely, let $\{\mu_J\}$ be a projective family of positive normed measures, defined over the spaces $(\mathbb{R}^J, \mathcal{R}^J)$. Kolmogoroff's theorem enables us to state that there exists a unique positive normed measure μ_ξ defined over the space $(\mathbb{R}^I, \mathcal{R}^I)$ which is projected onto the spaces $(\mathbb{R}^J, \mathcal{R}^J)$ in accordance with the measures μ_J (cf. Chapter 6, Section 3.2.3). The family $\{\mu_J\}$ thus defines a probability space $(\mathbb{R}^I, \mathcal{R}^I, \mu_\xi)$ with which it is possible to associate a real r.s. defined, in the sense of definition 1 (Section 1.1.1), by the identity mapping I_d from the space $(\mathbb{R}^I, \mathcal{R}^I, \mu_\xi)$ onto itself. This is illustrated in Fig. 4.6 (cf. Fig. 4.4).

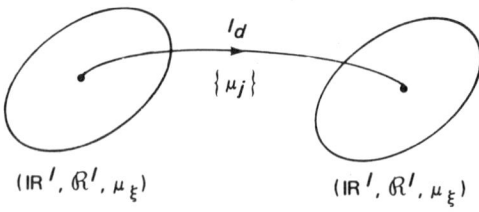

Fig. 4.6

The r.v.s ξ_t are then the composite mappings $\xi_t = \varphi_t \circ I_d : (\mathbb{R}^I, \mathcal{R}^I) \to (\mathbb{R}_t, \mathcal{R}_t)$. Let J be the set $J = \{t_1, \ldots, t_n\}$. The probability law of the $\overrightarrow{\text{r.v.}}$ $(\xi_1, \ldots, \xi_n)$ is the measure μ_J, which may be written

$$J = \{t_1, \ldots, t_n\} \qquad \mu_{\xi_1 \ldots \xi_n} = \mu_J \qquad (4.13)$$

This leads us to define in the following equivalent way a real r.s.

1.2.4. Definition 3

A projective family $\{\mu_J\}$ of positive normed measures defined on the measurable spaces $(\mathbb{R}^J, \mathcal{R}^J)$, associated with the family $\{J\}$ of finite parts of the set I, defines a probability space $(\mathbb{R}^I, \mathcal{R}^I, \mu_\xi)$ and a real r.s. $\xi(t)$, $t \in I$, defined, in the sense of definition 1, by the identity mapping I_d from the space $(\mathbb{R}^I, \mathcal{R}^I, \mu_\xi)$ onto itself.

It must be noted that this definition does not involve the realizations of the r.s. explicitly and that the latter are not specified, considered as functions of time. It enables us, however, to calculate the probabilities expressed by Eqns (4.9) and (4.12).

Example 1.4 Gaussian random signal

A gaussian r.s. $\xi(t)$, $t \in I$, is a real r.s. defined by a projective family $\{\mu_J\}$ of positive normed measures. The measure μ_J over the measurable space $(\mathbb{R}^J, \mathscr{R}^J)$ is defined by its density with respect to Lebesgue measure. It may be written

$$p_X(x_1, \ldots, x_n) = (2\pi)^{-n/2}(\det \Gamma_{XX})^{-1/2}$$
$$\times \exp\{-\tfrac{1}{2}(A - M_X)^T \Gamma_{XX}^{-1}(A - M_X)\} \quad (4.14)$$

where X is the random matrix associated with the $\vec{r.v.}$ $(\xi_1, \ldots, \xi_n)$ corresponding to the set J (cf. Eqn (2.2)) and A is a matrix such that $A^T = [x_1, \ldots, x_n]$.

The function p_X is the probability density of the $\vec{r.v.}$ $(\xi_1, \ldots, \xi_n)$ which is then gaussian (cf. Chapter 2, Example 1.6). It is a function of the matrix of means M_X and the covariance matrix Γ_{XX} which is symmetric since the r.s. ξ is real. These matrices are obtained from the marginal probability densities of order 1 and 2 of the r.v.s ξ_i and the $\vec{r.v.}$s (ξ_i, ξ_j). Since the densities of order 1 can be deduced by projection from those of order 2 (cf. Eqn (2.7)), the probability laws of arbitrary orders of a gaussian r.s. can be deduced from the marginal laws of order 2. In fact, a gaussian r.s. is therefore defined by the family of probability densities of the two-dimensional $\vec{r.v.}$s (ξ_i, ξ_j).

Comment: Gaussian r.s.s play a very important role in practice. In fact, many physical phenomena possessing the character of regularity, in contrast with point processes, are represented to a good approximation by a gaussian r.s. All the r.s.s resulting from the superposition of a large number of independent phenomena are gaussian (cf. Chapter 2, Section 2.5). This is the case in particular with thermal noise (cf. Chapter 5, Example 1.7). In addition, the gaussian probability law lends itself more easily than others to the solution of certain problems, especially those which involve non-linearities (cf. Chapter 5, Example 1.5).

1.3. PAIRS OF RANDOM SIGNALS

1.3.1. *General definition*

Consider a probability space $(\Omega, \mathscr{A}, P)$, two sets $I \subset \mathbb{R}$ and $I' \subset \mathbb{R}$ in which the variables t and t' take their values, and the measurable product space $(\mathbb{R}^I \times \mathbb{R}^{I'}, \mathscr{R}^I \otimes \mathscr{R}^{I'})$. Let ζ be a measurable mapping from the space $(\Omega, \mathscr{A})$ into the space $(\mathbb{R}^I \times \mathbb{R}^{I'}, \mathscr{R}^I \otimes \mathscr{R}^{I'})$ such that

$$\forall \omega \in \Omega \quad \zeta(\omega) = \{\xi(\omega), \eta(\omega)\} \quad \xi(\omega) \in \mathbb{R}^I \quad \eta(\omega) \in \mathbb{R}^{I'}$$

The mappings ξ and η are measurable and therefore define two real r.s.s $\xi(t)$, $t \in I$, and $\eta(t')$, $t' \in I'$. The mapping ζ defines the pair (ξ, η).

1.3.2. Probability law of the pair (ξ, η)

The probability law $\mu_{\xi\eta}$ of the pair (ξ, η) is the positive normed measure over the space $(\mathbb{R}^I \times \mathbb{R}^{I'}, \mathscr{R}^I \otimes \mathscr{R}^{I'})$, which is the image of the measure P through the measurable mapping ζ.

Consider two finite sets

$$J = \{t_1, \ldots, t_m) \subset I \quad J' = \{t'_1, \ldots, t'_n\} \subset I'$$

The measure $\mu_{\xi\eta}$ is projected onto the space $(\mathbb{R}^J \times \mathbb{R}^{J'}, \mathscr{R}^J \otimes \mathscr{R}^{J'})$ according to the probability law $\mu_{\xi_1 \ldots \xi_m \eta_1 \ldots \eta_n}$ of the $\overrightarrow{r.v.}$ $(\xi_1, \ldots, \xi_m, \eta_1, \ldots, \eta_n)$.

Consider the factor spaces $(\mathbb{R}_{t_i}, \mathscr{R}_{t_i})$ and $(\mathbb{R}_{t'_j}, \mathscr{R}_{t'_j})$ and the Borel sets

$$\forall i \in [1, m] \quad \beta_i \in \mathscr{R}_{t_i} \quad \forall j \in [1, n] \quad \beta'_j \in \mathscr{R}_{t'_j}$$

Finally, consider the Borel set

$$\beta_1 \times \ldots \times \beta_m \times \beta'_1 \times \ldots \times \beta'_n \in \mathscr{R}^J \otimes \mathscr{R}^{J'} \tag{4.15}$$

We get, reasoning as in Section 1.2.2,

$$\mu_{\xi_1 \ldots \xi_m \eta_1 \ldots \eta_n}(\beta_1 \times \ldots \times \beta_m \times \beta'_1 \times \ldots \times \beta'_n)$$
$$= P(\{\xi_1 \in \beta_1\} \cap \ldots \cap \{\xi_m \in \beta_m\} \cap \{\eta_1 \in \beta'_1\} \cap \ldots \cap \{\eta_n \in \beta'_n\}) \tag{4.16}$$

This is the probability that the r.s. ξ takes at the instants $t_1, \ldots, t_m$ values belonging to the Borel sets $\beta_1, \ldots, \beta_m$ respectively and that the r.s. η takes at the instants $t'_1, \ldots, t'_n$ values belonging to the Borel sets $\beta'_1, \ldots, \beta'_n$ respectively. This is illustrated in Fig. 4.7.

1.3.3. Independence of two random signals

Two real r.s.s ξ and η are independent if the probability law of the pair (ξ, η) is the tensor product of the probability laws μ_ξ and μ_η (cf. Chapter 6, Section 3.2). This is written

$$\mu_{\xi\eta} = \mu_\xi \otimes \mu_\eta \tag{4.17}$$

Hence, for a Borel set of the form of Eqn (4.15) we get

$$P(\{\xi_1 \in \beta_1\} \cap \ldots \cap \{\xi_m \in \beta_m\} \cap \{\eta'_1 \in \beta'_1\} \cap \ldots \cap \{\eta'_n \in \beta'_n\})$$
$$= P(\{\xi_1 \in \beta_1\} \cap \ldots \cap \{\xi_m \in \beta_m\}) P(\{\eta'_1 \in \beta'_1\} \cap \ldots \cap \{\eta'_n \in \beta'_n\}) \tag{4.18}$$

Example 1.5 Pair of mutually gaussian random signals

A pair (ξ, η) of mutually gaussian r.s.s $\xi(t)$ and $\eta(t)$, $t \in \mathbb{R}$, is defined by a projective family $\{\mu_{JJ'}\}$ of positive normed measures. The measure $\mu_{JJ'}$ over the measurable space $(\mathbb{R}^J \times \mathbb{R}^{J'}, \mathscr{R}^J \otimes \mathscr{R}^{J'})$ is defined by the probability density p_Z

Definitions and general properties 175

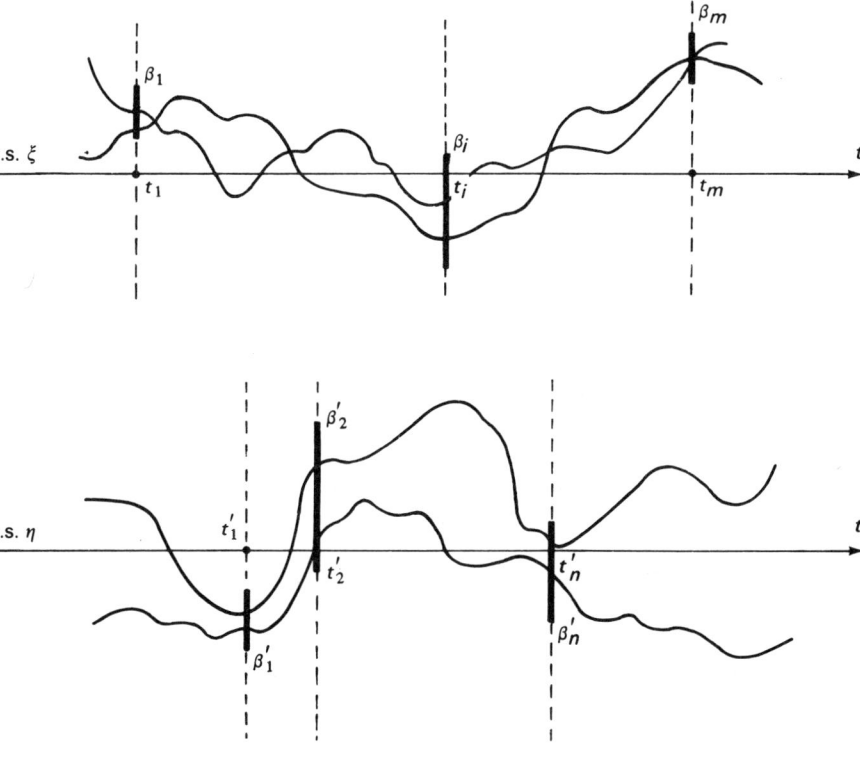

Fig. 4.7

of the $(m+n)$-dimensional $\vec{\text{r.v.}}$ $(\xi_1, \ldots, \xi_m, \eta_1, \ldots, \eta_n)$ associated with the sets J and J'. It can be written (cf. Eqn (4.14))

$$p_Z(x_1, \ldots, x_m, y_1, \ldots, y_n) = (2\pi)^{-(m+n)/2}(\det \Gamma_{ZZ})^{-1/2}$$
$$\times \exp\{-\tfrac{1}{2}(A - M_Z)^T \Gamma_{ZZ}^{-1}(A - M_Z)\} \quad (4.19)$$

where Z is a random matrix associated with the $\vec{\text{r.v.}}$ $(\xi_1, \ldots, \xi_m, \eta_1, \ldots, \eta_n)$ and A is a matrix such that $A^T = [x_1 \ldots x_m y_1 \ldots y_n]$.

Let X and Y be random matrices associated respectively with the $\vec{\text{r.v.}}$s $(\xi_1, \ldots, \xi_m)$ and $(\eta_1, \ldots, \eta_n)$. It follows that (cf. Chapter 2, Exercise 1.4)

$$Z = \begin{pmatrix} X \\ Y \end{pmatrix} \quad M_Z = \begin{pmatrix} M_X \\ M_Y \end{pmatrix} \quad \Gamma_{ZZ} = \begin{pmatrix} \Gamma_{XX} & \Gamma_{XY} \\ \Gamma_{YX} & \Gamma_{YY} \end{pmatrix}$$

It is immediately obvious that the r.s.s ξ and η are individually gaussian. Suppose that they are not correlated (cf. Section 1.6.3.3). The cross-correlation matrices Γ_{XY} and Γ_{YX} are then zero and it is easy to prove that we have

$$p_Z(x_1, \ldots, x_m, y_1, \ldots, y_n) = p_X(x_1, \ldots, x_m) p_Y(y_1, \ldots, y_n)$$

176 Random signals

It follows from this that the r.s.s ξ and η are independent since the probability law of the pair is the tensor product of the probability laws of the two r.s.s (cf. Eqn (4.17)). The mutually gaussian r.s.s therefore possess the very important property of being independent if they are not correlated (the converse is always true).

1.4. EQUIVALENCE OF TWO RANDOM SIGNALS

1.4.1. Weak equivalence

As we have seen, a projective family $\{\mu_J\}$ of positive normed measures defines a probability space $(\mathbb{R}^I, \mathcal{R}^I, \mu_\xi)$ and an r.s. $\xi(t)$, $t \in I$, which is the identity mapping from the space $(\mathbb{R}^I, \mathcal{R}^I, \mu_\xi)$ onto itself (cf. Section 1.2.4).

It is possible to define other r.s.s relative to the space $(\mathbb{R}^I, \mathcal{R}^I, \mu_\xi)$. Consider a probability space $(\Omega, \mathcal{A}, P)$ and a measurable mapping $\eta: (\Omega, \mathcal{A}) \to (\mathbb{R}^I, \mathcal{R}^I)$. The mapping η defines an r.s. $\eta(t)$, $t \in I$, whose probability law is the image measure μ_η. Suppose that we have (cf. Eqn (4.13))

$$\forall J \in \{J\} \qquad \mu_{\eta_1 \ldots \eta_n} = \mu_J = \mu_{\xi_1 \ldots \xi_n} \qquad (4.20)$$

The probability laws μ_ξ and μ_η are then identical by virtue of Kolmogoroff's theorem. We say that the r.s. η is **weakly equivalent** to the r.s. ξ. The realizations $\eta(\omega, t)$ of the r.s. η can be considered as representations of the r.s. ξ whose own realizations are not defined (cf. Section 1.2.4). This enables us to calculate the probabilities of Eqn (4.12) by arguing in terms of the realizations of the r.s. η as if they were those of the r.s. ξ and by writing

$$\mu_{\xi_1 \ldots \xi_n}(\beta_1 \times \ldots \times \beta_n) = P(\{\eta_1 \in \beta_1\} \cap \ldots \cap \{\eta_n \in \beta_n\}) \qquad (4.21)$$

1.4.2. Strong equivalence

Now let ξ and η be two r.s.s defined over the same probability space $(\Omega, \mathcal{A}, P)$. Suppose that the r.v.s $\xi(t)$ and $\eta(t)$ are almost surely (a.s.) equal for all $t \in I$, which may be written (cf. Chapter 1, Section 2.1.3)

$$\forall t \in I \qquad P\{\xi(t) \neq \eta(t)\} = 0 \qquad (4.22)$$

The r.s.s ξ and η are a.s. equal at every instant, which implies that their realizations a.s. pass through the same point at every instant. We say that the r.s.s ξ and η are **strongly equivalent.** It is easy to prove, reasoning as in Chapter 1, Example 2.4, that strong equivalence implies weak equivalence.

Consider the events

$$\forall t \in I \qquad A_t = \{\xi(t) \neq \eta(t)\}$$

The set

$$\bigcup_{t \in I} A_t$$

contains the trials ω for which the realizations $\xi(\omega, t)$ and $\eta(\omega, t)$ are not identical since they pass through two different points for at least one value of t. When the set I is not denumerable, the set

$$\bigcup_{t \in I} A_t$$

is not necessarily an event and, even if it is an event, its probability is not necessarily zero.

It is thus impossible to state, in the general case, that two strongly equivalent r.s.s have a.s. identical realizations even though these realizations pass a.s. through the same point at every instant. For this to be possible it is necessary that the r.s.s satisfy supplementary conditions called **separability** conditions, which the interested reader can find in the works cited in Neveu (1964) and Wong (1971). It can be shown that it is sufficient that the realizations of two strongly equivalent r.s.s be a.s. continuous to the right (or to the left) for them to be a.s. identical (cf. Exercise 1.11). This condition is sufficiently general that it is possible to accept, in practice, that it is always satisfied.

Example 1.6 Weakly equivalent polar representation

Let $\xi(t)$ and $\eta(t)$, $t \in \mathbb{R}$, be r.s.s defined over the same probability space $(\Omega, \mathscr{A}, P)$ by

$$\begin{aligned}\xi(t) &= x \cos(2\pi f_0 t + \varphi_0) + y \sin(2\pi f_0 t + \varphi_0) \\ \eta(t) &= \rho \cos(2\pi f_0 t + \varphi_0 + \vartheta)\end{aligned} \quad (4.23)$$

x, y, ρ and ϑ being real r.v.s defined over the space $(\Omega, \mathscr{A}, P)$. Suppose that the r.v.s x and y are gaussian, centred, with the same variance σ^2, and independent (cf. Chapter 1, Example 2.7). Suppose also that ρ is a Rayleigh r.v. with parameter a (cf. Chapter 1, Example 2.9) and that the r.v. ϑ is uniformly distributed over the interval $[-\pi, +\pi]$.

The r.s.s ξ and η are then weakly equivalent. We can write, in fact,

$$\xi(t) = (x^2 + y^2)^{1/2} \cos\left\{2\pi f_0 t + \varphi_0 - \arctan\left(\frac{y}{x}\right)\right\}$$

It follows from Eqn (2.59) that the r.v.s $\xi(t)$ and $\eta(t)$ have identical probability laws, for all $t \in I$. This does not imply, however, that the r.v.s $\xi(t)$ and $\eta(t)$ are a.s. equal for all $t \in I$ (cf. Chapter 1, Exercise 2.13). Thus we cannot state that the r.s.s ξ and η are strongly equivalent.

178 Random signals

1.5. EVENTS GENERATED BY A RANDOM SIGNAL

Let $\xi(t)$, $t \in I$, be an r.s. defined over a probability space $(\Omega, \mathscr{A}, P)$. The r.v. $\xi(t)$ generates in the space $(\Omega, \mathscr{A}, P)$ a sub-Borel field which we shall denote by $\mathscr{B}_{\xi(t)}$ (cf. Chapter 1, Section 2.1.1.2). It contains the events generated by the r.s. ξ at the instant t.

Consider the family of r.v.s $\{\xi(s) | s \leq t, s \in I\}$. It generates in the space $(\Omega, \mathscr{A}, P)$ a sub-Borel field which we shall denote by $\mathscr{B}_\xi(-\infty, t)$. This is the smallest sub-Borel field containing the sub-Borel fields $\mathscr{B}_{\xi(s)}$ and is denoted (cf. Chapter 6, Section 1.1.4 and Exercise 1.11)

$$\mathscr{B}_\xi(-\infty, t) = \widetilde{\bigcup_{\substack{s \leq t \\ s \in I}}} \mathscr{B}_{\xi(s)} \tag{4.24}$$

It contains the events generated by the r.s. ξ prior to the instant t. We define in the same way the sub-Borel field $\mathscr{B}_\xi(t, +\infty)$ of events generated by the r.s. ξ subsequent to the instant t.

It is immediately obvious that the sub-Borel fields defined above satisfy

$$t_1 \leq t_2 \Rightarrow \begin{cases} \mathscr{B}_\xi(-\infty, t_1) \subset \mathscr{B}_\xi(-\infty, t_2) \\ \mathscr{B}_\xi(t_2, +\infty) \subset \mathscr{B}_\xi(t_1, +\infty) \end{cases} \tag{4.25}$$

Example 1.7 Markov random signal

Definition 1
Consider an r.s. $\xi(t)$, $t \in \mathbb{R}$, a pair of instants (t_1, t_2) such that $t_1 \leq t_2$, and the sub-Borel fields $\mathscr{B}_\xi(-\infty, t_1)$ and $\mathscr{B}_\xi(t_2, +\infty)$. The r.s. ξ is **markovian** when

$$\forall A \in \mathscr{B}_\xi(t_2, +\infty) \qquad E[I_A | \mathscr{B}_\xi(-\infty, t_1)] \underset{\text{a.s.}}{=} E[I_A | \mathscr{B}_{\xi(t_1)}] \tag{4.26}$$

The a.s. equality of the conditional expectation r.v.s $E[I_A | \mathscr{B}_{\xi(t_1)}]$ and $E[I_A | \mathscr{B}_\xi(-\infty, t_1)]$ means that conditioning of an event generated by the r.s. ξ after the instant t_2 on the sub-Borel field of events generated before the instant t_1 may be reduced to conditioning on the sub-Borel field of events generated at the instant t_1. In other words, only events generated in the most recent past, at the instant t_1, occur in the conditioning on events generated before the instant t_1. The memory of a markovian r.s. does not go further back than the nearest past. This is illustrated schematically in Fig. 4.8.

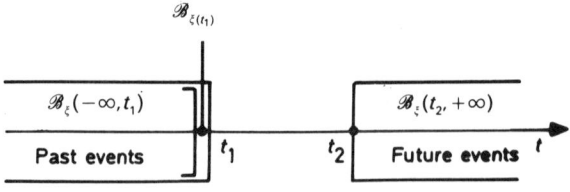

Fig. 4.8

Definitions and general properties   179

Definition 2
An equivalent definition is the subject of Exercise 1.4. It is expressed by the relation (4.54) which makes it clear that two events generated respectively before and after an instant t are conditionally independent with respect to the sub-Borel field $\mathscr{B}_{\xi(t)}$. The past and the present are thus symmetrical with respect to the present.

Property
Consider a markovian r.s. $\xi(t)$, $t \in \mathbb{R}$, and a sequence $(t_1, \ldots, t_n)$ of instants such that $t_1 \leqslant t_2 \leqslant \ldots \leqslant t_n$; denote $\forall i \in [1, n]$ $\xi(t_i) = \xi_i$. The sub-Borel field $\mathscr{B}_{\xi(t_{n-1})}$ is contained within the sub-Borel field generated by the r.v. $(\xi_1, \ldots, \xi_{n-1})$ and the latter is contained within the sub-Borel field $\mathscr{B}_{\xi}(-\infty, t_{n-1})$. From this, by applying Eqn (1.108), we obtain the a.s. equality

$$E[\xi_n | \xi_1, \ldots, \xi_{n-1}] \underset{\text{a.s.}}{=} E[\xi_n | \xi_{n-1}] \qquad (4.27)$$

When the probability law of the r.v. $(\xi_1, \ldots, \xi_n)$ is continuous and defined by a density $p_{\xi_1 \ldots \xi_n}$, Eqn (4.27) implies the equality of the conditional probability densities (cf. Chapter 2, Section 1.2.4):

$$p_{\xi_n | \xi_1 \ldots \xi_{n-1}}(x_n | x_1, \ldots, x_{n-1}) = p_{\xi_n | \xi_{n-1}}(x_n | x_{n-1}) \qquad (4.28)$$

Finally, by applying Eqn (2.14), we obtain the fundamental relation

$$p_{\xi_1 \ldots \xi_n}(x_1, \ldots, x_n) = p_{\xi_1}(x_1) p_{\xi_2 | \xi_1}(x_2 | x_1) \ldots p_{\xi_n | \xi_{n-1}}(x_n | x_{n-1}) \qquad (4.29)$$

Conversely when, according to the case, Eqn (4.27) or (4.28) is satisfied for all finite ordered sequences of instants, the r.s. ξ is markovian. This enables us to give the following equivalent definition. It is much more useful than the last two.

Definition 3
Let $\xi(t)$, $t \in \mathbb{R}$, be an r.s. and let $\{J\}$ be the family of finite ordered sequences of instants. The r.s. ξ is markovian when, according to the situation, one or other of relations (4.27) and (4.28) is satisfied for all sequences of the family $\{J\}$.

A markovian r.s. having a continuous probability law is thus entirely defined by

- the family of marginal probability densities of the r.v.s $\xi(t)$ for all instants t
- the family of conditional probability densities $p_{\xi_n | \xi_{n-1}}$ for all ordered pairs (t_{n-1}, t_n)

The probability densities $p_{\xi_n | \xi_{n-1}}$ are also called **transition probabilities**.

The Chapman–Kolmogoroff equation
The conditional probability densities $p_{\xi_n | \xi_{n-1}}$ are not arbitrary. Consider, in fact, instants t_1, t_2, t_3 such that $t_1 \leqslant t_2 \leqslant t_3$. We write

$$p_{\xi_1 \xi_2 \xi_3}(x_1, x_2, x_3) = p_{\xi_1}(x_1) p_{\xi_2 | \xi_1}(x_2 | x_1) p_{\xi_3 | \xi_2}(x_3 | x_2)$$

180 *Random signals*

It follows from Eqn (2.7) that

$$p_{\xi_1\xi_3}(x_1, x_3) = \int_{\mathbb{R}} p_{\xi_1\xi_2\xi_3}(x_1, x_2, x_3)\,dx_2$$

$$= p_{\xi_1}(x_1) \int_{\mathbb{R}} p_{\xi_2|\xi_1}(x_2|x_1) p_{\xi_3|\xi_2}(x_3|x_2)\,dx_2$$

Hence, by dividing both sides by $p_{\xi_1}(x_1)$ and applying Eqn (2.12), we obtain

$$p_{\xi_3|\xi_1}(x_3|x_1) = \int_{\mathbb{R}} p_{\xi_2|\xi_1}(x_2|x_1) p_{\xi_3|\xi_2}(x_3|x_2)\,dx_2 \tag{4.30}$$

This is the **Chapman–Kolmogoroff equation**. It connects the conditional transition probabilities of a markovian r.s. It must be noted that an r.s. satisfying the Chapman–Kolmogoroff equation is not necessarily markovian.

1.6. MOMENTS ASSOCIATED WITH A PAIR OF RANDOM SIGNALS

Consider two real r.s.s $\xi(t)$ and $\eta(t)$, $t \in I$, defined over one and the same probability space. The probability law $\mu_{\xi\eta}$ defines all the properties of the pair (ξ, η). In particular it enables us to calculate the mean values of r.v.s defined by the two r.s.s as well as the covariances of pairs of these r.v.s.

In practice, the probability law $\mu_{\xi\eta}$ is not in general known except when we are concerned with gaussian r.s.s. Moreover, when we are interested only in the energy properties of the r.s., as is often the case, knowledge of this probability law is not necessary. We content ourselves then with knowledge of the means and covariances, which are the moments of order 1 and 2 associated with the pair (ξ, η). The object of this section is to introduce the moments of order 1 and 2 associated with a pair of complex r.s.s. It must be noted that all that follows applies to r.s.s in continuous time as well as to r.s.s in discrete time.

1.6.1. *Mean value and variance*

(a) The mean value at instant $t \in I$ of a complex r.s. ξ is the mean value $E[\xi(t)]$ of the r.v. $\xi(t)$. It is a complex function, defined over the set I (cf. Chapter 1, Section 2.4.1.1). The r.s. ξ is **centred**, or of mean zero, when $E[\xi(t)] = 0$ for all $t \in I$.

(b) The variance at instant $t \in I$ of the r.s. ξ is the variance $V[\xi(t)]$ of the r.v. $\xi(t)$. It is a real function, defined over the set I. The variance of an r.s. of finite mean power is always finite (cf. Example 1.1 and Chapter 1, Section 2.4.2).

1.6.2. Autocorrelation function and coefficient

1.6.2.1. Autocorrelation function
The **autocorrelation function** $R_{\xi\xi}$ of the r.s. ξ is the complex function defined over the set I^2 by

$$R_{\xi\xi}(t_1, t_2) = E[\xi_1 \xi_2^*] \tag{4.31}$$

where we write $\xi(t_1) = \xi_1$ and $\xi(t_2) = \xi_2$. It obviously satisfies the relation

$$R_{\xi\xi}(t_1, t_2) = R_{\xi\xi}^*(t_2, t_1) \tag{4.32}$$

It therefore has hermitian symmetry. It follows immediately that

$$R_{\xi\xi}(t, t) = E[|\xi(t)|^2] = P_\xi(t) \tag{4.33}$$

The value taken by the autocorrelation function for $t_1 = t_2 = t$ is the mean power of the r.s. ξ at the instant t (cf. Eqn (4.1)).

Let g be an arbitrary complex function defined over I and let $t_1, \ldots, t_n$ be instants belonging to I. Let η be the r.v.

$$\eta = \sum_{i=1}^{n} g(t_i) \xi(t_i)$$

Its mean square value is the positive or zero quantity

$$E[|\eta|^2] = \sum_{i=1}^{n} \sum_{j=1}^{n} g(t_i) g^*(t_j) R_{\xi\xi}(t_i, t_j) \geq 0 \tag{4.34}$$

The function $R_{\xi\xi}$ is therefore non-negative definite.

1.6.2.2. Autocorrelation coefficient
The **autocorrelation coefficient** $r_{\xi\xi}$ of the r.s. ξ is defined by

$$r_{\xi\xi}(t_1, t_2) = \frac{R_{\xi\xi}(t_1, t_2) - E[\xi_1]E[\xi_2^*]}{(V[\xi_1]V[\xi_2])^{1/2}} = \frac{C[\xi_1, \xi_2]}{(V[\xi_1]V[\xi_2])^{1/2}} \tag{4.35}$$

It satisfies (cf. Eqns (2.24), (2.27) and (2.28))

$$r_{\xi\xi}(t_1, t_2) = r_{\xi\xi}^*(t_2, t_1) \quad |r_{\xi\xi}(t_1, t_2)| \leq 1 \quad r_{\xi\xi}(t, t) = 1 \tag{4.36}$$

Example 1.8 Autocorrelation function of a product
The product of two measurable mappings being measurable, the product $\zeta = \xi\eta$ defines an r.s. $\zeta(t)$, $t \in I$, over the same probability space as the r.s.s ξ and η (cf. Chapter 6, Section 2.2.4). Its autocorrelation function is defined by

$$R_{\zeta\zeta}(t_1, t_2) = E[\zeta_1 \zeta_2^*] = E[\xi_1 \eta_1 \xi_2^* \eta_2^*]$$

When the r.s.s ξ and η are independent, it follows from Eqns (4.17) and (2.47) that

$$R_{\zeta\zeta} = R_{\xi\xi} R_{\eta\eta} \tag{4.37}$$

182 Random signals

The autocorrelation function of the product of two independent r.s.s is the product of the autocorrelation functions of the two r.s.s.

1.6.3. Cross-correlation function and coefficient

1.6.3.1. Cross-correlation function
The **cross-correlation function** $R_{\xi\eta}$ of the r.s.s ξ and η is the complex function defined over the set I^2 by

$$R_{\xi\eta}(t_1, t_2) = E[\xi_1 \eta_2^*] \qquad (4.38)$$

where we have written $\xi(t_1) = \xi_1$ and $\eta(t_2) = \eta_2$. It obviously satisfies the relation

$$R_{\xi\eta}(t_1, t_2) = R_{\eta\xi}^*(t_2, t_1) \qquad (4.39)$$

When the r.s.s ξ and η are independent, it follows from Eqns (4.17) and (2.47) that

$$R_{\xi\eta}(t_1, t_2) = E[\xi_1] E[\eta_2^*] \qquad (4.40)$$

1.6.3.2. Cross-correlation coefficient
The **cross-correlation coefficient** $r_{\xi\eta}$ of the r.s.s ξ and η is defined by (cf. Eqn (2.24))

$$r_{\xi\eta}(t_1, t_2) = \frac{R_{\xi\eta}(t_1, t_2) - E[\xi_1]E[\eta_2^*]}{(V[\xi_1]V[\eta_2])^{1/2}} = \frac{C[\xi_1, \eta_2]}{(V[\xi_1]V[\eta_2])^{1/2}} \qquad (4.41)$$

It satisfies (cf. Eqns (2.26) and (2.27))

$$r_{\xi\eta}(t_1, t_2) = r_{\eta\xi}^*(t_2, t_1) \qquad |r_{\xi\eta}(t_1, t_2)| \leq 1 \qquad (4.42)$$

1.6.3.3. Non-correlation
The r.s.s ξ and η are **uncorrelated** when their cross-correlation coefficient is the null function, written

$$\forall (t_1, t_2) \in I^2 \qquad r_{\xi\eta}(t_1, t_2) = 0 \qquad (4.43)$$

It follows from Eqn (4.40) that two independent r.s.s are uncorrelated. The converse is not in general true. It is true, however, in the case (which is very important in practice) when the r.s.s considered are mutually gaussian. Independence and non-correlation are then equivalent properties (cf. Example 1.5).

Example 1.9 Autocorrelation function of a sum
The sum of two measurable mappings being a measurable mapping, the sum $\zeta = \xi + \eta$ defines an r.s. $\zeta(t)$, $t \in I$, over the same probability space as the r.s.s ξ

and η (cf. Chapter 6, Section 2.2.4). Its autocorrelation function may be written

$$R_{\zeta\zeta}(t_1, t_2) = E[(\xi_1 + \eta_1)(\xi_2^* + \eta_2^*)]$$
$$= R_{\xi\xi}(t_1, t_2) + R_{\eta\eta}(t_1, t_2) + R_{\xi\eta}(t_1, t_2) + R_{\eta\xi}(t_1, t_2) \quad (4.44)$$

When the cross-correlation function of the r.s.s ξ and η is zero, $R_{\xi\eta} = R_{\eta\xi} = 0$ and the autocorrelation function of the sum is the sum of the autocorrelation functions:

$$R_{\xi\eta} = R_{\eta\xi} = 0 \Rightarrow R_{\zeta\zeta} = R_{\xi\xi} + R_{\eta\eta} \quad (4.45)$$

This is the case in particular when the r.s.s ξ and η are independent and when at least one of them is centred, since then

$$R_{\xi\eta}(t_1, t_2) = E[\xi_1]E[\eta_2^*] = 0$$

Exercises

Exercise 1.1 Poisson process

Let $(T_n, n \in \mathbb{N}^*)$ be a sequence defining a Poisson process and let $N(t)$, $t \in \mathbb{R}_+$, be a Poisson r.s. with parameter ϑ (cf. Example 1.3). We construct a new process by selecting the point events of the above process, independently of each other, with probability p, as Fig. 4.9 shows. Let $M(t)$, $t \in \mathbb{R}_+$, be the r.s. defined by the counting function of the new process.

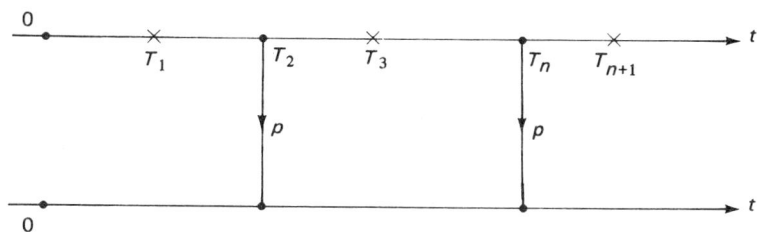

Fig. 4.9

(1) Calculate the conditional probability

$$\forall n \in \mathbb{N}^* \quad \forall k \leqslant n \quad P\{M(t) = k \mid N(t) = n\}$$

(2) Hence deduce that the r.s. $M(t)$, $t \in \mathbb{R}_+$, is poissonian and that it has parameter $p\vartheta$.

Exercise 1.2 Poisson process

Let $(T_1, \ldots, T_n)$ be a sequence of r.v.s defining a Poisson process with parameter ϑ (cf. Example 1.3), let $t_1, \ldots, t_n, t$, be instants such that

184 Random signals

$0 < t_1 < \ldots < t_n < t$ and let β_i be infinitesimal disjoint intervals such that

$$\forall i \in \{1, \ldots, n\} \qquad \beta_i =]t_i, t_i + \Delta t_i]$$

Consider the events

$$A_n = \{N(t) = n\} \qquad B_n = \bigcap_{i=1}^{n} \{T_i \in \beta_i\}$$

(1) Prove that

$$P(A_n \cap B_n) = \vartheta^n \exp(-\vartheta t) \; I_{\{0 < t_1 < \ldots < t_n < t\}}(t_1, \ldots, t_n) \Delta t_1 \ldots \Delta t_n \quad (4.46)$$

$I_{\{0 < t_1 < \ldots < t_n < t\}}$ being the indicator function of the set $\{(t_1, \ldots, t_n) | 0 < t_1 < \ldots < t_n < t\}$. This is the probability that the point events $T_1, \ldots, T_n$ occur respectively in the intervals $\beta_1, \ldots, \beta_n$ and that the point event T_{n+1} occurs outside the interval $]0, t]$. This is illustrated by

```
        T₁                              Tₙ                Tₙ₊₁
────●───×──────────────────────────×────●──────×──────────
    0  t₁  t₁+Δt₁                 tₙ  tₙ+Δtₙ  t
```

(Integrate the probability density of the $(n+1)$-dimensional $\vec{r.v.}$ $(T_1, \ldots, T_n, T_{n+1})$ over the set $\beta_1 \times \ldots \times \beta_n \times]0, t]$, using Eqn (4.7).)

(2) Hence deduce that the conditional probability density of the $\vec{r.v.}$ $(T_1, \ldots, T_n)$, given that the event A_n is realized, is written

$$P_{T_1 \ldots T_n | A_n}(t_1, \ldots, t_n) = \frac{n!}{t^n} I_{\{0 < t_1 < \ldots < t_n < t\}}(t_1, \ldots, t_n) \quad (4.47)$$

(Use Eqns (4.46) and (4.8) to start to calculate $P(B_n | A_n)$.)

(3) Deduce from Eqn (4.47) the conditional probability

$$\forall s \leq t \quad \forall k \leq n \quad P\{N(s) = k | N(t) = n\} = \binom{n}{k} \left(\frac{s}{t}\right)^k \left(1 - \frac{s}{t}\right)^{n-k} \quad (4.48)$$

The conditional law is therefore a binomial law (cf. Chapter 1, Example 2.6).

Exercise 1.3 Poisson process and Poisson random signal

Let $(T_n, n \in \mathbb{N}^*)$ be a sequence of r.v.s defining a Poisson process with parameter ϑ and let $N(t), t \in \mathbb{R}_+^*$, be the Poisson r.s. associated with it (cf. Example 1.3).

(1) Let $Q(t, \tau) = N(t + \tau) - N(t)$ be the number of point events realized in the interval $]t, t + \tau]$, $\tau \in \mathbb{R}_+^*$. This is illustrated by

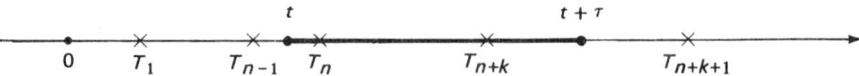

Prove that

$$\forall k \in \mathbb{N} \quad P\{Q(t,\tau) = k\} = \exp(-\vartheta\tau)\frac{(\vartheta\tau)^k}{k!} \quad (4.49)$$

The number of point events realized in an arbitrary interval of duration τ is therefore a Poisson r.v. with parameter $\vartheta\tau$ (cf. Chapter 1, Example 2.5). (Start from Eqn (4.48) and remove the conditioning by using Eqn (4.8) and summing over $\mathbb{N}$.)

(2) Deduce from the above that

$$P\{Q(t, \Delta t) = 1\} = \vartheta\Delta t + o(\Delta t)$$
$$P\{Q(t, \Delta t) = 0\} = 1 - \vartheta\Delta t + o(\Delta t) \quad (4.50)$$
$$P\{Q(t, \Delta t) > 1\} = o(\Delta t)$$

These are specific properties of the Poisson process with parameter ϑ.

(3) Consider now an interval $]t, t + \xi]$, ξ being a positive r.v. with arbitrary law. Consider the positive r.v. $\eta = N(t + \xi) - N(t)$. It represents the number of point events realized during the random interval ξ. This is illustrated by

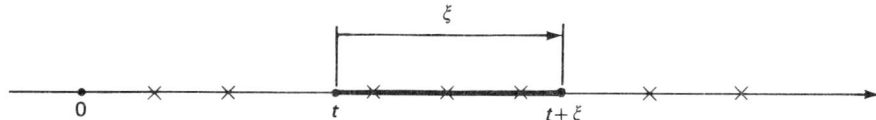

Show that the mean and variance of the r.v. η are respectively

$$E[\eta] = \vartheta E[\xi] \quad V[\eta] = \vartheta E[\xi] + \vartheta^2 V[\xi] \quad (4.51)$$

(4) Consider two disjoint intervals $]t_1, t_1 + \tau_1]$ and $]t_2, t_2 + \tau_2]$ and the events

$$A_1 = \{Q(t_1, \tau_1) = k_1\} \quad A_2 = \{Q(t_2, \tau_2) = k_2\}$$

Show that

$$P(A_1 \cap A_2) = \exp(-\vartheta\tau_1)\frac{(\vartheta\tau_1)^{k_1}}{k_1!} \exp(-\vartheta\tau_2)\frac{(\vartheta\tau_2)^{k_2}}{k_2!} = P(A_1)P(A_2)$$

(4.52)

The events A_1 and A_2 are thus independent and the Poisson r.s. has independent increments. (Integrate the probability density of the r.v.

$$(T_1, \ldots, T_{m+k_1+n+k_2+1})$$

which is given by Eqn (4.7), so as to write that the numbers of point events realized in the intervals $]0, t_1],]t_1, t_1 + \tau_1],]t_1 + \tau_1, t_2]$ and $]t_2, t_2 + \tau_2]$ are respectively m, k_1, n and k_2. Then remove the conditioning on m and n by

186 *Random signals*

summing twice over $\mathbb{N}$. Use the result

$$\int_a^b dt_1 \int_{t_1}^b dt_2 \ldots \int_{t_{n-1}}^b dt_n = \frac{(b-a)^n}{n!} \quad (4.53)$$

for this purpose.)

Exercise 1.4 Markov random signal

Let $\xi(t)$, $t \in \mathbb{R}$, be an r.s., t_1, t, t_2 be instants such that $t_1 \leqslant t \leqslant t_2$, and $A_1 \in \mathscr{B}_\xi(-\infty, t_1)$ and $A_2 \in \beta_\xi(t_2, +\infty)$ be events (cf. Eqn (4.25)).

(1) Suppose that the r.s. ξ is markovian. Prove that

$$E[I_{A_1} I_{A_2} | \mathscr{B}_{\xi(t)}] \underset{\text{a.s.}}{=} E[I_{A_1} | \mathscr{B}_{\xi(t)}] E[I_{A_2} | \mathscr{B}_{\xi(t)}] \quad (4.54)$$

$\mathscr{B}_{\xi(t)}$ being the sub-Borel field generated by the r.v. $\xi(t)$. (Show that Eqn (4.26) implies Eqn (4.54) by applying Eqns (1.108) and (1.120) and using the property that a conditional expectation r.v. $E[\xi | \mathscr{B}]$ is measurable with respect to the sub-Borel field $\mathscr{B}$ (cf. Chapter 1, Section 2.6.1(a)).)

This means that, for a markovian r.s., past events (instant t_1) and future events (instant t_2) are conditionally independent with respect to the sub-Borel field of present events (instant t) (cf. Eqn (2.48)).

(2) Prove, conversely, that, if Eqn (4.54) is satisfied for all instants t_1, t, t_2 such that $t_1 \leqslant t \leqslant t_2$, the r.s. ξ is markovian. (Prove, by taking $t_1 = t$ and applying Eqns (1.120), (4.54) and (1.93), that the r.v. $E[I_{A_2} | \mathscr{B}_{\xi(t)}]$ is also the conditional expectation of the r.v. I_{A_2} with respect to the sub-Borel field $\mathscr{B}_\xi(-\infty, t_1)$.)

Exercise 1.5 Markov random signal

Let $(\xi_n, n \in \mathbb{N}^*)$ be an independent sequence of r.v.s. Show that the r.s. in discrete time $(\xi(t), t \in T\mathbb{N}^*)$ with

$$\xi(nT) = \sum_{i=1}^n \xi_i$$

is markovian.

Exercise 1.6 Martingale

Let (Ω, A, P) be a probability space and let $(\mathscr{B}_t, t \in I)$ be a family of sub-Borel fields of the Borel field $\mathscr{A}$, and $\xi(t), t \in I$, be an r.s. defined over the space $(\Omega, \mathscr{A}, P)$ such that, for all $t \in I$, the r.v. $\xi(t)$ is measurable with respect to the sub-Borel field $\mathscr{B}_t$ and has finite mean value. The r.s. ξ is a **martingale** with respect to the family $(\mathscr{B}_t)$ when

$$\forall t \in I \quad \forall s \in I \quad s \leqslant t \quad E[\xi(t) | \mathscr{B}_s] \underset{\text{a.s.}}{=} \xi(s) \quad (4.55)$$

(1) Let $\xi(t), t \in T\mathbb{N}^*$, be an r.s. in discrete time, and let $(\mathscr{B}_n, n \in \mathbb{N}^*)$ be a sequence of sub-Borel fields, $\mathscr{B}_n$ being the sub-Borel field generated by the $\overrightarrow{\text{r.v.}}$

$(\xi(T), \ldots, \xi(nT))$. Suppose that the r.v.s $\xi(nT)$ are centred and put

$$S(nT) = \sum_{i=1}^{n} \xi(nT)$$

Prove that the r.s. in discrete time $S(t), t \in T\mathbb{N}^*$, is a martingale with respect to the family $(\mathscr{B}_n)$.

(2) Let $(\mathscr{B}_t, t \in \mathbb{R})$ be an increasing family of sub-Borel fields, ξ be a centred r.v., and $\xi(t)$ be the r.s. $\xi(t) = E[\xi | \mathscr{B}_t]$, $t \in \mathbb{R}$. Prove that the r.s. ξ is a martingale with respect to the family $(\mathscr{B}_t)$. (Use Eqn (1.103).)

Exercise 1.7 Binary random signal

Let $\xi(t), t \in \mathbb{R}_+^*$, be a binary r.s., taking values 0 or 1, such that $\xi(0) = 0$ and defined by two independent and mutually independent sequences of positive r.v.s $(X_n, n \in \mathbb{N}^*)$ and $(Y_n, n \in \mathbb{N}^*)$. A realization of this r.s. is represented in Fig. 4.10. The r.s. ξ takes the value 0 during the time intervals X_n and the value 1 during the intervals Y_n. We shall suppose that the r.v.s X_n have the same probability density p_X and that the r.v.s Y_n have the same probability density p_Y. We shall suppose, in addition, that the mean values $E[X]$ and $E[Y]$ are finite.

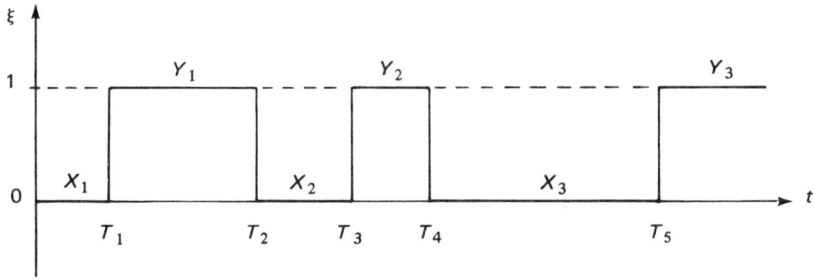

Fig. 4.10

The r.s. ξ can be considered as the mathematical model of the state of an arbitrary system which is in operation (in working order) during the intervals Y_n and is at rest (broken down) during the intervals X_n. The sequence of positive r.v.s $(T_{2n-1}, n \in \mathbb{N}^*)$ defines the instants of transition from the state of rest to the operating state.

(1) Let $p_1(t)$ be the function $p_1(t) = P\{\xi(t) = 1\}$, $t \in \mathbb{R}_+^*$, and let $\mathscr{L}p_1$ be its Laplace transform. Prove that

$$\mathscr{L}p_1(p) = \frac{\mathscr{L}p_X(p)\{1 - \mathscr{L}p_Y(p)\}}{p\{1 - \mathscr{L}p_X(p)\mathscr{L}p_Y(p)\}} \quad (4.56)$$

(Write
$$\forall t > 0 \quad P\{\xi(t) = 1\} = P\left(\bigcup_{n \in \mathbb{N}^*} \{T_{2n-1} \leq t \leq T_{2n-1} + Y_n\}\right)$$

Apply Eqn (2.69) and take the Laplace transform using Appendix 3.)

(2) Hence deduce, by applying the final value theorem (Dupraz, 1977), that

$$\lim_{t \to \infty} p_1(t) = \lim_{p \to 0} p\mathscr{L}p_1(p) = \frac{E[Y]}{E[X] + E[Y]} \quad (4.57)$$

The probability that the system considered is in operation at the instant t thus tends to a constant value when t tends to infinity.

(3) Verify that the sequence $(T_{2n-1}, n \in \mathbb{N}^*)$ tends a.s. to infinity when n tends to infinity (cf. Chapter 2, Exercise 2.1).

(4) Consider the sequences $(U_n, n \in \mathbb{N}^*)$ and $(V_n, n \in \mathbb{N}^*)$ of r.v.s defined by

$$U_n = \sum_{i=1}^{n} \frac{Y_i}{T_{2n-1}} \quad V_n = \frac{n}{T_{2n-1}} \quad (4.58)$$

The r.v. U_n represents the proportion of time that the system is in operation during the interval $[0, T_{2n-1}]$, whereas the r.v. V_n represents the relative frequency of the 'rest → operation' transitions during the same interval. Prove that we have

$$\lim_{n \to \infty} U_n \underset{\text{a.s.}}{=} \lim_{n \to \infty} E[Y]V_n \underset{\text{a.s.}}{=} \frac{E[Y]}{E[X] + E[Y]} \quad (4.59)$$

(Apply the strong law of large numbers to the sequences (X_n) and (Y_n) (cf. Chapter 2, Section 2.4).)

Comment: In the case when the system considered is a telephone line, the quantity

$$E[Y] \lim_{n \to \infty} V_n$$

is the **mean traffic** carried by the line in the interval $]0, +\infty[$. This is the product of the mean duration of telephone calls and the relative frequency of calls. It is a dimensionless number which is usually expressed in **erlangs**.

Formulae (4.57) and (4.59) express the fact that the mean traffic, the proportion of time that the line is occupied, and the probability that the line is occupied at a sufficiently large time t are quantities which are a.s. equal.

Exercise 1.8 Gaussian random signal

Let $\xi(t), t \in I$, be a gaussian r.s., a be a real function defined over the set I, $b: I \to I$ be a mapping, and $\eta(t)$ be the r.s.

$$\eta(t) = a(t)\xi\{b(t)\}$$

Show that the r.s. $\eta(t), t \in I$, is gaussian. (Use Chapter 2, Examples 1.6 and 1.7.)

Exercise 1.9 Renewal process

Let $(T_n, n \in \mathbb{N}^*)$ be a sequence of r.v.s defining a renewal process and let $(X_n, n \in \mathbb{N}^*)$ be the sequence of positive r.v.s such that, $\forall n \in \mathbb{N}^*$, $X_n = T_{n+1} - T_n$. A realization of the process is represented by

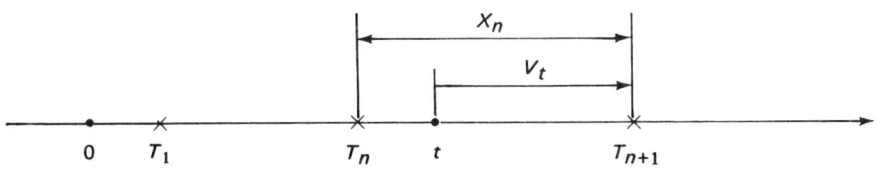

Let p_{T_1} be the probability density of the r.v. T_1 and let p_X be the common probability density of the r.v.s $X_n, n \in \mathbb{N}^*$. Finally, let $m(t) = E[N(t)], t \in \mathbb{R}_+^*$, be the mean value of the r.v. $N(t)$ defined by Eqn (4.3) (cf. Example 1.3).

(1) Prove that the Laplace transforms $\mathscr{L}m$, $\mathscr{L}p_{T_1}$ and $\mathscr{L}p_X$ are connected by the relation

$$\mathscr{L}m(p) = \frac{\mathscr{L}p_{T_1}(p)}{p\{1 - \mathscr{L}p_X(p)\}} \qquad (4.60)$$

(Starting with Eqn (4.3), write the distribution function F_{T_n} of the r.v. T_n in the form of a convolution $F_{T_n} = Y * p_{T_n}$; then take Laplace transforms using Appendix 3.)

(2) We impose the supplementary condition

$$m(t) = Y(t)at \qquad a \in \mathbb{R}_+^*$$

Deduce directly from Eqn (4.60) that

$$E[X] = \frac{1}{a} \qquad E[T_1] = \frac{E[X^2]}{2E[X]} \qquad (4.61)$$

and then that the probability densities p_{T_1} and p_X are connected by

$$p_{T_1}(x) = aY(x) * \{\delta(x) - p_X(x)\}$$

(Develop the generating functions in the neighbourhood of the origin and apply Eqn (1.90) to obtain Eqn (4.61).)

(3) Verify that the Poisson process studied in Example 1.3 fulfils the preceding conditions.

(4) Let $t \in \mathbb{R}_+^*$ be an instant and V_t the positive r.v. representing the time which elapses between the instant t and the instant of realization of the first point event occurring after the instant t. This is illustrated by

190 Random signals

(4.1) Put, $\forall x \in \mathbb{R}_+^*$, $\beta =]x, x + \Delta x]$. Verify that

$$\{V_t \in \beta\} = \{T_1 - t \in \beta\} \cup \ldots \cup (\{T_n < t\} \cap \{T_{n+1} - t \in \beta\}) \cup \ldots$$

(4.2) Hence deduce, when $m(t) = Y(t)at$, $a \in \mathbb{R}_+^*$, that

$$p_{V_t}(x) = p_{T_1}(t + x) + a \int_0^t p_X(u + x) \, du \tag{4.62}$$

(Use the result of Question (4.1), take the Laplace transform of the right-hand side with respect to the variable t, x being fixed, and apply Eqn (4.60).)

(4.3) Deduce from the preceding relation, always supposing that $m(t) = at$, that

$$p_{V_t}(x) = p_{T_1}(x) \tag{4.63}$$

The probability law of the residual r.v. V_t is then identical with that of the r.v. T_1. The renewal process is **stationary** and its origin has no significance. (Use Eqn (4.60) to express p_X as a function of p_{T_1}.)

Exercise 1.10 Renewal process

The object of this exercise is to give a more general definition of a renewal process than that which is given in Example 1.3.

Let $]t, t + \tau]$, $t \in \mathbb{R}$, $\tau \in \mathbb{R}_+^*$, be an interval and let $Q(t, \tau)$ be the number of point events realized in the interval considered. Consider a positive measure over the space $(\mathbb{R}, \mathscr{R})$ which is absolutely continuous with respect to Lebesgue measure and has as density a locally integrable function λ.

The point process is defined by the following properties.

(i) The r.v.s $Q(t_1, \tau_1)$ and $Q(t_2, \tau_2)$ are independent when the intervals $]t_1, t_1 + \tau_1]$ and $]t_2, t_2 + \tau_2]$ are disjoint.

(ii) The probability law of the r.v. $Q(t, \tau)$ is such that

$$P\{Q(t, \Delta t) = 1\} = \lambda(t) \, \Delta t + o(\Delta t)$$
$$P\{Q(t, \Delta t) = 0\} = 1 - \lambda(t) \, \Delta t + o(\Delta t) \tag{3.64}$$
$$P\{Q(t, \Delta t) > 1\} = o(\Delta t)$$

This is illustrated by

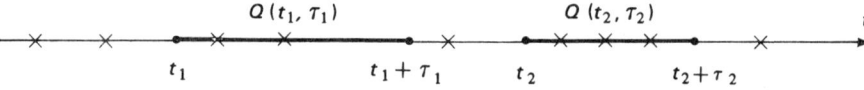

(1) Prove that the characteristic function of the r.v. $Q(t, \tau)$ is

$$\Theta_Q(u) = \exp[q(t, \tau)\{\exp(iu) - 1\}] \qquad q(t, \tau) = \int_t^{t+\tau} \lambda(u) \, du \tag{3.65}$$

Hence deduce that the probability law of the r.v. $Q(t, \tau)$ is a Poisson law with parameter $q(t, \tau)$ and that consequently

$$P\{Q(t, \tau) = k\} = \exp\{-q(t, \tau)\} \frac{\{q(t, \tau)\}^k}{k!} \quad (4.66)$$

(Divide the interval $]t, t + \tau]$ into disjoint infinitesimal intervals of the same length Δt, use the independence of the corresponding r.v.s Q, and then pass to the limit as Δt tends to zero. Apply Eqns (2.49) and (1.115).)

(2) Let $]t, t + \tau]$ be an interval and $t_1, \ldots, t_n$ be n instants such that $t < t_1 < \ldots < t_n < t + \tau$. Prove that the probability that n point events occur in the interval $]t, t + \tau]$, in the disjoint infinitesimal intervals $]t_i, t_i + \Delta t_i]$ respectively, is

$$\exp\{-q(t, \tau)\} \lambda(t_1) \ldots \lambda(t_n) I_{[t,\tau]}(t_1, \ldots, t_n) \Delta t_1 \ldots \Delta t_n \quad (4.67)$$

where we denote by $I_{[t,\tau]}$ the indicator function of the set $\mathbb{R}^n$:

$$\{(t_1, \ldots, t_n) \mid t < t_1 < \ldots < t_n < t + \tau\}$$

This is illustrated by

(3) Suppose that a point event occurs at the instant t and let V_t be the interval of time which elapses until the realization of the following point event. Prove that the r.v. V_t has as probability density the function defined by

$$p_{V_t}(v) = Y(v)\lambda(v) \exp\left\{-\int_t^{t+v} \lambda(x)\,dx\right\} \quad (4.68)$$

Hence deduce that the locally integrable function λ must be non-integrable for the function p_{X_t} to define a probability density, which implies that at least one point event occurs after the instant t.

(4) Prove that, when the function λ is constant and equal to ϑ, the process defined is a stationary Poisson process with parameter ϑ (cf. Example 1.3 and Exercises 1.2 and 1.3).

Exercise 1.11 Almost surely zero random signal

(1) Let g be a real function defined over $\mathbb{R}$ and continuous to the right (or the left).

(1.1) Prove that we have, for all $t \in \mathbb{N}^*$,

$$\sup_{t \in I} |g(t)| \leq \frac{1}{n} \Leftrightarrow \forall t \in \mathbb{Q} \quad g(t) \leq \frac{1}{n}$$

(The implication $\Rightarrow$ is obvious. Use *reductio ad absurdum* to prove the implication $\Leftarrow$, using the property that the set $\mathbb{Q}$ is dense everywhere in $\mathbb{R}$.)

(1.2) Deduce from the above that g is the zero function if and only if

$$\forall n \in \mathbb{N}^* \quad \forall t \in \mathbb{Q} \quad |g(t)| \leq \frac{1}{n}$$

(Use

$$g \equiv 0 \Leftrightarrow \forall n \in \mathbb{N}^* \quad \sup_{t \in \mathbb{R}} |g(t)| \leq \frac{1}{n}$$

for this.)

(2) Let $\xi(t)$, $t \in \mathbb{R}$, be a real r.s., defined over a probability space $(\Omega, \mathcal{A}, P)$, such that

$$\forall t \in \mathbb{R} \quad P\{\omega \,|\, \xi(\omega, t) \neq 0\} = 0$$

which means that the r.s. ξ is a.s. zero for all t. Suppose moreover that the realizations $\xi(\omega, t)$ are a.s. continuous to the right (or to the left).

(2.1) Prove that the set of trials corresponding to realizations which are not the zero function can be defined by

$$\{\omega \,|\, \xi(\omega, t) \not\equiv 0\} = \bigcup_{n \in \mathbb{N}^*} \bigcup_{t \in \mathbb{Q}} \left\{ \omega \,\Big|\, \xi(\omega, t) > \frac{1}{n} \right\}$$

(Use the result of Question (1.2).)

(2.2) Hence deduce that

$$P\{\omega \,|\, \xi(\omega, t) \not\equiv 0\} = 0$$

This means that the set of trials corresponding to realizations which are not the zero function is an event of zero probability.

(Use the property that the set $\mathbb{Q}$ is denumerable to establish that the set of trials considered is indeed an event belonging to the Borel field $\mathcal{A}$.)

(2.3) Then conclude that, when an r.s. is a.s. zero at every instant and its realizations are a.s. continuous to the right (or the left), these are a.s. zero functions.

2. Stationarity

2.1. STRONG STATIONARITY

Let $\xi(t)$ and $\eta(t)$, $t \in I$, be two r.s.s, I being either the set $\mathbb{R}$, when the r.s.s are in continuous time, or the set $T\mathbb{Z}$, $T \in \mathbb{R}^*_+$, when they are in discrete time.

(a) Let $t_1, \ldots, t_n$ be instants belonging to the set I and let τ be a number $\tau \in I$. The r.s. ξ is **strongly stationary** or **strictly stationary** if the n-dimensional $\overrightarrow{\text{r.v.}}$s $\{\xi(t_1), \ldots, \xi(t_n)\}$ and $\{\xi(t_1 + \tau), \ldots, \xi(t_n + \tau)\}$ have identical probability laws for all $n \in \mathbb{N}^*$ and for all $\tau \in I$.

(b) Let $t_1, \ldots, t_m, t'_1, \ldots, t'_n$ be instants belonging to I and let τ be a number $\tau \in I$. The r.s.s ξ and η are mutually strongly stationary if they are individually strongly stationary and if the $(m+n)$-dimensional $\vec{r.v.}$s $\{\xi(t_1), \ldots, \xi(t_m), \eta(t'_1), \ldots, \eta(t'_n)\}$ and $\{\xi(t_1 + \tau), \ldots, \xi(t_m + \tau), \eta(t'_1 + \tau), \ldots, \eta(t'_n + \tau)\}$ have identical probability laws for all $m \in \mathbb{N}^*$, for all $n \in \mathbb{N}^*$, and for all $\tau \in I$.

2.2. WEAK STATIONARITY

It is very difficult if not impossible in practice to use the notion of strong stationarity. Thus we usually content ourselves with using the notion of weak stationarity which is much easier to operate and which is adequate for a large number of problems.

Weak stationarity is also called second-order stationarity for it involves only the moments of order 1 and 2 associated with the r.s. in question.

2.2.1. *Weakly stationary random signals*

2.2.1.1. Definition

Let $\xi(t)$, $t \in I$, be an r.s. with $I = \mathbb{R}$ or $I = T\mathbb{Z}$, $T \in \mathbb{R}_+^*$. The r.s. ξ is **weakly stationary** when its mean value is constant and when its autocorrelation function is a function only of the difference $\tau = t_1 - t_2$. This may be written

$$\forall t \in I \quad E[\xi(t)] = m_\xi$$
$$\forall t \in I \quad \forall \tau \in I \quad E[\xi(t)\xi^*(t-\tau)] = R_{\xi\xi}(\tau) \tag{4.69}$$

on adopting the same notation $R_{\xi\xi}$ to denote the function of the pair (t_1, t_2) defined by Eqn (4.31) and the function of the variable τ defined by Eqn (4.69).

2.2.1.2. Properties

It follows from Eqn (4.69) that the mean power of the r.s. ξ is the constant $P_\xi = R_{\xi\xi}(0)$ (cf. Eqn (4.33)) and that its variance is the constant (cf. Eqn (1.70))

$$\sigma_\xi^2 = R_{\xi\xi}(0) - |m_\xi|^2 \tag{4.70}$$

It follows also that the covariance of the r.v.s $\xi(t)$ and $\xi(t - \tau)$ is a function of the variable τ which can be written (cf. Eqn (4.35))

$$C_{\xi\xi}(\tau) = R_{\xi\xi}(\tau) - |m_\xi|^2 = \sigma_\xi^2 r_{\xi\xi}(\tau) \tag{4.71}$$

It follows immediately that strong stationarity implies weak stationarity. The converse is not in general true. It is true when the r.s. in question is gaussian (cf. Example 2.1).

194 Random signals

Example 2.1 Stationary gaussian random signals

Let $\xi(t), t \in \mathbb{R}$, be a gaussian r.s. in continuous time. It is defined by the family of probability densities p_X of the $\vec{\text{r.v.}}$s $(\xi_1, \ldots, \xi_n)$ (cf. Example 1.4).

When the r.s. ξ is weakly stationary, the matrix M_X is constant and the covariance matrix Γ_{XX} is a function only of the differences $t_j - t_i$. The probability density p_X is then invariant for every translation of the time origin. A weakly stationary gaussian r.s. is therefore strongly stationary. It follows from Eqn (4.71) that its probability law is completely defined by its mean value m_ξ and its autocorrelation function $R_{\xi\xi}$.

Example 2.2 Weakly stationary random signals

Let $(\Omega, \mathcal{A}, P)$ be a probability space, ξ be a complex r.v., s be a function $s: \mathbb{R} \to \mathbb{C}$, and $\eta(t), t \in \mathbb{R}$, be the complex r.s. whose realizations can be written

$$\eta(\omega, t) = \xi(\omega) s(t)$$

Let us find conditions under which the r.s. η is weakly stationary.

It follows immediately that

$$E[\eta(t)] = E[\xi] s(t)$$

It is thus necessary that the mean value of the r.v. ξ be zero for the mean value of the r.s. η to be independent of time (excluding the trivial case in which the function s is zero).

It then follows that

$$R_{\eta\eta}(t, t - \tau) = E[|\xi|^2] s(t) s^*(t - \tau)$$

In order that the function $R_{\eta\eta}$ be independent of t, it is necessary that this is the case for $\tau = 0$. For this it is necessary that the modulus of the function s be constant. Then let

$$s(t) = r \exp\{i\varphi(t)\} \qquad r \in \mathbb{C}$$

where φ is a function $\mathbb{R} \to \mathbb{R}$. It follows that

$$R_{\eta\eta}(t, t - \tau) = E[|\xi|^2] |r|^2 \exp[i\{\varphi(t) - \varphi(t - \tau)\}]$$

It is then necessary that the function $\varphi(t) - \varphi(t - \tau)$ does not depend on t, which implies that

$$\frac{d\varphi}{dt} = \frac{d\varphi(t - \tau)}{dt} = \lambda \qquad \lambda \in \mathbb{R}$$

The function φ is therefore of the form $\varphi(t) = \lambda t + \vartheta$.

It is immediately obvious that the above necessary conditions are also sufficient. They can be written

$$E[\xi] = 0 \qquad s(t) = c \exp(i\lambda t) \qquad c \in \mathbb{C} \qquad \lambda \in \mathbb{R} \qquad (4.72)$$

2.2.2. Pair of mutually weakly stationary random signals

2.2.2.1. Definition
Let $\xi(t)$ and $\eta(t)$, $t \in I$, be two r.s.s, I being one of the sets $\mathbb{R}$ or $T\mathbb{Z}$, $T \in \mathbb{R}_+^*$. They are **mutually weakly stationary** if they are individually weakly stationary and if their cross-correlation function is a function only of the difference $\tau = t_1 - t_2$. This may be written

$$\forall t \in I \quad \forall \tau \in I \quad E[\xi(t)\eta^*(t-\tau)] = R_{\xi\eta}(\tau) \quad (4.73)$$

adopting the same notation $R_{\xi\eta}$ to denote the function of the pair (t_1, t_2) defined by Eqn (4.38) and the function of τ defined by Eqn (4.73).

2.2.2.2. Properties
The covariance of the r.v.s $\xi(t)$ and $\eta(t-\tau)$ is a function of the variable τ which is written (cf. Eqn (4.41))

$$C_{\xi\eta}(\tau) = R_{\xi\eta}(\tau) - m_\xi m_\eta^* = \sigma_\xi \sigma_\eta r_{\xi\eta}(\tau) \quad (4.74)$$

Example 2.3 Sinusoidal random signals
Consider the sinusoidal r.s. (cf. Example 1.2)

$$\xi(t) = A \cos(2\pi f_0 t + \varphi_0) \quad \eta(t) = B \sin(2\pi f_0 t + \varphi_0) \quad (4.75)$$

We suppose that the r.v. φ_0 is uniformly distributed over the interval $[-\pi, +\pi]$ and is independent of the real r.v.s A and B.

The mean value of the r.v. ξ is obtained by applying Eqn (1.71). We obtain

$$E[\xi(t)] = E[A] \frac{1}{2\pi} \int_{-\pi}^{+\pi} \cos(2\pi f_0 t + u) \, du = 0 \quad (4.76)$$

The mean value of the r.s. η is also zero.

The autocorrelation function of the r.s. ξ is obtained by applying Eqn (4.31). We obtain

$$R_{\xi\xi}(t, t-\tau) = E[\xi(t)\xi^*(t-\tau)]$$
$$= E[A^2]\tfrac{1}{2}E[\cos(2\pi f_0 \tau) + \cos(4\pi f_0 t - 2\pi f_0 \tau + 2\varphi_0)]$$

The term containing $2\varphi_0$ vanishes because of the uniform distribution of the r.v. φ_0, and so the autocorrelation function depends only on τ. In the same way we calculate the autocorrelation function of the r.s. η. We get

$$R_{\xi\xi}(\tau) = \tfrac{1}{2}E[A^2] \cos(2\pi f_0 t)$$
$$R_{\eta\eta}(\tau) = \tfrac{1}{2}E[B^2] \cos(2\pi f_0 t) \quad (4.77)$$

The r.s.s ξ and η are thus centred and weakly stationary.

The cross-correlation function of the r.s.s ξ and η is obtained by applying

196 Random signals

Eqn (4.39). We get

$$R_{\xi\eta}(t, t-\tau) = E[\xi(t)\eta^*(t-\tau)]$$
$$= E[AB]\tfrac{1}{2}E[-\sin(2\pi f_0 \tau) + \sin(4\pi f_0 t - 2\pi f_0 \tau + 2\varphi_0)]$$

The term containing $2\varphi_0$ vanishes as before. In the same way we calculate the cross-correlation function of the r.s.s η and ξ. We get

$$R_{\xi\eta}(\tau) = -R_{\eta\xi}(\tau) = -\tfrac{1}{2}E[AB]\sin(2\pi f_0 \tau) \tag{4.78}$$

The r.s.s ξ and η are thus mutually weakly stationary.

Example 2.4 Periodic random signal

Definitions
A periodic r.s. $\xi(t)$, $t \in \mathbb{R}$, of period T, is defined by its realizations which are written

$$\xi(\omega, t) = s(t) * \delta\{t - \vartheta(\omega)\} = s\{t - \vartheta(\omega)\} \tag{4.79}$$

where s is a periodic d.s. of period T (cf. Chapter 3, Section 3.4) and ϑ is an r.v. uniformly distributed over the period interval $[-T/2, +T/2]$.

The d.s. s can be expanded in a Fourier series in the space $\mathscr{S}'$. The same is true for each realization of the r.s. ξ. This enables us to write (suppressing the ω), without seeking to give a rigorous meaning to the equality,

$$\xi(t) = \sum_{n \in \mathbb{Z}} \xi_n(t) \qquad \xi_n(t) = c_n \exp\left(-in\frac{2\pi\vartheta}{T}\right)\exp\left(in\frac{2\pi t}{T}\right) \tag{4.80}$$

where c_n is the Fourier coefficient of order n of the d.s. s.

The r.s. ξ_n is a complex sinusoidal r.s., of frequency n/T and of amplitude $|c_n|$. It is easy to verify, since the r.v. ϑ is uniformly distributed over the period interval, that we have

$$E[\xi_0(t)] = c_0$$
$$\forall n \in \mathbb{Z}^* \qquad E[\xi_n(t)] = 0$$
$$\forall (m,n) \in \mathbb{Z}^2 \quad m \neq n \qquad E[\xi_m(t_1)\xi_n^*(t_2)] = 0$$

It follows from this that two r.s.s ξ_m and ξ_n, $m \neq n$, are not correlated (cf. Section 1.6.3.3).

A periodic r.s. can thus be considered as a sum of complex sinusoidal r.s.s, pairwise uncorrelated.

Properties
It is immediate, from the above, that the mean value of the r.s. ξ is constant:

$$E[\xi(t)] = m_\xi = c_0 \tag{4.81}$$

Then the autocorrelation function may be written

$$R_{\xi\xi}(t, t-\tau) = E[\xi(t)\xi^*(t-\tau)]$$

$$= \sum_{m \in \mathbb{Z}} \sum_{n \in \mathbb{Z}} c_m c_n^* E\left[\exp\left\{-i(m-n)2\pi\frac{\vartheta}{T}\right\}\right]$$

$$\times \exp\left\{i(m-n)2\pi\frac{t}{T}\right\} \exp\left(in2\pi\frac{\tau}{T}\right)$$

The only non-zero terms are those for which $m \neq n$. It follows that

$$R_{\xi\xi}(\tau) = \sum_{n \in \mathbb{Z}} |c_n|^2 \exp\left(in2\pi\frac{\tau}{T}\right) \tag{4.82}$$

A periodic r.s. ξ is thus weakly stationary.

The r.s. ξ is of finite mean power if and only if the series with general term $|c_n|^2$ converges. The d.s. s thus has finite mean power and its (temporal) mean power P_s is equal to the (statistical) mean power P_ξ of the r.s. ξ (cf. Eqn (3.58)):

$$P_\xi = \sum_{n \in \mathbb{Z}} |c_n|^2 = \frac{1}{T} \int_{-T/2}^{T/2} |s(t)|^2 \, dt = P_s \tag{4.83}$$

Example 2.5 Ideal sampling signal

The ideal sampling r.s. is the periodic r.s. defined formally (we have not defined random distributions) by

$$\xi(\omega, t) = \sum_{n \in \mathbb{Z}} \delta\{t - nT - \vartheta(\omega)\} \tag{4.84}$$

Since the Fourier coefficients c_n are all equal to $1/T$ (cf. Appendix 1, case 23), it is possible to write the expansion

$$\xi(\omega, t) = \frac{1}{T} \sum_{n \in \mathbb{Z}} \exp\left\{-in2\pi\frac{\vartheta(\omega)}{T}\right\} \exp\left(in2\pi\frac{t}{T}\right) \tag{4.85}$$

The mean value and the autocorrelation function of the r.s. may be obtained from Eqns (4.81) and (4.82) respectively. We get

$$E[\xi] = \frac{1}{T} \qquad R_{\xi\xi}(\tau) = \frac{1}{T^2} \sum_{n \in \mathbb{Z}} \exp\left(in2\pi\frac{\tau}{T}\right) \tag{4.86}$$

These results are very useful in practice when we are dealing with sampled r.s.s (cf. Chapter 5, Section 2.4).

3. Random signals of finite mean power

R.s.s of finite mean power are defined in Example 1.1. They play a very important part in practice in view of the fact that an r.s. having concrete

198 *Random signals*

existence always has finite mean power inasmuch as it is physically measurable. Therefore it is necessary to devote a special section to them.

3.1. GENERALITIES

3.1.1. Definition

A complex r.s. $\xi(t)$, $t \in I$, defined over a probability space $(\Omega, \mathcal{A}, P)$ has finite mean power when (cf. Example 1.1)

$$\forall t \in I \qquad E[|\xi(t)|^2] = P_\xi(t) < +\infty \qquad (4.87)$$

The power $P_\xi(t)$ is the mean value of the instantaneous power developed by the r.s. ξ in a normalized resistance of $1\,\Omega$.

The complex r.v. $\xi(t)$ belongs then, for every $t \in I$, to the Hilbert space $L^2(\Omega, \mathcal{A}, P)$ and the instantaneous mean power $P_\xi(t)$ is the square of the norm of the r.v. $\xi(t)$ (cf. Chapter 1, Section 2.4.2). Hence it follows that an r.s. of finite mean power satisfies particular properties, which form the subject of this section.

3.1.2. Properties

The autocorrelation function of the r.s. ξ is defined by Eqn (4.31). Its value at the point $(t_1, t_2) \in I^2$ is the scalar product in the space $L^2(\Omega, \mathcal{A}, P)$ of the r.v.s $\xi(t_1)$ and $\xi(t_2)$ (cf. Eqn (6.69)). Its value at the point $(t, t) \in I^2$ is the instantaneous mean power $P_\xi(t)$.

It is easy to prove that the r.s. ξ has finite mean power if and only if its autocorrelation function is bounded: this may be written

$$\forall (t_1, t_2) \in I^2 \qquad |R_{\xi\xi}(t_1, t_2)| < +\infty \qquad (4.88)$$

It is immediately obvious that the condition is sufficient. It is necessary since Schwarz's inequality enables us to write (cf. Eqn (6.70))

$$\forall (t_1, t_2) \in I^2 \qquad |R_{\xi\xi}(t_1, t_2)| \leq (P_{\xi_1} P_{\xi_2})^{1/2} \qquad (4.89)$$

Formula (4.88) gives the criterion used to verify whether an r.s. has finite mean power.

3.2. CONTINUITY IN MEAN SQUARE

3.2.1. Definitions

An r.s. $\xi(t)$, $t \in \mathbb{R}$, in continuous time, with finite mean power, is **continuous in mean square** (m.s.) at the point t when the r.v. $\xi(t + \tau)$ tends in the space

$L^2(\Omega, \mathscr{A}, P)$ to the r.v. $\xi(t)$ when τ tends to zero. This may be written

$$\lim_{\tau \to 0} E[|\xi(t+\tau) - \xi(t)|^2] = 0 \qquad (4.90)$$

The r.s. ξ is continuous in m.s. when it is continuous in m.s. for every $t \in \mathbb{R}$.

3.2.2. Criterion for continuity in mean square

3.2.2.1. Theorem

For an r.s. $\xi(t)$, $t \in \mathbb{R}$, of finite mean power to be continuous in m.s. at the point t it is necessary and sufficient that its autocorrelation function be continuous at the point (t, t). For it to be continuous for all points $t \in \mathbb{R}$ it is necessary and sufficient that its autocorrelation function be continuous for all points $(t_1, t_2) \in \mathbb{R}^2$.

3.2.2.2. Proof

The proof of this theorem is left to the reader. It uses the continuity properties of the scalar product (cf. Chapter 6, Exercises 4.4 and 4.5). It should be noted that continuity of the autocorrelation function on its principal diagonal implies its continuity everywhere in $\mathbb{R}^2$.

3.2.2.3. Property

It follows from the above theorem that the instantaneous mean power of an m.s. continuous r.s. is a continuous function of time.

3.3. DIFFERENTIABILITY IN MEAN SQUARE

3.3.1. Definitions

An r.s. $\xi(t)$, $t \in \mathbb{R}$, with finite mean power is **differentiable in m.s.** at the instant t and has as derivative the r.v. $\xi'(t)$ when the r.v. $X_\tau(t)$ defined by

$$X_\tau(t) = \frac{\xi(t+\tau) - \xi(t)}{\tau} \qquad (4.91)$$

tends in the space $L^2(\Omega, \mathscr{A}, P)$ to the r.v. $\xi'(t)$ when τ tends to zero. This may be written

$$\lim_{\tau \to 0} E[|X_\tau(t) - \xi'(t)|^2] = 0 \qquad (4.92)$$

The r.s. ξ is differentiable in m.s. when it is differentiable in m.s. for every $t \in \mathbb{R}$. It then has as derivative the r.s. $\xi'(t)$, $t \in \mathbb{R}$.

3.3.2. Criterion for differentiability in mean square

3.3.2.1. Theorem

For an r.s. $\xi(t)$, $t \in \mathbb{R}$, of finite mean power to be differentiable in m.s. at the point t it is necessary and sufficient that its autocorrelation function satisfy

$$\lim_{\substack{\tau \to 0 \\ \tau' \to 0}} \frac{1}{\tau \tau'} \{R_{\xi\xi}(t+\tau, t+\tau') - R_{\xi\xi}(t, t+\tau') - R_{\xi\xi}(t+\tau, t) + R_{\xi\xi}(t, t)\} < +\infty \tag{4.93}$$

3.3.2.2. Proof

The proof of this theorem is left to the reader. It uses the continuity properties of the scalar product (cf. Chapter 6, Exercise 4.5). It is sufficient to note that the expression appearing in Eqn (4.93) is the scalar product of the r.v.s $X_\tau(t)$ and $X_{\tau'}(t)$ defined by Eqn (4.91).

3.3.3. Property

3.3.3.1. Statement

An r.s. $\xi(t)$, $t \in \mathbb{R}$, of finite mean power differentiable in m.s. at the instant t, is continuous in m.s. at the same instant.

3.3.3.2. Proof

We can write, using the properties of the expectation E and taking account of Eqn (4.91),

$$E[|\xi(t+\tau) - \xi(t)|^2] = \tau^2 E[|X_\tau(t) - \xi'(t) + \xi'(t)|^2]$$
$$\leq \tau^2 E[|X_\tau(t) - \xi'(t)|^2] + \tau^2 E[|\xi'(t)|^2]$$

The property follows immediately from this by making τ tend to zero.

3.3.4. Mean value of the mean-square derivative random signal

Let $\xi(t)$, $t \in \mathbb{R}$, be an r.s. with finite mean power, differentiable in m.s., and let $\xi'(t)$, $t \in \mathbb{R}$, be the m.s. derivative r.s. Schwarz's inequality enables us to write

$$\left| \frac{E[\xi(t+\tau)] - E[\xi(t)]}{\tau} - E[\xi'(t)] \right| = |E[X_\tau(t) - \xi'(t)]|$$
$$\leq E[|X_\tau(t) - \xi'(t)|^2]$$

It follows from this, by making τ tend to zero, that

$$\frac{dE[\xi(t)]}{dt} = E[\xi'(t)] \tag{4.94}$$

Random signals of finite mean power 201

The mean value of the m.s. derivative r.s. ξ' is thus the derivative of the mean value of the r.s. ξ.

3.3.5. *Correlation functions*

Let $\xi(t), t \in \mathbb{R}$, be an r.s. with finite mean power, differentiable in m.s., and let $\xi'(t), t \in \mathbb{R}$, be the m.s. derivative r.s. The cross-correlation functions of the r.s.s ξ and ξ' and the autocorrelation function of the r.s. ξ' are respectively the derivatives

$$R_{\xi'\xi}(t_1, t_2) = \frac{\partial R_{\xi\xi}(t_1, t_2)}{\partial t_1}$$

$$R_{\xi\xi'}(t_1, t_2) = \frac{\partial R_{\xi\xi}(t_1, t_2)}{\partial t_2}$$
(4.95)

$$R_{\xi'\xi'}(t_1, t_2) = \frac{\partial^2 R_{\xi\xi}(t_1, t_2)}{\partial t_1 \, \partial t_2}$$
(4.96)

We shall prove the first relation. The others may be obtained in the same way. The r.v. $\xi'(t_1)$ is the limit of the r.v. $X_\tau(t_1)$ when τ tends to zero. It is therefore possible to write, using the property of continuity of the scalar product (cf. Chapter 6, Exercise 4.4),

$$R_{\xi'\xi}(t_1, t_2) = E[\xi'(t_1)\xi^*(t_2)]$$

$$= \lim_{\tau \to 0} E[X_\tau(t_1)\xi^*(t_2)]$$

$$= \lim_{\tau \to 0} \frac{R_{\xi\xi}(t_1 + \tau, t_2) - R_{\xi\xi}(t_1, t_2)}{\tau}$$

The stated result is thus proved.

Example 3.1 Wiener–Levy random signal

Definition
A Wiener–Levy (WL) r.s. is a causal gaussian r.s. $\xi(t), t \in \mathbb{R}_+$, defined by the condition $P\{\xi(0) = 0\} = 1$ and by the probability density

$$p_X(x_1, \ldots, x_n) = \frac{1}{(2\pi b^2 t_1)^{1/2}} \exp\left(-\frac{x_1^2}{2b^2 t_1}\right) \times \cdots$$

$$\times \frac{1}{\{2\pi b^2(t_n - t_{n-1})\}^{1/2}} \exp\left\{-\frac{(x_n - x_{n-1})^2}{2b^2(t_n - t_{n-1})}\right\} \quad (4.97)$$

The realizations of the r.s. ξ start a.s. from the origin and p_X is the probability density of the $\overrightarrow{\text{r.v.}}$ $(\xi_1, \ldots, \xi_n)$, where we put $\xi(t_i) = \xi_i$. The WL r.s. is thus defined by a family of projective probability measures (cf. Section 1.2.4).

Property
The WL r.s. is markovian.

It is in fact easy to verify, using Eqns (2.7) and (2.12), that the conditional probabilities $p_{\xi_n|\xi_1\ldots\xi_{n-1}}$ and $p_{\xi_n|\xi_{n-1}}$ satisfy

$$p_{\xi_n|\xi_1\ldots\xi_{n-1}}(x_n|x_1,\ldots,x_{n-1}) = p_{\xi_n|\xi_{n-1}}(x_n|x_{n-1})$$

$$= \frac{1}{\{2\pi b^2(t_n - t_{n-1})\}^{1/2}} \exp\left\{-\frac{(x_n - x_{n-1})^2}{2b^2(t_n - t_{n-1})}\right\} \quad (4.98)$$

The r.s. ξ is thus markovian (cf. Eqn (4.28)) and the probability density p_X can be written

$$p_X(x_1,\ldots,x_n) = p_{\xi_1}(x_1)p_{\xi_2|\xi_1}(x_2|x_1)\ldots p_{\xi_n|\xi_{n-1}}(x_n|x_{n-1})$$

Property
The WL r.s. is gaussian.

Let $(\eta_1,\ldots,\eta_n)$ be the $\vec{r.v.}$ obtained by the linear transformation of the $\vec{r.v.}$ $(\xi_1,\ldots,\xi_n)$ defined by

$$\eta_1 = \xi_1 \quad \eta_2 = \xi_2 - \xi_1 \quad \cdots \quad \eta_n = \xi_n - \xi_{n-1} \quad (4.99)$$

The probability density p_Y of the new $\vec{r.v.}$ is obtained by change of variable, starting from the probability density p_X. It follows, since the jacobian of the inverse transformation is equal to unity (cf. Chapter 2, Section 1.2.3), that

$$p_Y(y_1,\ldots,y_n) = \frac{1}{(2\pi b^2 t_1)^{1/2}} \exp\left(-\frac{y_1^2}{2b^2 t_1}\right) \times \cdots$$

$$\times \frac{1}{\{2\pi b^2(t_n - t_{n-1})\}^{1/2}} \exp\left\{-\frac{y_n^2}{2b^2(t_n - t_{n-1})}\right\} \quad (4.100)$$

It is easy to verify that the density p_Y is of the form of Eqn (4.14) with

$$M_Y = \begin{bmatrix} 0 \\ \vdots \\ 0 \end{bmatrix} \quad \Gamma_{YY} = \begin{bmatrix} b^2 t_1 & & 0 \\ & \ddots & \\ 0 & & b^2(t_n - t_{n-1}) \end{bmatrix} \quad (4.101)$$

The $\vec{r.v.}$ $(\eta_1,\ldots,\eta_n)$ is thus gaussian and consequently the $\vec{r.v.}$ $(\xi_1,\ldots,\xi_n)$ is also gaussian (cf. Chapter 2, Example 1.7). The WL r.s. is therefore gaussian.

Property
The WL r.s. has independent increments.

In fact, the r.v.s $\eta_1,\ldots,\eta_n$ represent the increments of the r.s. ξ during the intervals $]0, t_1],\ldots,]t_{n-1}, t_n]$. It follows from Eqn (4.101) that the r.v.s η_n are centred and pairwise uncorrelated. The gaussian sequence $(\eta_n, n \in \mathbb{N}^*)$ is thus independent (cf. Chapter 2, Example 1.10). For this reason we say that the WL r.s. has independent increments.

Properties

The WL r.s. is centred and has finite mean power. It is continuous in m.s. but is not differentiable in m.s.

The r.v. $\xi(t), t \in \mathbb{R}_+$, is the increment of the r.s. ξ during the interval $[0, t]$. Its mean value is zero (cf. Eqn (4.101)). The WL r.s. is therefore centred.

Let us calculate the autocorrelation function of the r.s. ξ. It follows that, for $t_1 < t_2$, since the increments ξ_1 and $\xi_2 - \xi_1$ are independent and centred,

$$R_{\xi\xi}(t_1, t_2) = E[\xi_1 \xi_2] = E[\xi_1(\xi_2 - \xi_1 + \xi_1)] = V[\xi_1] = b^2 t_1$$

It follows likewise, when $t_2 < t_1$, that

$$R_{\xi\xi}(t_1, t_2) = b^2 t_2$$

The autocorrelation function can thus be written

$$R_{\xi\xi}(t_1, t_2) = b^2 \min(t_1, t_2) \tag{4.102}$$

The function $R_{\xi\xi}$ is bounded. The r.s. ξ thus has finite mean power (cf. Eqn (4.88)). The function $R_{\xi\xi}$ is continuous at the point (t, t). The r.s. ξ is therefore continuous in m.s. (cf. Section 3.2.2). The function $R_{\xi\xi}$ is not twice differentiable. In fact

$$\frac{\partial^2 R_{\xi\xi}(t_1, t_2)}{\partial t_1 \, \partial t_2} = \frac{\partial}{\partial t_2} \{b^2 Y(t_2 - t_1)\} \tag{4.103}$$

and the Heaviside function is not differentiable at the origin (at least in terms of functions). The r.s. ξ is therefore not differentiable in m.s. since the necessary condition (4.96) is not satisfied.

Physical interpretation

The WL r.s. is the mathematical model of **brownian motion**. The object of this section is to show that it can be defined, by passage to the limit, starting from an r.s. called a **random walk**.

Let $(X_n, n \in \mathbb{N}^*)$ be an independent sequence of real binary r.v.s taking the values Δ and $-\Delta$ and having the same probability law. Put

$$\forall n \in \mathbb{N}^* \quad P\{X_n = \Delta\} = p \quad P\{X_n = -\Delta\} = q = 1 - p \tag{4.104}$$

Consider the real causal r.s. in discrete time $\xi(t), t \in \tau\mathbb{N}, \tau \in \mathbb{R}_+^*$, defined by

$$\xi(0) = 0 \quad \forall n \in \mathbb{N}^* \quad \xi(n\tau) = \sum_{i=1}^{n} X_i \tag{4.105}$$

The r.s. ξ is a one-dimensional random walk. In fact the r.v. $\xi(n\tau)$ can represent the position, at the instant $n\tau$, of a body moving along an axis, stepping forward or backward at each instant $k\tau$ always by an amount with the same amplitude Δ. The mean value and variance of the r.v. $\xi(n\tau)$ are respectively

$$E[\xi(n\tau)] = n \Delta (p - q) \quad V[\xi(n\tau)] = n \Delta^2 4pq \tag{4.106}$$

Fix $t = n\tau$ and put

$$\Delta = b\tau^{1/2} + o(\tau^{1/2}) \qquad p = q = \tfrac{1}{2} + o(\tau^{1/2}) \qquad (4.107)$$

It follows then that

$$E[\xi(t)] = o(\tau^{1/2}) \qquad V[\xi(t)] = b^2 t + o(\tau^{1/2}) \qquad (4.108)$$

If we make τ tend to zero, the displacement Δ tends to zero, the speed of displacement tends to infinity, and the probabilities p and q tend to $\tfrac{1}{2}$. Moreover, t being fixed, n tends to infinity. The central limit theory enables us to state, taking account of Eqn (4.108), that the probability law of the r.v. $\xi(t)$ tends to a centred gaussian law with variance $b^2 t$ (cf. Chapter 2, Section 2.5).

We can define thus, by passage to the limit, a causal r.s. $\xi(t), t \in \mathbb{R}_+$. Let $t_1, \ldots, t_n$ be instants such that $0 < t_1 < \ldots < t_n$. It follows immediately, by construction and by using Eqn (4.108), that the r.v.s $\xi_1, \ldots, \xi_n - \xi_{n-1}$ are centred, gaussian and independent, and that their variances are equal to those of the r.v.s defined by Eqn (4.99).

It follows from this that the r.s. defined by passage to the limit of the random walk is the WL r.s. defined by Eqn (4.97).

3.4. INTEGRABILITY IN MEAN SQUARE

3.4.1. Definition

Let $\xi(u), u \in \mathbb{R}$, be an r.s. with finite mean power defined over a probability space $(\Omega, \mathscr{A}, P)$, let $[a, b] \subset \mathbb{R}$ be a bounded interval, and consider a subdivision of the interval $[a, b]$ into n disjoint segments:

$$a = u_0 < u_1 < \ldots < u_{n-1} < u_n = b$$

Put $\Delta_n = \max |u_i - u_{i-1}|$ and consider points $\vartheta_i \in [u_{i-1}, u_i]$. Let $g(t, u)$ be a complex function defined over the set $I \times \mathbb{R} \subset \mathbb{R}^2$ which, for all $t \in I$, is a bounded function of the variable $u \in \mathbb{R}$. Finally, t being fixed, let $X_n(t)$ be the complex r.v.

$$X_n(t) = \sum_{i=1}^{n} g(t, \vartheta_i) \xi(\vartheta_i)(u_i - u_{i-1}) \qquad (4.109)$$

It belongs to the space $L^2(\Omega, \mathscr{A}, P)$. Suppose that, when n tends to infinity and Δ_n tends to zero, it tends to a limit independent of the subdivision chosen. The limit is the Riemann integral

$$X(t) = \lim_{n \to \infty} X_n(t) = \int_a^b g(t, u)\xi(u)\, du \qquad (4.110)$$

It defines an r.s. $X(t), t \in I$, with finite mean power obtained by integration in m.s. from the r.s. ξ. As for the Riemann integral, the definition (4.110) and the

Random signals of finite mean power 205

condition (4.111) can be generalized when one of the limits tends to infinity. When the r.s. ξ is gaussian, the r.v. $X_n(t)$ is gaussian (cf. Exercise 1.8). We shall accept that this property is preserved by passage to the limit and that the integrated r.s. X is also gaussian.

3.4.2. Criterion for integrability in mean square

3.4.2.1. Theorem
For the r.s. defined by the integral (4.110) to exist it is necessary and sufficient that the following Riemann integral exist:

$$\int_a^b \int_a^b g(t, u_1)g^*(t, u_2)R_{\xi\xi}(u_1, u_2)\, du_1\, du_2 < +\infty \quad (4.111)$$

3.4.2.2. Proof
The proof of this theorem is left to the reader. It uses the continuity properties of scalar products (cf. Chapter 6, Exercise 4.5).

3.4.3. Mean value of the mean-square integrated random signal

Consider the r.s. $X(t), t \in I$, defined by Eqn (4.110); let us write, interchanging the passage to the limit and the expectation (cf. Chapter 6, Exercise 4.6),

$$E[X(t)] = E\left[\lim_{n \to \infty} X_n(t)\right]$$

$$= \lim_{n \to \infty} E[X_n(t)]$$

$$= \lim_{n \to \infty} \left\{ \sum_{i=1}^{n} g(t, \vartheta_i) E[\xi(\vartheta_i)](u_i - u_{i-1}) \right\}$$

It follows from this that

$$\forall t \in I \quad E[X(t)] = \int_a^b g(t, u) E[\xi(u)]\, du \quad (4.112)$$

3.4.4. Correlation functions

Let $\xi(t), t \in U$, be an r.s. with finite power, integrable in m.s., and let $X(t), t \in I$, be the r.s. defined by Eqn (4.110). The cross-correlation function of the r.s.s ξ and X and the autocorrelation function of the r.s. X are defined respectively by

206 Random signals

the integrals

$$R_{\xi X}(t_1, t_2) = \int_a^b g^*(t_2, u) R_{\xi\xi}(t_1, u)\, du \qquad (4.113)$$

$$R_{XX}(t_1, t_2) = \int_a^b \int_a^b g(t_1, u_1) g^*(t_2, u_2) R_{\xi\xi}(u_1, u_2)\, du_1\, du_2 \qquad (4.114)$$

The proof is left to the reader. It uses the properties of continuity of the scalar product (cf. Chapter 6, Exercise 4.4).

Example 3.2 Integration and differentiation in mean square

Let $\xi(t)$ and $\eta(t)$, $t \in \mathbb{R}$, be two r.s.s continuous in m.s. over $\mathbb{R}$. It follows from this that the autocorrelation functions $R_{\xi\xi}$ and $R_{\eta\eta}$ are continuous over $\mathbb{R}^2$ (cf. Section 3.2.2.1). They are then integrable over the set $[0, t] \times [0, t]$ and it follows from Eqn (4.111) that the r.s.s ξ and η are integrable in m.s. over the interval $[0, t]$. This enables us to define the integrated r.s.s:

$$X(t) = \int_0^t \xi(u)\, du \qquad Y(t) = \int_0^t \eta(v)\, dv \qquad (4.115)$$

The mean value and the autocorrelation function of the r.s. X are obtained by applying Eqns (4.112) and (4.114) respectively. We get

$$E[X(t)] = \int_0^t E[\xi(u)]\, du \qquad (4.116)$$

$$R_{XX}(t_1, t_2) = \int_0^{t_1} du_1 \int_0^{t_2} R_{\xi\xi}(u_1, u_2)\, du_2 \qquad (4.117)$$

The cross-correlation function of the r.s.s X and Y is obtained by arguing as in the preceding section. We obtain

$$R_{XY}(t_1, t_2) = \int_0^{t_1} du_1 \int_0^{t_2} R_{\xi\eta}(u_1, u_2)\, du_2 \qquad (4.118)$$

It is easy to verify, by applying Eqn (4.93), that the r.s. X is differentiable in m.s. The autocorrelation function of the differentiated r.s. X' can be written (cf. Eqn (4.96))

$$R_{X'X'}(t_1, t_2) = R_{\xi\xi}(t_1, t_2)$$

The continuity property of the scalar product enables us to conclude that the differentiated r.s. X' is the r.s. ξ (cf. Chapter 6, Exercise 4.4).

Exercises

Exercise 3.1 Wiener–Levy random signal
Let $\xi(t), t \in \mathbb{R}$, be a gaussian r.s., centred, having an autocorrelation function such that

$$\forall t_1 \leqslant t_2 \quad R_{\xi\xi}(t_1, t_2) = b^2 u(t_1) v(t_2)$$

u and v being two real functions such that the function $g = u/v$ is a continuous mapping, bijective and increasing, from an interval $I \subset \mathbb{R}$ into $\mathbb{R}_+$.
Consider the r.s. $X(t)$, $t \in \mathbb{R}_+$, defined by

$$\forall t \in \mathbb{R}_+ \quad \eta(t) = \frac{\xi\{g^{-1}(t)\}}{v\{g^{-1}(t)\}} \quad X(t) = \eta(t) - \eta(0)$$

(1) Prove that the autocorrelation function of the r.s. X can be written

$$R_{XX}(t_1, t_2) = b^2 \min(t_1, t_2)$$

(2) Hence deduce that the r.s. X is a WL r.s. with parameter b. (Show that its probability density is of the form of Eqn (4.97).)

(3) Suppose that the autocorrelation function of the r.s. ξ can be written

$$R_{\xi\xi}(t_1, t_2) = b^2 \exp(-a|t_2 - t_1|) \quad a \in \mathbb{R}_+^*$$

Hence deduce that the r.s. X can be written

$$X(t) = t^{1/2} \xi\left(\frac{1}{2a} \ln t\right)$$

Exercise 3.2 Wiener–Levy random signal
Let $\xi(t), t \in \mathbb{R}_+$, be a WL r.s. and let $u(t)$ and $v(t)$, $t \in \mathbb{R}_+$, be the r.s.s defined by

$$\forall t \in \mathbb{R}_+ \quad u(t) = t^{1/2} \xi\left(\frac{1}{t}\right) \quad v(t) = \frac{1}{t^{1/2}} \xi(t)$$

Show that the r.s.s u and v are weakly equivalent. (Verify that they are gaussian, centred, and with the same autocorrelation function (cf. Exercise 1.8).)

Exercise 3.3 Differentiation in mean square
Let $\xi(t), t \in \mathbb{R}$, be an r.s. of finite mean power, differentiable in m.s. at the instant t. Prove that

$$\lim_{\tau \to 0} E[|X_\tau(t)|^2] = E[|\xi'(t)|^2] \qquad (4.119)$$

$X_\tau(t)$ being the r.v. defined by Eqn (4.91). (Write $X_\tau(t) = X_\tau(t) - \xi'(t) + \xi'(t)$ and apply the Schwarz inequality (6.70).)

Exercise 3.4 Poisson random signal

Consider a Poisson process with parameter ϑ and let $N(t)$, $t \in \mathbb{R}_+^*$, be its associated Poisson r.s. (cf. Example 1.3). Consider an independent sequence $(X_n, n \in \mathbb{N}^*)$ of square integrable r.v.s having identical probability laws and the causal r.s. $\xi(t)$, $t \in \mathbb{R}_+^*$, defined by

$$\forall t \in \mathbb{R}_+^* \qquad \xi(t) = \sum_{i=1}^{N(t)} X_i \qquad (4.120)$$

A realization of the r.s. ξ is represented in Fig. 4.11.

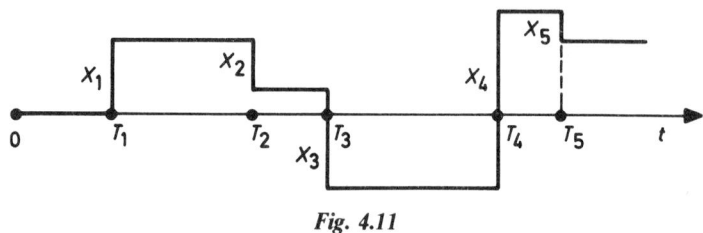

Fig. 4.11

(1) Prove that

$$E[\xi(t)] = E[X]\,\vartheta t \qquad (4.121)$$

(Argue conditionally with respect to the r.v. $N(t)$ and then remove the conditioning by applying Eqns (4.8) and (1.116).)

(2) Prove that

$$R_{\xi\xi}(t_1, t_2) = E[|X|^2]\vartheta \min(t_1, t_2) + |E[X]|^2 \vartheta^2 t_1 t_2 \qquad (4.122)$$

(Argue conditionally with respect to the r.v.s $N(t_1)$ and $N(t_2)$ and then remove the conditioning by applying Eqns (4.8), (4.49) and (1.116).)

(3) Deduce from the above that the r.s. ξ is continuous in m.s., although its realizations are not continuous. Deduce also that the r.s. ξ is not differentiable in m.s. and is integrable in m.s. over every interval $[0, t]$. (Apply Eqns (4.88), (4.96) and (4.111).)

(4) Verify that the above properties apply to the Poisson r.s. $N(t)$ whose autocorrelation function can be written

$$R_{NN}(t_1, t_2) = \vartheta \min(t_1, t_2) + \vartheta^2 t_1 t_2 \qquad (4.123)$$

(Note that the r.s. N is identical with the r.s. ξ when the r.v.s ξ_i are a.s. equal to 1.)

References

BLANC-LAPIERRE A. and PICINBONO B. (1981): *Fonctions Aléatoires*, Masson, Paris.

COX D.R. and MILLER H.D. (1967): *The Theory of Stochastic Processes*, Methuen, London, p. 146.
DUPRAZ J. (1977): *La Théorie des Distributions et ses Applications*, Cepadues, Toulouse, p. 125.
KLEINROCK L. (1975): *Queuing Systems*, Vol. I, Wiley, New York, p. 10.
NEVEU J. (1964): *Bases Mathématiques du Calcul des Probabilités*, Masson, Paris.
WONG, E. (1971): *Stochastic Processes in Information and Dynamical Systems*, McGraw-Hill, New York.

Chapter 5
Weakly stationary random signals

Weakly stationary random signals of finite mean power are extremely important in practice. Thus it is necessary to devote a special chapter to them, as a continuation of the previous chapter in which they are defined in a general manner.

The properties of random signals (r.s.s) in continuous and in discrete time, particularly spectral properties and linear filtering, are studied in parallel, the connection between the two types of signal being achieved by sampling. Non-linear filtering is treated as in the case of gaussian signals.

1. Random signals in continuous time

1.1. PROPERTIES OF THE CORRELATION FUNCTIONS

1.1.1. *Autocorrelation function*

Let $\xi(t)$, $t \in \mathbb{R}$, be a complex r.s. in continuous time, weakly stationary and with finite mean power. Its autocorrelation function $R_{\xi\xi}$ has the following properties.

1.1.1.1. Symmetry and parity

It follows immediately from Eqns (4.69) and (4.31) that

$$\forall \tau \in \mathbb{R} \qquad R_{\xi\xi}(\tau) = R_{\xi\xi}{}^*(-\tau) \qquad (5.1)$$

The function $R_{\xi\xi}$ thus has hermitian symmetry.

When the r.s. ξ is real, the function $R_{\xi\xi}$ is real and it follows from Eqn (5.1) that it is an even function.

1.1.1.2. Value at the origin

Since the r.s. ξ is of finite mean power, it follows from Eqn (4.33) that

$$\forall t \in \mathbb{R} \qquad R_{\xi\xi}(t, t) = R_{\xi\xi}(0) = P_\xi$$

Random signals in continuous time

The **instantaneous mean power** of the r.s. ξ is a constant P_ξ. It is equal to the value at the origin of the autocorrelation function. It follows in addition from Eqn (4.89) that

$$\forall \tau \in \mathbb{R} \qquad |R_{\xi\xi}(\tau)| \leq R_{\xi\xi}(0) = P_\xi \tag{5.2}$$

The function $R_{\xi\xi}$ is thus bounded by its value at the origin.

1.1.1.3. Continuity

Using Schwarz's inequality, we can write (cf. Eqn (6.70))

$$\forall t \in \mathbb{R} \quad \forall \tau \in \mathbb{R} \quad |R_{\xi\xi}(\tau) - R_{\xi\xi}(0)| = |E[\xi(t)\{\xi^*(t+\tau) - \xi^*(t)\}]|$$
$$\leq (E[|\xi(t)|^2] E[|\xi(t+\tau) - \xi(t)|^2])^{1/2}$$

In addition, we have the result on applying Eqn (5.1)

$$E[|\xi(t+\tau) - \xi(t)|^2] = 2 \operatorname{Re}\{R_{\xi\xi}(0) - R_{\xi\xi}(\tau)\}$$
$$\leq 2|R_{\xi\xi}(0) - R_{\xi\xi}(\tau)| \tag{5.3}$$

It follows from these two inequalities and from Eqn (4.90) that the r.s. ξ is continuous in mean square (m.s.) if and only if the function $R_{\xi\xi}$ is continuous at the origin. It is then easy to verify that the function $R_{\xi\xi}$ is everywhere continuous (cf. Exercise 1.1).

1.1.1.4. Non-negative definiteness

The autocorrelation function is non-negative definite (cf. Eqn (4.34)). This enables us to write here

$$\forall (c_1, \ldots, c_n) \in \mathbb{C}^n \qquad \forall (t_1, \ldots, t_n) \in \mathbb{R}^n$$

$$\sum_{i=1}^{n} \sum_{j=1}^{n} c_i c_j^* R_{\xi\xi}(t_j - t_i) \geq 0 \tag{5.4}$$

1.1.2. Cross-correlation function

Let $\xi(t)$ and $\eta(t)$, $t \in \mathbb{R}$, be two complex r.s.s in continuous time, weakly stationary and with finite mean power. Their cross-correlation functions $R_{\xi\eta}$ and $R_{\eta\xi}$ have the following properties which result immediately from Eqns (4.40) and (4.73):

$$\forall \tau \in \mathbb{R} \qquad R_{\xi\eta}(\tau) = R_{\eta\xi}^*(-\tau) \qquad |R_{\xi\eta}(\tau)| \leq (P_\xi P_\eta)^{1/2} \tag{5.5}$$

The functions $R_{\xi\eta}$ and $R_{\eta\xi}$ are therefore bounded.

1.2. SPECTRAL PROPERTIES

1.2.1. Spectrum of a random signal in continuous time

Let $\xi(t)$, $t \in \mathbb{R}$, be a complex r.s. in continuous time, weakly stationary and with finite mean power. Its spectrum can be introduced in two ways. The first is an immediate application of Bochner's theorem. It has the disadvantage of being purely mathematical. The second is the subject of the Wiener–Kintchine theorem. It involves explicitly the complex spectra of realizations of the r.s. under consideration and is for this reason more meaningful physically.

1.2.1.1. First method

The function $R_{\xi\xi}$, being bounded, defines a tempered distribution. It therefore has a Fourier transform $G_{\xi\xi}$ which is a tempered distribution (Dupraz, 1977, p. 29).

Suppose that the r.s. ξ is continuous in m.s. The function $R_{\xi\xi}$ is then continuous and, since it is non-negative definite, Bochner's theorem can be applied to it (cf. Chapter 6, Section 4.3.4). It follows from this that the distribution $G_{\xi\xi}$ is a bounded positive measure over the space $(\mathbb{R}, \mathscr{R})$. It is connected with the autocorrelation function by the integral (cf. Eqn (6.81))

$$R_{\xi\xi}(\tau) = \int_{\mathbb{R}} \exp(i2\pi\tau f)\, dG_{\xi\xi}(f) \tag{5.6}$$

We have then

$$P_\xi = R_{\xi\xi}(0) = \int_{\mathbb{R}} dG_{\xi\xi}(f) \tag{5.7}$$

This relation shows that the measure $G_{\xi\xi}$ defines the distribution of the power P_ξ over the frequency space. We say more simply that it is the **spectrum** of the r.s. ξ. When the measure $G_{\xi\xi}$ is absolutely continuous with respect to Lebesgue measure its density (which we shall write in the same way) is a positive integrable function (with respect to Lebesgue measure). It is the **spectral density** of the r.s. ξ. In the general case, the spectrum is the sum of a positive integrable function and a combination of Dirac distributions.

To summarize, we can write the fundamental relation which in the space $\mathscr{S}'$ connects the autocorrelation function $R_{\xi\xi}$ and the spectrum $G_{\xi\xi}$ as follows:

$$R_{\xi\xi} \underset{\mathscr{F}}{\overset{\mathscr{F}}{\rightleftarrows}} G_{\xi\xi} \tag{5.8}$$

1.2.1.2. Second method

Suppose that the realizations of the r.s. ξ are locally square integrable functions and let us define the time average power $\langle |\xi(\omega, t)|^2 \rangle_{2T}$ of the realization $\xi(\omega, t)$ truncated to the interval $[-T, +T]$ as shown in Fig. 5.1.

Random signals in continuous time 213

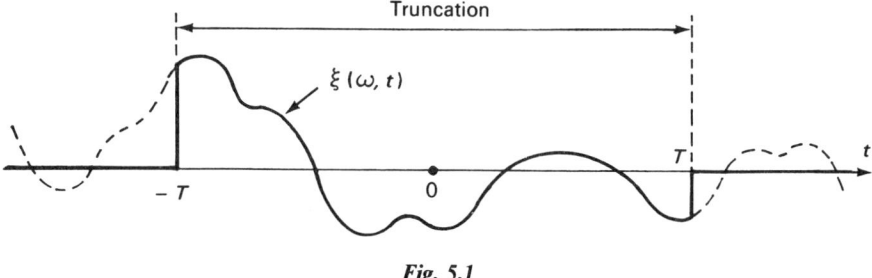

Fig. 5.1

Then

$$\forall \omega \in \Omega \quad \langle |\xi(\omega, t)|^2 \rangle_{2T} = \frac{1}{2T} \int_{-T}^{T} |\xi(\omega, t)|^2 \, dt \quad (5.9)$$

Let $X_{2T}(\omega, f)$ be the complex spectrum of the truncated realization. It is defined by the integral

$$X_{2T}(\omega, f) = \int_{-T}^{T} \xi(\omega, t) \exp(-i2\pi ft) \, dt \quad (5.10)$$

This is a continuous function which is square integrable and Parseval's theorem enables us to write (cf. Eqn (3.32))

$$\langle |\xi(\omega,t)|^2 \rangle_{2T} = \frac{1}{2T} \int_{\mathbb{R}} |X_{2T}(\omega, f)|^2 \, df$$

It follows from this that the positive function $(1/2T)|X_{2T}(\omega, f)|^2$ is the spectral density of the time average power of the truncated realization. For all $f \in \mathbb{R}$ it is a positive random variable (r.v.) which may be written

$$\frac{1}{2T}|X_{2T}(\omega, f)|^2 = \frac{1}{2T} \int_{-T}^{T} \int_{-T}^{T} \xi(\omega, u) \xi^*(\omega, v) \exp\{-i2\pi f(u - v)\} \, du \, dv$$

Its mean value may be written (cf. Eqn (4.114))

$$E\left[\frac{1}{2T}|X_{2T}(\omega, f)|^2\right] = \frac{1}{2T} \int_{-T}^{T} \int_{-T}^{T} R_{\xi\xi}(u - v) \exp\{-i2\pi f(u - v)\} \, du \, dv$$

We note that, since the r.s. ξ has finite mean power, the function $R_{\xi\xi}$ is bounded and consequently the above integral exists. Let us put $\tau = u - v$ and leave the variable v unchanged. The change is illustrated in Fig. 5.2.

214 Weakly stationary random signals

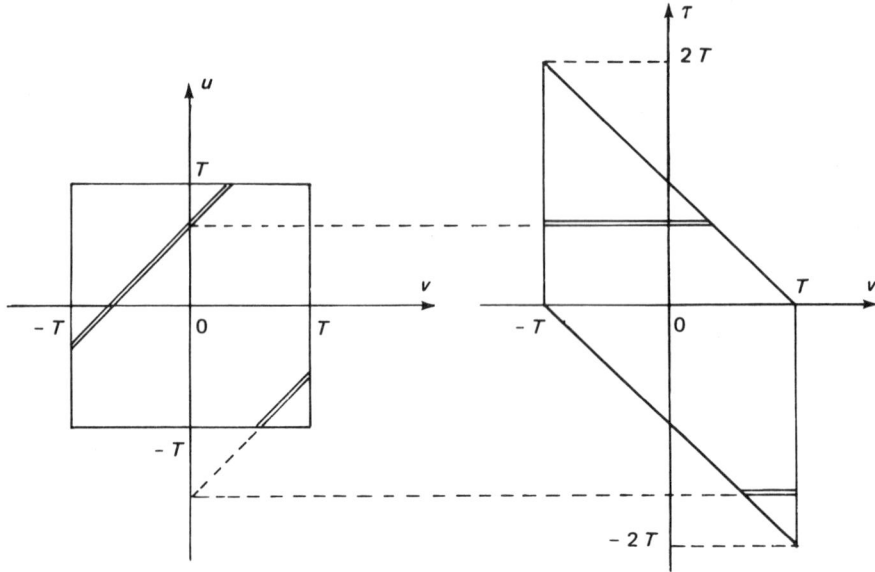

Fig. 5.2

It follows that

$$E\left[\frac{1}{2T}|X_{2T}(f)|^2\right] = \int_{-2T}^{2T} R_{\xi\xi}(\tau)\left(1 - \frac{|\tau|}{2T}\right)\exp(-i2\pi f\tau)\,d\tau$$

$$= \mathscr{F}\left\{R_{\xi\xi}(\tau)\left(1 - \frac{|\tau|}{2T}\right)\mathring{I}_{4T}(\tau)\right\}(f)$$

$$= G_{\xi\xi}(f) * \mathscr{F}\left\{\left(1 - \frac{|\tau|}{2T}\right)\mathring{I}_{4T}(\tau)\right\}(f)$$

where we denote by $\mathring{I}_{4T}$ the indicator function of the interval $[-2T, 2T]$ and apply Eqn (3.5). Let us make T tend to infinity. The second term of the convolution tends to the Dirac distribution $\delta(f)$ and the continuity of the convolution in the space $\mathscr{S}'$ enables us to write

$$\lim_{T \to \infty} E\left[\frac{1}{2T}|X_{2T}(f)|^2\right] = G_{\xi\xi}(f) \qquad (5.11)$$

The spectrum $G_{\xi\xi}$ is thus the limit, when T tends to infinity, of the ensemble mean of the spectral density of the time average power of the truncated realizations of the r.s. ξ. This is the Wiener–Kintchine theorem (Blanc-Lapierre and Picinbono, 1981).

1.2.1.3. Band-limited random signal
A weakly stationary r.s. ξ is band limited when its spectrum $G_{\xi\xi}$ has bounded

support. When the r.s. ξ satisfies certain supplementary conditions, we can show that the supports of the complex spectra of the realizations $\xi(\omega, t)$ are almost surely (a.s.) included in the support of $G_{\xi\xi}$ (cf. Example 1.13). These conditions are generally fulfilled in practice.

An r.s. ξ having concrete existence is not in general band limited. Nevertheless, for such an r.s. we can define a positive frequency f_c such that the mean power of the r.s. outside the spectral band $[-f_c, +f_c]$ is negligible. This may be written

$$\int_{-f_c}^{f_c} dG_{\xi\xi}(f) \approx \int_{\mathbb{R}} dG_{\xi\xi}(f) = P_\xi$$

We can then accept that the r.s. ξ is practically band limited by making the approximation

$$\text{supp } G_{\xi\xi} \approx [-f_c, +f_c] \tag{5.12}$$

The frequency f_c is the (upper) **cut-off frequency** of the r.s. ξ. The value chosen depends on the level of approximation which we are prepared to accept. This is poorer the lower the value applied.

1.2.2. Cross-correlation spectrum of two random signals

Let $\xi(t)$ and $\eta(t)$, $t \in \mathbb{R}$, be two r.s.s in continuous time, weakly stationary and of finite mean power. Their cross-correlation functions $R_{\xi\eta}$ and $R_{\eta\xi}$ are bounded. They therefore define tempered distributions and have Fourier transforms $G_{\xi\eta}$ and $G_{\eta\xi}$ which are tempered distributions. These are the **cross-correlation spectra** of the r.s.s ξ and η. It follows from Eqn (5.5) (cf. Appendix 2) that

$$G_{\xi\eta} = G_{\eta\xi}^* \tag{5.13}$$

Unlike the spectra $G_{\xi\xi}$ and $G_{\eta\eta}$, the cross-correlation spectra are not necessarily real and for this reason do not have physical significance. They can be used, however, as intermediaries in a calculation, to take advantage of the properties of the Fourier transform.

Example 1.1 Spectrum of a real random signal
Let $\xi(t)$, $t \in \mathbb{R}$, be a real r.s. in continuous time, weakly stationary and having finite mean power P_ξ. Since the function $R_{\xi\xi}$ is even, the spectrum $G_{\xi\xi}$ is an even distribution (cf. Appendix 2). Its support therefore extends to negative frequencies and we must take precautions if we wish the frequency to be preserved as a physical quantity.

Let f_1 and f_2 therefore be two frequencies such that $0 < f_1 < f_2$. The mean power contained within the spectral band $[f_1, f_2]$ is given by the integral

$$P_\xi[f_1, f_2] = 2 \int_{f_1}^{f_2} dG_{\xi\xi}(f) \tag{5.14}$$

216 Weakly stationary random signals

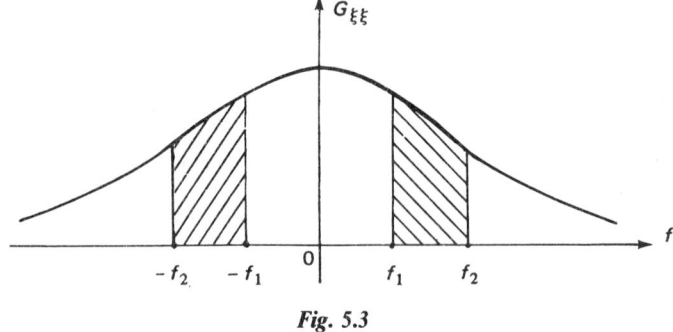

Fig. 5.3

This is illustrated in Fig. 5.3 for the case in which the spectrum is a function. We must not forget in calculating the integral (5.14) to take into account the Dirac distributions which could be centred on the frequencies f_1 and f_2.

Example 1.2 Spectrum of a periodic random signal

Let $\xi(t)$, $t \in \mathbb{R}$, be the periodic r.s. defined in Chapter 4, Example 2.4. Its spectrum can be obtained by taking Fourier transforms of the two sides of Eqn (4.82). We obtain

$$G_{\xi\xi}(f) = \sum_{n \in \mathbb{Z}} |c_n|^2 \delta(f - n/T) \qquad (5.15)$$

The spectrum of a periodic r.s. is therefore discrete. We say also that it is a **line spectrum**. When the r.s. ξ is real, we know that $c_{-n} = c_n^*$, which implies that $|c_n|^2 = |c_{-n}|^2$ and that the spectrum is even. We must note that the amplitude of the lines tends to zero at infinity since the r.s. ξ has finite mean power (cf. Eqn (4.83)). The behaviour of the spectrum is represented in Fig. 5.4 for the case in which the r.s. ξ is real.

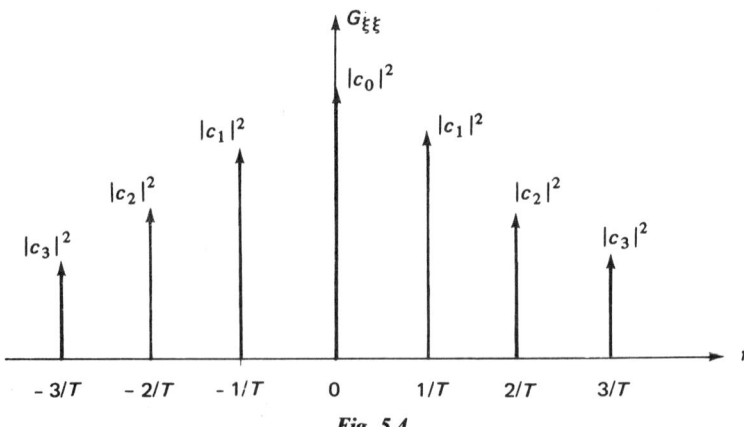

Fig. 5.4

Random signals in continuous time 217

We should note that the line at frequency n/T is the spectrum of the complex sinusoidal r.s. ξ_n defined by Eqn (4.80). Note also that (cf. Eqn (4.81))

$$G_{\xi\xi}(\{0\}) = |c_0|^2 = |m_\xi|^2$$

Hence a periodic r.s. is ergodic in the mean (cf. Section 1.6.2).

Example 1.3 Spectrum of an ideal sampling random signal

Consider the ideal sampling r.s. defined in Chapter 4, Example 2.5. Its spectrum can be obtained from Eqn (4.86) by applying Eqn (5.15). It follows that

$$\xi(t) = \sum_{n\in\mathbb{Z}} \delta(t - nT - \vartheta) \qquad G_{\xi\xi}(f) = \frac{1}{T^2} \sum_{n\in\mathbb{Z}} \delta\left(f - \frac{n}{T}\right) \qquad (5.16)$$

It should be noted that the ideal sampling signal and its spectrum are both trains of Dirac distributions. They are shown in Fig. 5.5.

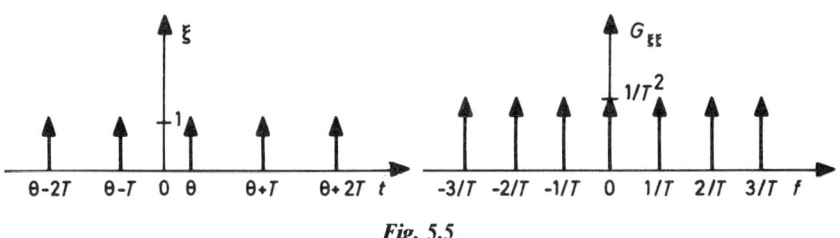

Fig. 5.5

Example 1.4 Spectrum of a markovian gaussian random signal

Let $\xi(t)$, $t \in \mathbb{R}$, be a centred r.s., weakly stationary, gaussian and markovian. Let t_1, t_2, t_3 be instants such that $t_1 \leq t_2 \leq t_3$. Since the r.s. ξ is centred and gaussian, it follows from Eqn (2.18) that

$$\forall i, \forall j \in \{1, 2, 3\} \qquad E[\xi_i | \xi_j] \underset{\text{a.s.}}{=} r_{ij} \frac{\sigma_i}{\sigma_j} \xi_j \qquad (5.17)$$

The r.s. ξ is markovian and it follows from Eqn (4.27) that

$$E[\xi_3 | \xi_1, \xi_2] \underset{\text{a.s.}}{=} E[\xi_3 | \xi_2]$$

Moreover, Eqn (1.108) enables us to write

$$E[E[\xi_3 | \xi_1, \xi_2] | \xi_1] \underset{\text{a.s.}}{=} E[\xi_3 | \xi_1]$$

It then follows from Eqn (5.17) that

$$E[\xi_3 | \xi_1] \underset{\text{a.s.}}{=} E[E[\xi_3 | \xi_2] | \xi_1] \underset{\text{a.s.}}{=} r_{23} \frac{\sigma_3}{\sigma_2} E[\xi_2 | \xi_1]$$

and then

$$r_{13} = r_{12}r_{23}$$

Put $t_1 = t$, $t_2 = t + \tau$, $t_3 = t + \tau + \Delta\tau$. The above relation may then be written, since the r.s. ξ is weakly stationary (it is actually strongly stationary),

$$\forall \tau \in \mathbb{R} \quad \forall \Delta\tau \in \mathbb{R} \quad r_{\xi\xi}(\tau + \Delta\tau) = r_{\xi\xi}(\tau)r_{\xi\xi}(\Delta\tau) \tag{5.18}$$

Suppose, in addition, that the r.s. ξ is continuous in m.s. The autocorrelation coefficient $r_{\xi\xi}$ is then a continuous function and it can be shown by classical methods that Eqn (5.18) implies

$$\exists a \in \mathbb{R}_+^* : r_{\xi\xi}(\tau) = \exp(-a|\tau|) \tag{5.19}$$

It follows from this that the autocorrelation function and the spectrum of a centred r.s., weakly stationary, continuous in m.s., gaussian and markovian, are respectively of the form

$$R_{\xi\xi}(\tau) = \sigma_\xi^2 \exp(-a|\tau|)$$
$$G_{\xi\xi}(f) = \frac{2a\sigma_\xi^2}{a^2 + 4\pi^2 f^2} \quad a \in \mathbb{R}_+^* \tag{5.20}$$

Conversely, a centred r.s., weakly stationary, gaussian, and having a correlation coefficient of the form of Eqn (5.19), is markovian. In fact, it follows immediately that the correlation coefficient satisfies Eqn (5.18). Hence it follows that the probability density of Eqn (4.19) satisfies the relation (4.28), which implies that the r.s. ξ is markovian.

In contrast, a centred r.s., weakly stationary and markovian, can have a correlation coefficient of the form of Eqn (5.19) without being gaussian.

Example 1.5 Spectrum of a random signal function of a gaussian random signal

Consider an r.s. $\xi(t)$, $t \in \mathbb{R}$, weakly stationary, gaussian, centred, with variance σ^2 and autocorrelation coefficient $r(\tau)$, and a Borel function $g: \mathbb{R} \to \mathbb{C}$. Consider the complex r.s. $\eta(t)$, $t \in \mathbb{R}$, defined by

$$\eta(t) = g\{\xi(t)\} \tag{5.21}$$

and determine its characteristics as a function of those of the r.s. ξ.

Mean value of the random signal η

The mean value of the r.v. $\eta(t)$ is obtained by applying Eqn (1.71). We get

$$\forall t \in \mathbb{R} \quad E[\eta(t)] = \int_\mathbb{R} g(x) p_\xi(x) \, dx \tag{5.22}$$

The mean value of the r.s. η is constant since the probability density p_ξ does not depend on t. When the function g is odd this constant is zero since the function p_ξ is even (cf. Eqn (1.46)). The r.s. η is then centred.

Autocorrelation function of the random signal η

The autocorrelation function $R_{\eta\eta}$ is obtained by applying Eqn (4.69). We get

$$R_{\eta\eta}(\tau) = E[\eta(t)\eta^*(t-\tau)]$$

$$= \iint_{\mathbb{R}^2} g(x_1)g^*(x_2)p_X(x_1, x_2, \tau)\,dx_1\,dx_2 \qquad (5.23)$$

p_X being the probability density of the two-dimensional gaussian random vector (r.v.) $\{\xi(t), \xi(t-\tau)\}$. The function $R_{\eta\eta}$ depends only on τ since the function p_X depends only on τ (cf. Chapter 4, Example 2.1). The r.s. η is therefore weakly stationary.

The function $R_{\eta\eta}$ can also be written by the duality of the Fourier transform (Dupraz, 1977, p. 88)

$$R_{\eta\eta}(\tau) = \iint_{\mathbb{R}^2} \mathscr{F}g(x_1)\mathscr{F}g^*(x_2)\bar{\mathscr{F}}p_X(x_1, x_2, \tau)\,dx_1\,dx_2$$

It then follows, since the r.s. ξ is gaussian and centred (cf. Eqns (2.39), (2.42) and (4.71)), that

$$\bar{\mathscr{F}}p_X(x_1, x_2) = \Theta_X(2\pi x_1, 2\pi x_2)$$

$$= \exp[-2\pi^2\sigma^2\{x_1^2 + 2x_1x_2r(\tau) + x_2^2\}] \qquad (5.24)$$

It follows from this that

$$R_{\eta\eta}(\tau) = \iint_{\mathbb{R}^2} \mathscr{F}g(x_1)\mathscr{F}g^*(x_2)$$
$$\times \exp[-2\pi^2\sigma^2\{x_1^2 + 2x_1x_2r(\tau) + x_2^2\}]\,dx_1\,dx_2$$

Finally, by expanding part of the exponential and introducing the autocorrelation function $R_{\xi\xi}(\tau) = \sigma^2 r(\tau)$ of the r.s. ξ, we have

$$R_{\eta\eta}(\tau) = \sum_{n\in\mathbb{N}} \frac{1}{n!}(2\pi)^{2n}|h_n|^2 R_{\eta\eta}^n(\tau) \qquad (5.25)$$

the coefficient h_n being defined by

$$h_n = \int_{\mathbb{R}} x^n \mathscr{F}g(x)\exp(-2\pi^2\sigma^2 x^2)\,dx \qquad (5.26)$$

When the transform $\mathscr{F}g$ is not a function, the integral (5.26) must be taken in the sense of distributions. It can be written in another way if we note that

$$x^n\mathscr{F}g(x) = \frac{1}{(i2\pi)^n}\mathscr{F}(g^{(n)})(x)$$

where the derivative $g^{(n)}$ is taken in the space of tempered distributions (cf. Eqns (3.2) and (3.5) and Appendix 2). It follows then, through the duality of the

220 Weakly stationary random signals

Fourier transform, that

$$h_n = \frac{1}{(i2\pi)^n} \int_\mathbb{R} \mathscr{F}(g^{(n)})(x) \exp(-2\pi^2\sigma^2 x^2)\,dx$$

$$= \frac{1}{(i2\pi)^n} \int_\mathbb{R} g^{(n)}(x) \mathscr{F}\{\exp(-2\pi^2\sigma^2 x^2)\}(x)\,dx$$

Finally (cf. Appendix 2) we have

$$h_n = \frac{1}{(i2\pi)^n} \frac{1}{(2\pi\sigma^2)^{1/2}} \int_\mathbb{R} g^{(n)}(x) \exp\left(-\frac{x^2}{2\sigma^2}\right) dx \qquad (5.27)$$

The integral (5.27) is to be taken in the sense of distributions when $g^{(n)}$ is not a function.

The coefficients h_n can be calculated by using Eqn (5.26) or (5.27), as the case may be. It is interesting to apply Eqn (5.27) when the derivatives $g^{(n)}$ vanish beyond a certain order. This is the case when g is a quadratic function (cf. Exercise 1.2).

Spectrum of the random signal η

The spectrum $G_{\eta\eta}$ is obtained by taking the Fourier transform of the expansion (5.25). We get

$$G_{\eta\eta}(f) = |h_0|^2 \delta(f) + \sum_{n\in\mathbb{N}^*} \frac{1}{n!} (2\pi)^{2n} |h_n|^2 G_{\xi\xi}^{*n}(f) \qquad (5.28)$$

In the most general case the spectrum $G_{\eta\eta}$ contains an infinity of terms. The term $n = 0$ is a line of power $|h_0|^2$ at the frequency 0. From Eqn (5.22) we have

$$h_0 = E[\eta] = m_\eta \qquad (5.29)$$

It follows from this that the r.s. η is ergodic on average (cf. Eqn (5.67)).

When the function g is linear all the terms of order $n \geq 2$ vanish and the spectrum $G_{\eta\eta}$ reduces to the first two terms. Its support is identical with that of the spectrum $G_{\xi\xi}$ and the r.s. η occupies the same spectral band as the r.s. ξ.

When g is a non-linear function, the term of order $n > 2$ is proportional to the n-fold convolution of the spectrum $G_{\xi\xi}$ with itself. Its support is the vector sum of the supports of each of the terms. It is therefore n times wider than the support of $G_{\xi\xi}$. It must be concluded from this that the spectral band occupied by the r.s. η is much wider than that occupied by the r.s. ξ. We usually say that a non-linearity broadens the spectral band.

This is illustrated in Fig. 5.6 for the particular case in which the spectrum $G_{\xi\xi}$ is centred on the frequency 0 and is a function with bounded support. Only the first three terms of the expansion are represented.

Random signals in continuous time 221

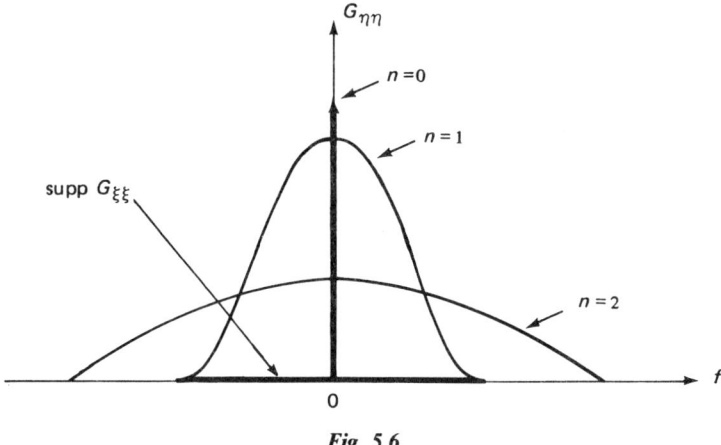

Fig. 5.6

1.3. WEAKLY EQUIVALENT REPRESENTATION IN (x, y) OF A RANDOM SIGNAL

(a) Let $\xi(t)$, $t \in \mathbb{R}$, be a real r.s., weakly stationary and centred, whose spectrum $G_{\xi\xi}$ is represented, in an intentionally schematic manner, in Fig. 5.7.

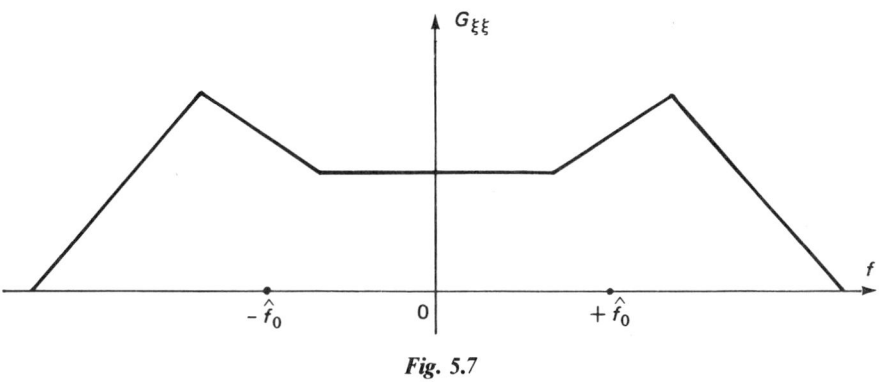

Fig. 5.7

The object of this section is to introduce an r.s. $\eta(t)$, $t \in \mathbb{R}$, real, weakly stationary, centred, and having the same autocorrelation function and consequently the same spectrum as the r.s. ξ. It follows from this, when the r.s.s ξ and η are gaussian, that the r.s. η is a weakly equivalent representation of the r.s. ξ (cf. Chapter 4, Section 1.4.2).

(b) Let $\hat{f}_0$ be an arbitrary positive frequency, not necessarily belonging to the support of the spectrum $G_{\xi\xi}$, and let us define the spectra $G_{\xi\xi}^+$ and $G_{\xi\xi}^-$ such that

$$G_{\xi\xi}(f) = G_{\xi\xi}^+(f) * \delta(f - \hat{f}_0) + G_{\xi\xi}^-(f) * \delta(f + \hat{f}_0) \qquad (5.30)$$

222 Weakly stationary random signals

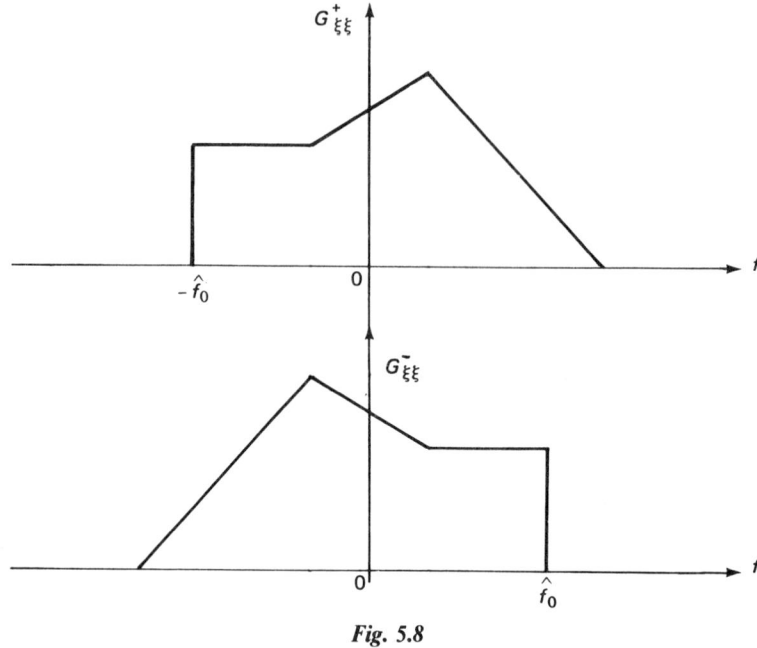

Fig. 5.8

They are represented in Fig. 5.8. Since the r.s. ξ is real, the spectrum $G_{\xi\xi}$ is even, which implies that

$$\forall f \in \mathbb{R} \qquad G_{\xi\xi}^{-}(f) = G_{\xi\xi}^{+}(-f) \qquad (5.31)$$

Put

$$R_{\xi\xi}^{+} = \bar{\mathscr{F}} G_{\xi\xi}^{+} \qquad R_{\xi\xi}^{-} = \bar{\mathscr{F}} G_{\xi\xi}^{-} \qquad \hat{\omega}_0 = 2\pi \hat{f}_0 \qquad (5.32)$$

The autocorrelation function $R_{\xi\xi}$ may then be written

$$R_{\xi\xi}(\tau) = \{R_{\xi\xi}^{+}(\tau) + R_{\xi\xi}^{-}(\tau)\}\cos(\hat{\omega}_0\tau)$$
$$+ i\{R_{\xi\xi}^{+}(\tau) - R_{\xi\xi}^{-}(\tau)\}\sin(\hat{\omega}_0\tau) \qquad (5.33)$$

(c) Consider the real r.s. $\eta(t)$, $t \in \mathbb{R}$, defined by

$$\eta(t) = x(t)\cos(\hat{\omega}_0 t) + y(t)\sin(\hat{\omega}_0 t)$$

where $x(t)$ and $y(t)$, $t \in \mathbb{R}$, are mutually weakly stationary r.s.s, real and centred.

It is immediately obvious that the r.s. η is centred. Its autocorrelation function may be written

$$R_{\eta\eta}(t, t-\tau) = \tfrac{1}{2}\{R_{xx}(\tau) + R_{yy}(\tau)\}\cos(\hat{\omega}_0\tau)$$
$$- \tfrac{1}{2}\{R_{xy}(\tau) - R_{yx}(\tau)\}\sin(\hat{\omega}_0\tau)$$
$$+ \tfrac{1}{2}\{R_{xx}(\tau) - R_{yy}(\tau)\}\cos(2\hat{\omega}_0 t - \hat{\omega}_0\tau)$$
$$+ \tfrac{1}{2}\{R_{yx}(\tau) + R_{xy}(\tau)\}\sin(2\hat{\omega}_0 t - \hat{\omega}_0\tau)$$

The r.s. η is therefore not weakly stationary in the general case. It becomes so if we require that the r.s.s x and y are such that
$$R_{xx} = R_{yy} \qquad R_{xy} = -R_{yx}$$
and its autocorrelation function is written
$$R_{\eta\eta}(\tau) = R_{xx}(\tau)\cos(\hat{\omega}_0\tau) - R_{xy}(\tau)\sin(\hat{\omega}_0\tau) \tag{5.34}$$
Impose on the r.s.s x and y the supplementary condition
$$\begin{aligned} R_{xx} &= R_{yy} = R_{\xi\xi}^+ + R_{\xi\xi}^- \\ R_{xy} &= -R_{yx} = -i(R_{\xi\xi}^+ - R_{\xi\xi}^-) \end{aligned} \tag{5.35}$$
The autocorrelation functions of the r.s.s ξ and η are then identical and the spectra of the r.s.s x and y, and their intercorrelation spectra, are functions of the spectra $G_{\xi\xi}^+$ and $G_{\xi\xi}^-$. It follows from Eqn (5.35) that
$$\begin{aligned} G_{xx} &= G_{yy} = G_{\xi\xi}^+ + G_{\xi\xi}^- \\ G_{xy} &= -G_{yx} = -i(G_{\xi\xi}^+ - G_{\xi\xi}^-) \end{aligned} \tag{5.36}$$
The r.s.s x and y have identical spectra and are in general correlated. It is immediately obvious that
$$E[x^2] = E[y^2] = E[\xi^2] \tag{5.37}$$
The r.s.s x and y thus have the same mean power as the r.s. ξ.

(d) The spectra $G_{\xi\xi}^+$ and $G_{\xi\xi}^-$ depend on the choice of the frequency $\hat{f}_0$. Suppose that this frequency is such that
$$G_{\xi\xi}^+ = G_{\xi\xi}^- \tag{5.38}$$
This implies, as shown in Figs 5.7 and 5.8, that
$$\text{supp } G_{\xi\xi}^+ = \text{supp } G_{\xi\xi}^- \subset [-\hat{f}_0, \hat{f}_0] \qquad \text{supp } G_{\xi\xi} \subset [-2\hat{f}_0, 2\hat{f}_0] \tag{5.39}$$
The r.s. ξ is thus necessarily band limited. Moreover, it follows from Eqn (5.31) that the spectra $G_{\xi\xi}^+$ and $G_{\xi\xi}^-$ are even. It then follows from Eqns (5.35) and (5.36) that
$$R_{xy} = R_{yx} = 0 \qquad G_{xx} = G_{yy} = 2G_{\xi\xi}^+ \tag{5.40}$$
The r.s.s x and y are therefore uncorrelated since they are centred. In addition, they are band limited, by virtue of Eqn (5.39).

The autocorrelation function and the spectrum of the r.s. ξ can then be written respectively
$$R_{\xi\xi}(\tau) = 2R_{\xi\xi}^+(\tau)\cos(\hat{\omega}_0\tau) \tag{5.41}$$
and
$$G_{\xi\xi}(f) = G_{\xi\xi}^+(f) * \{\delta(f+\hat{f}_0) + \delta(f-\hat{f}_0)\} \tag{5.42}$$
(e) Suppose finally that the r.s. ξ is gaussian and let us require that the r.s.s x

and y be mutually gaussian. The r.s. η is then gaussian (cf. Chapter 4, Exercise 1.8). Under these conditions, the r.s.s ξ and η are gaussian and centred and have equal autocorrelation functions. Their probability laws are therefore identical and they are weakly equivalent (cf. Chapter 4, Section 1.4). This enables us, in practice, to substitute the r.s. η for the r.s. ξ, writing

$$\xi(t) = x(t)\cos(\hat{\omega}_0 t) + y(t)\sin(\hat{\omega}_0 t) \qquad (5.43)$$

The r.s. ξ can then be represented by the Fresnel diagram in Fig. 5.9.

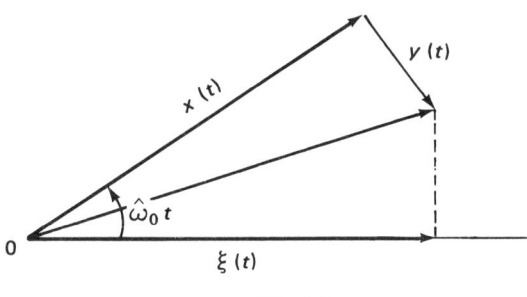

Fig. 5.9

In the general case the r.s.s x and y are not independent. They are independent, however, when the condition (5.38) is satisfied since they are then uncorrelated and mutually gaussian. It is for this reason that it is interesting to choose a suitable frequency $\hat{f}_0$. Now, to be completely rigorous, condition (5.38) cannot be satisfied for an r.s. having a concrete existence since the support of its spectrum is not bounded. However, when the width of the spectral band occupied by the r.s. ξ is small in comparison with the frequency $\hat{f}_0$, i.e. when the r.s. ξ is **narrow band**, a frequency $\hat{f}_0$ exists for which condition (5.38) is approximately satisfied. The r.s.s x and y are then effectively independent. This is demonstrated by the problem in Exercise 1.11.

The representation (5.43) is called the representation in (x, y) of the r.s. ξ. Another representation of the same type is the subject of Exercise 1.4. It is very widely used in the theory of the reception of radioelectric r.s.s to represent gaussian white noise, filtered by a bandpass filter. The case in which the bandpass filter is ideal is studied in Exercise 1.12.

1.4. DIFFERENTIABILITY IN MEAN SQUARE

1.4.1. Criterion for differentiability

Let $\xi(t)$, $t \in \mathbb{R}$, be a weakly stationary r.s. with finite mean power. It is differentiable in m.s. if and only if its autocorrelation function is twice differentiable at the origin.

Random signals in continuous time 225

The proof of this property uses the criterion (4.93). It is left to the reader. The same criterion enables us to prove that, if the r.s. ξ is differentiable in m.s. at one point, it is differentiable in m.s. at every point (cf. Example 1.6). We can thus write $\xi'(t)$, $t \in \mathbb{R}$, for the r.s. differentiated in m.s.

1.4.2. Mean value of the derivative random signal

It follows immediately from Eqn (4.94), since the mean value of the r.s. ξ is constant, that

$$\forall t \in \mathbb{R} \quad E[\xi'(t)] = 0 \tag{5.44}$$

The r.s. ξ' is therefore centred.

1.4.3. Correlation functions

We obtain from Eqns (4.95) and (4.96)

$$\forall t \in \mathbb{R} \quad \forall \tau \in \mathbb{R} \quad R_{\xi'\xi}(t, t-\tau) = -R_{\xi\xi}(t, t-\tau) = -R'_{\xi\xi}(\tau)$$

$$\forall t \in \mathbb{R} \quad \forall \tau \in \mathbb{R} \quad R_{\xi'\xi'}(t, t-\tau) = -R''_{\xi\xi}(\tau)$$

It follows from this that the r.s.s ξ and ξ' are mutually weakly stationary and that

$$R_{\xi'\xi} = -R_{\xi\xi'} = -R'_{\xi\xi} \tag{5.45}$$

$$R_{\xi'\xi'} = -R''_{\xi\xi} \tag{5.46}$$

1.4.4. Spectrum of the derivative random signal

The spectrum $G_{\xi'\xi'}$ of the r.s. ξ' is obtained from Eqn (5.46) (cf. Appendix 2):

$$G_{\xi'\xi'}(f) = 4\pi^2 f^2 G_{\xi\xi}(f) \tag{5.47}$$

It is thus necessary, for the r.s. ξ to be differentiable in m.s., that its spectrum be such that the function f^2 is integrable with respect to the measure $G_{\xi\xi}$. The integral is then the mean power $P_{\xi'}$ of the r.s. ξ':

$$\int_\mathbb{R} 4\pi^2 f^2 \, dG_{\xi\xi}(f) = P_{\xi'} < +\infty \tag{5.48}$$

Example 1.6 Derivative random signal of a real random signal

Let $\xi(t)$, $t \in \mathbb{R}$, be an r.s. differentiable in m.s. We know that (cf. Chapter 4, Exercise 3.3)

$$\forall t \in \mathbb{R} \quad \lim_{\tau \to 0} E[|X_\tau(t)|^2] = E[|\xi'(t)|^2]$$

When the r.s. ξ is real and weakly stationary, we can write (cf. Eqn (5.3))

$$\forall t \in \mathbb{R} \qquad E[|X_\tau(t)|^2] = \frac{2}{\tau^2}\{R_{\xi\xi}(0) - R_{\xi\xi}(\tau)\}$$

It follows from this that

$$R_{\xi\xi}(\tau) = R_{\xi\xi}(0) - \frac{\tau^2}{2} E[|\xi'|^2] + o(\tau^2) \qquad (5.49)$$

The autocorrelation function $R_{\xi\xi}$ is thus twice differentiable at the origin and

$$R'_{\xi\xi}(0) = 0 \qquad R''_{\xi\xi}(0) = -E[|\xi'|^2] \qquad (5.50)$$

It follows then from Eqns (5.44) and (5.45) that

$$\forall t \in \mathbb{R} \qquad r_{\xi\xi'}(t, t) = 0 \qquad (5.51)$$

The r.v.s $\xi(t)$ and $\xi'(t)$ are therefore uncorrelated and the r.s.s ξ and ξ' are uncorrelated at the same instant.

Example 1.7 White noise and thermal noise
(a) A **white noise** in continuous time is an r.s. $n(t)$, $t \in \mathbb{R}$, weakly stationary, centred, having as autocorrelation 'function'

$$R_{nn}(\tau) = \frac{N_0}{2} \delta(\tau) \qquad N_0 \in \mathbb{R}_+^* \qquad (5.52)$$

Its spectral 'density' is constant since (cf. Appendix 2)

$$G_{nn}(f) = \frac{N_0}{2} \qquad (5.53)$$

The r.s. n is called white noise by analogy with white light which is a superposition of colours of all frequencies and of the same intensity.

(b) Let f be a positive frequency. The mean power contained within the spectral band $[f, f + \Delta f]$ is obtained from Eqn (5.14). We get

$$P_n[f, f + \Delta f] = N_0 \Delta f \qquad (5.54)$$

N_0 being expressed in watts per hertz (or in joules). It must be noted that N_0 is not a variance.

It is obvious that white noise has no concrete existence. In fact, its mean power is infinite and its autocorrelation 'function' is not a true function. If we consider two instants t_1 and t_2, the r.v.s $n(t_1)$ and $n(t_2)$ are never correlated, even if t_2 is very close to t_1. White noise therefore has no memory, which cannot be the case for an r.s. having concrete existence. However, white noise is a mathematical model which is very widely used.

(c) It is possible to suppose, formally, that a white noise is gaussian even though we cannot write down its probability density because the variances are infinite (cf. Chapter 4, Example 1.4).

Consider the Wiener–Levy (WL) r.s. ξ defined in Chapter 4, Example 3.1. It is not differentiable in m.s. since the differentiation in Eqn (4.103) is not possible. The differentiation can be carried out, however, in the sense of distributions, in the space $\mathscr{S}'$. We obtain then

$$\forall t_1, \forall t_2 \in \mathbb{R}_+^* \qquad R_{\xi'\xi'}(t_1, t_2) = b^2 \delta(t_2 - t_1)$$

This enables us to define formally an r.s. that is the 'derivative' in m.s. of a WL r.s. which should have been in existence since $-\infty$. It is centred since the WL r.s. is centred and its correlation 'function' may be written

$$R_{\xi'\xi'}(\tau) = b^2 \delta(\tau)$$

It is gaussian since the WL r.s. is gaussian.

Gaussian white noise with spectral density $N_0/2$ can therefore be obtained formally by 'differentiation' in m.s. of a WL r.s. with parameter $b = (N_0/2)^{1/2}$.

(d) **Thermal noise**: Consider the ideal situation in which a resistance R is in thermal equilibrium at absolute temperature T. The resistance radiates energy and it can be regarded as a voltage generator consisting of an electromotive force $e(t)$ in series with an internal resistance R, supposed to be in equilibrium at temperature $T = 0$. The electromotive force results from the thermal agitation of the electrons in the interior of the resistance. This is a stationary centred r.s., and the central limit theorem enables us to assert that it is gaussian (cf. Chapter 2, Section 2.5). This is illustrated in Fig. 5.10.

Let $n(t)$, $t \in \mathbb{R}$, be the r.s. such that $n(t) = e(t)/2R^{1/2}$. The instantaneous mean maximal power available at the terminals of the resistance R is the mean power P_n of the r.s. n:

$$P_n = \frac{E[e^2(t)]}{4R} = E[n^2(t)] \qquad (5.55)$$

This is the mean power dissipated in a resistance equal to R. The laws of thermodynamics enable us to prove that the spectral density of this mean power is the function (Gagliardi and Karp, 1976)

$$G_{nn}(f) = \frac{h|f|}{2\{\exp(h|f|/kT) - 1\}} = \frac{kT}{2}\left(1 - \frac{h|f|}{2kT} + \ldots\right) \qquad (5.56)$$

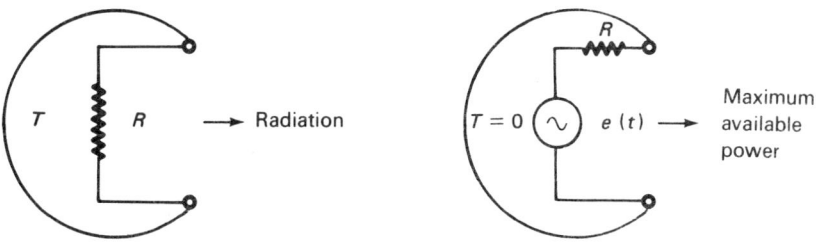

Fig. 5.10

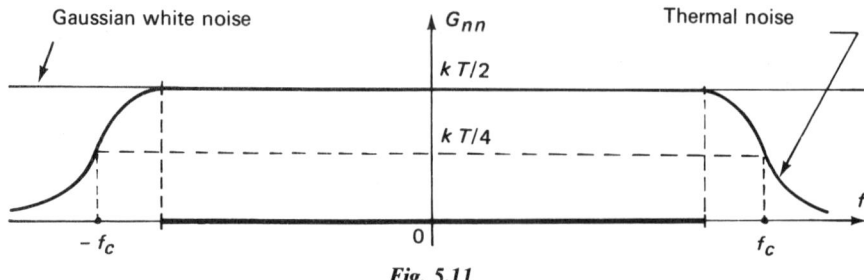

Fig. 5.11

h being Planck's constant and k Boltzmann's constant. It is represented in Fig. 5.11.

The spectral density G_{nn} tends to zero at infinity and its cut-off frequency at half-value is $f_c \approx 2.6 \times 10^{10} T$. It is sufficiently high, in practice, for us to be able to accept that the interval $[-f_c, f_c]$ contains the supports of the spectra of all the other r.s.s with which we are likely to be concerned. This enables us to accept that the spectral density is constant and equal to $kT/2$. Within the limits of this approximation, thermal noise is gaussian white noise. This is illustrated in Fig. 5.11.

Example 1.8 Zero crossings of a random signal

General case

Let $\xi(t)$, $t \in \mathbb{R}$, be a real r.s., weakly stationary and of finite mean power, and let the real r.s. $\eta(t)$, $t \in \mathbb{R}$, be defined by

$$\forall t \in \mathbb{R} \qquad \eta(t) = Y\{\xi(t)\}$$

Y being the Heaviside function (cf. Eqn (6.22)). The r.s. η is obviously of finite mean power. Suppose that the r.s. ξ is differentiable in m.s. and let us write, differentiating (formally) in the space $\mathscr{S}'$,

$$\eta'(t) = \delta\{\xi(t)\}\xi'(t)$$

The situation is illustrated in Fig. 5.12.

It is possible to state, still formally, that the number $N[0, T]$ of passages of the r.s. ξ through zero in the interval $[0, T]$ is the integral

$$N[0, T] = \int_0^T |\eta'(t)| \, dt$$

This is a real r.v. whose mean value is obtained by applying Eqn (4.116). We get, the probability density $p_{\xi\xi'}$ being independent of t,

$$E[N[0, T]] = \int_0^T dt \left\{ \iint_{\mathbb{R}^2} p_{\xi\xi'}(x, y)\delta(x)|y| \, dx \, dy \right\}$$

$$= T \int_{\mathbb{R}} p_{\xi\xi'}(0, y)|y| \, dy$$

Random signals in continuous time 229

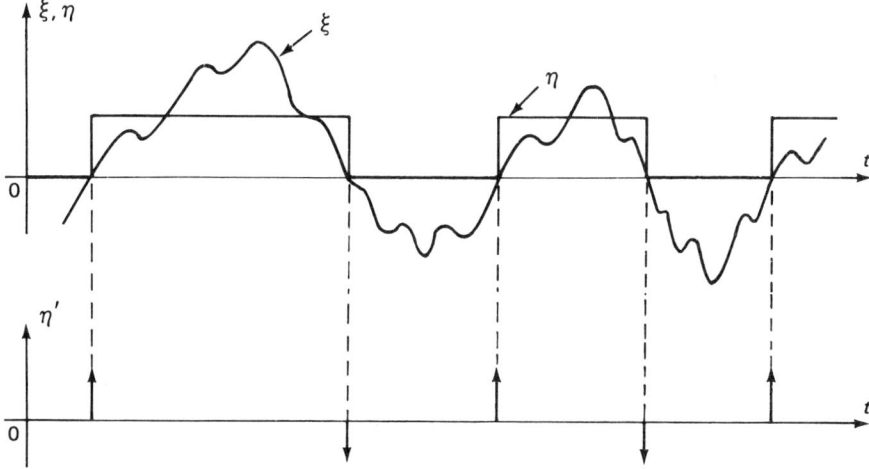

Fig. 5.12

The mean number of passages through zero per unit of time is the apparent frequency f_0^* of the r.s. ξ. It follows that

$$f_0^* = \int_R p_{\xi\xi'}(0, y)|y|\,dy \tag{5.57}$$

Case of a gaussian random signal

Suppose now that the r.s. is gaussian and let m_ξ be its mean value, σ_ξ^2 its variance and $R_{\xi\xi}$ its autocorrelation function. The r.s. ξ' derived in m.s. is centred and its variance is written (cf. Eqn (5.50))

$$\sigma_{\xi'}^2 = -R''_{\xi\xi}(0) \tag{5.58}$$

The gaussian r.v.s $\xi(t)$ and $\xi'(t)$ are independent since they are uncorrelated (cf. Eqn (5.51)). Hence it follows that

$$p_{\xi\xi'}(0, y) = p_\xi(0)p_{\xi'}(y)$$

$$= \frac{1}{(2\pi\sigma_\xi^2)^{1/2}} \exp\left(-\frac{m_\xi^2}{2\sigma_\xi^2}\right) \frac{1}{\{-2\pi R''_{\xi\xi}(0)\}^{1/2}} \exp\left\{\frac{y^2}{2R''_{\xi\xi}(0)}\right\}$$

It follows on calculating the integral (5.57) that

$$f_0^* = \frac{1}{\pi\sigma_\xi^2} \exp\left(-\frac{m_\xi^2}{2\sigma_\xi^2}\right)\{-R''_{\xi\xi}(0)\}^{1/2} \tag{5.59}$$

1.5. INTEGRABILITY IN MEAN SQUARE

1.5.1. Definition and existence

Let $\xi(t)$, $t \in \mathbb{R}$, be a weakly stationary r.s., of finite mean power, and let $X(t)$, $t \in [a, +\infty[$, be the r.s. defined by the integral

$$X(t) = \int_a^t \xi(u)\,du \qquad (5.60)$$

It is immediately obvious that the r.s. X is of the form of Eqn (4.110) with $g(t, u) = Y(t-u)$ and $b = +\infty$.

The integral (4.111) may be written

$$J(a, t) = \int_a^t \int_a^t R_{\xi\xi}(u_1 - u_2)\,du_1\,du_2$$

Put $u_1 - u_2 = \tau$. It follows, by reasoning as in Section 1.2.1.2 and denoting by $\overset{\circ}{I}_{2(t-a)}$ the indicator function of the interval $[-t+a, t-a]$, that

$$J(a, t) = \int_\mathbb{R} R_{\xi\xi}(\tau)(t - a - |\tau|)\overset{\circ}{I}_{2(t-a)}(\tau)\,d\tau$$

It then follows, by using Eqn (5.7) and interchanging the integration signs, which is permissible in view of Fubini's theorem (cf. Chapter 6, Section 4.2), that

$$J(a, t) = \int_\mathbb{R} dG_{\xi\xi}(f) \int_\mathbb{R} \overset{\circ}{I}_{2(t-a)}(\tau)(t - a - |\tau|)\exp(i2\pi f\tau)\,d\tau$$

$$= \int_\mathbb{R} \frac{\sin^2[\pi(t-a)f]}{\pi^2 f^2}\,dG_{\xi\xi}(f)$$

$$\leq \int_\mathbb{R} \frac{1}{\pi^2 f^2}\,dG_{\xi\xi}(f) \qquad (5.61)$$

It follows from this, by applying the integrability criterion (cf. Chapter 4, Section 3.4.2), that the r.s. ξ is integrable in m.s. over every interval $[a, t]$ when the function $1/f^2$ is integrable with respect to the measure $G_{\xi\xi}$. We shall assume that this condition is satisfied in what follows, which implies that the mean power of the r.s. X is finite.

Remark: When a tends to $-\infty$, $J(a, t)$ does not necessarily have a limit and consequently the r.s. ξ is not necessarily integrable in m.s. over the interval $]-\infty, t]$ and the r.s.

$$X(t) = \int_{-\infty}^t \xi(u)\,du$$

is not necessarily defined.

1.5.2. Mean value of the integrated random signal

The mean value of the r.s. X may be obtained by applying Eqn (4.112). Now, the mean value $E[\xi]$ is zero since the function $1/f^2$ is assumed to be integrable with respect to the measure $G_{\xi\xi}$ (cf. Exercise 1.16). Hence

$$\int_{\mathbb{R}} \frac{1}{f^2} dG_{\xi\xi}(f) < +\infty \Rightarrow E[X(t)] = 0 \tag{5.62}$$

The r.s. X is thus centred.

1.5.3. Instantaneous mean power of an integrated random signal

The instantaneous mean power of the r.s. X at the instant t can be obtained from Eqn (4.114) by taking $t_1 = t_2 = t$. It is a function of a and t which is written

$$E[|X(t)|^2] = \sigma_X^2(t) = J(a, t)$$

The r.s. X is thus not weakly stationary.

1.5.4. Time average power of the integrated random signal

The time average value of the instantaneous mean power of the r.s. X is defined by the limit

$$\langle E[|X(t)|^2] \rangle = \langle \sigma_X^2(t) \rangle = \lim_{T \to \infty} \frac{1}{2T} \int_{-T}^{T} E[|X(t)|^2] dt \tag{5.63}$$

$$= \lim_{T \to \infty} \frac{1}{2T} \int_{a}^{T} E[|X(t)|^2] dt$$

It follows from this, by using Eqn (5.61) and changing the order of integration, that

$$\langle \sigma_X^2(t) \rangle = \frac{1}{4\pi^2} \lim_{T \to \infty} \int_{\mathbb{R}} \frac{1}{f^2} \left(1 - \frac{a}{T}\right) \left[1 - \frac{\sin\{2\pi(T-a)f\}}{2\pi(T-a)f}\right] dG_{\xi\xi}$$

The theorem of dominated convergence (cf. Chapter 6, Section 4.1.3.5) may be applied since we are assuming that the function $1/f^2$ is integrable with respect to the measure $G_{\xi\xi}$. Hence we have the result

$$\langle \sigma_X^2(t) \rangle = \int_{\mathbb{R}} \frac{1}{4\pi^2 f^2} dG_{\xi\xi}(f) \tag{5.64}$$

We can say, in these conditions, that the r.s. $X(t), t \in [a, +\infty[$, although it is not weakly stationary, has as spectrum the positive bounded measure G_{XX}

defined by

$$G_{XX}(f) = \frac{1}{4\pi^2 f^2} G_{\xi\xi}(f) \qquad (5.65)$$

This is the spectral distribution of the time average power $P_X = \langle \sigma^2(t) \rangle$.

1.6. ERGODICITY (Parzen, 1962, p. 72)

1.6.1. Generalities

The mean value, the autocorrelation function and, in general, the moments of an r.s. are defined by ensemble means (stochastic means) which take into account all possible realizations of the r.s.

More precisely, a moment of any order is defined by an integral with respect to the marginal probability law of the same order. Now in practice the latter is not necessarily known whereas it is always possible, at least potentially, to observe and record a finite number of realizations of the r.s. in question during an *a priori* undetermined period.

We are thus inevitably led to ask what information relative to the set of all its realizations can be obtained from observation of a single realization of the r.s. The answer is that there exists a class of r.s.s called **ergodic** for which observation of a single realization for all instants belonging to the indexation set I enables us to determine all the moments of the r.s. by means of temporal averages.

In general, ergodicity implies the possibility of obtaining ensemble means from temporal means taken over an arbitrary realization; this has great practical importance. We shall confine ourselves, however, to stating and proving the simplest of the ergodic theorems, which relates to ergodicity in the mean.

1.6.2. Ergodicity in the mean

1.6.2.1. Theorem

Let $\xi(t), t \in \mathbb{R}$, be a weakly stationary r.s. of finite mean power having a mean value m_ξ. The necessary and sufficient condition for

$$\lim_{T \to \infty} E\left[\left| \frac{1}{2T} \int_{-T}^{T} \xi(t) \, dt - m_\xi \right|^2 \right] = 0 \qquad (5.66)$$

to be true is that the measure $G_{\xi\xi}$ satisfies

$$G_{\xi\xi}(\{0\}) = |m_\xi|^2 \qquad (5.67)$$

1.6.2.2. Proof
To prove this, we write

$$J_T = \left| \frac{1}{2T} \int_{-T}^{T} \xi(t)\,dt - m_\xi \right|^2$$

$$= \frac{1}{4T^2} \int_{-T}^{T} \int_{-T}^{T} \xi(u)\xi^*(v)\,du\,dv$$

$$- m_\xi \frac{1}{2T} \int_{-T}^{T} \xi^*(t)\,dt - m_\xi^* \frac{1}{2T} \int_{-T}^{T} \xi(t)\,dt + |m_\xi|^2$$

It follows, by taking the mean values, that

$$E[J_T] = \frac{1}{4T^2} \int_{-T}^{T} \int_{-T}^{T} R_{\xi\xi}(u - v)\,du\,dv - |m_\xi|^2$$

and then, by reasoning in exactly the same way as in Section 1.2.1.2, that

$$E[J_T] = \int_{\mathbb{R}} \left\{ \frac{\sin(2\pi T f)}{2\pi T f} \right\}^2 dG_{\xi\xi}(f) - |m_\xi|^2$$

The theorem of dominated convergence may be applied since the r.s. has finite mean power (cf. Chapter 6, Section 4.1.3.5). We then get, where $I_{\{0\}}$ denotes the indicator function of the singleton $\{0\}$,

$$\lim_{T \to \infty} E[J_T] = \int_{\mathbb{R}} I_{\{0\}}(f)\,dG_{\xi\xi}(f) - |m_\xi|^2$$

$$= G_{\xi\xi}(\{0\}) - |m_\xi|^2$$

The theorem follows immediately from this.

When condition (5.67) is satisfied, it follows (cf. Chapter 6, Exercise 4.6) that

$$E\left[\left| \lim_{T \to \infty} \frac{1}{2T} \int_{-T}^{T} \xi(t)\,dt - m_\xi \right|^2 \right] = 0$$

which implies that the r.v.

$$\lim_{T \to \infty} \frac{1}{2T} \int_{-T}^{T} \xi(t)\,dt$$

is a.s. equal to the constant r.v. m_ξ (cf. Eqn (1.39)). In other words, if we consider an arbitrary realization $\xi(\omega, t)$ of the r.s. ξ, the probability that the time average

$$\lim_{T \to \infty} \frac{1}{2T} \int_{-T}^{T} \xi(\omega, t)\,dt$$

differs from the ensemble mean m_ξ is zero.

1.7. ANALOGUE LINEAR FILTERING

1.7.1. *Generalities*

Let $\xi(t), t \in \mathbb{R}$, be an r.s. in continuous time, weakly stationary, defined over a probability space $(\Omega, \mathscr{A}, P)$. Consider the block diagram of Fig. 5.13. The r.s. ξ is filtered by a filter having impulse response h to generate the r.s. in continuous time $\eta(t), t \in \mathbb{R}$. In all that follows, in the absence of any statement to the contrary, the r.s. and the impulse responses can be complex.

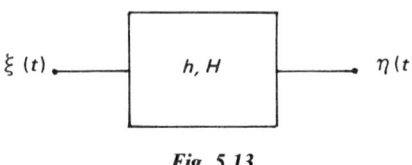

Fig. 5.13

Each realization of the r.s. η is obtained by filtering from a realization of the r.s. ξ. We can therefore write (cf. Eqn (3.74))

$$\forall \omega \in \Omega \quad \eta(\omega, t) = \xi(\omega, t) * h(t) \tag{5.68}$$

or again, suppressing from now on the variable ω and writing the explicit form of the convolution,

$$\eta(t) = \int_{\mathbb{R}} \xi(u) h(t - u) \, du \tag{5.69}$$

The r.s. η is thus defined by an integral of the form of Eqn (4.110). The only difference is that here, the integral being taken over $\mathbb{R}$, we must pass to the limit by making the limits a and b tend to $-\infty$ and $+\infty$ respectively. We shall accept that the existence criterion (4.111) is still valid in these conditions.

When the impulse response is causal, it follows from Eqn (5.69) that the value taken by the r.v. $\eta(t)$ depends on the values taken by the non-denumerable infinity of r.v.s belonging to the family $\{\xi(s) | s \leqslant t\}$. In other words, the value taken by the r.s. η at an instant t is conditioned by the events generated by the r.s. ξ prior to the instant t. This makes it clear that it is impossible, in the general case, to obtain the probability law even of the r.v. $\eta(t)$ and *a fortiori* the probability law of the r.s. η. We must content ourselves in most cases with characterizing the filtered r.s. by its mean value and autocorrelation function, which can be obtained, by contrast, very easily.

When the r.s. ξ is gaussian, the r.s. η, which is obtained from it by a linear transformation, is also gaussian. Its probability law is then well defined since it can be expressed as a function of its mean value and autocorrelation function. This is the only case in which it is possible to determine the probability law of the filtered r.s.

1.7.2. *Mean value of the filtered random signal*

The mean value of the r.s. η is obtained by applying Eqn (4.112). We get

$$\forall t \in \mathbb{R} \qquad E[\eta(t)] = E[\xi] \int_{\mathbb{R}} h(t-u)\,du$$

and then, by bringing in the transfer function $\mathscr{F}h = H$,

$$E[\eta] = E[\xi]H(0) \qquad (5.70)$$

The mean value of the filtered r.s. is therefore constant. It is the product of the mean value of the r.s. to be filtered and the value at zero frequency of the transfer function of the filter.

1.7.3. *Autocorrelation function and spectrum of the filtered random signal*

The autocorrelation function of the r.s. η is obtained by applying Eqn (4.69). We get

$$E[\eta(t)\eta^*(t-\tau)] = E\left[\int_{\mathbb{R}} \xi(u)\xi^*(v)h(t-u)h^*(t-\tau-v)\,du\,dv\right]$$

and then, by putting $u - v = x$ and introducing the autocorrelation function of the r.s. ξ and that of the filter h (cf. Eqn (3.25)),

$$R_{\eta\eta}(\tau) = \int_{\mathbb{R}} R_{\xi\xi}(x)R_{hh}(\tau-x)\,dx = R_{\xi\xi}(\tau) * R_{hh}(\tau) \qquad (5.71)$$

The autocorrelation function of the filtered r.s. does not depend on t. Since its mean value is constant, it is weakly stationary.

The spectrum of the filtered r.s. is obtained from Eqn (5.71) by taking account of Eqn (3.26). We get

$$G_{\eta\eta}(f) = G_{\xi\xi}(f)|H(f)|^2 \qquad (5.72)$$

This is the product of the spectrum of the r.s. to be filtered and the square of the modulus of the transfer function of the filter. This result is extremely important in practice.

1.7.4. *Cross-correlation function of a random signal and the filtered random signal*

The cross-correlation function of the r.s.s η and ξ is obtained from Eqn (4.73). We get

$$E[\eta(t)\xi^*(t-\tau)] = E\left[\xi^*(t-\tau)\int_{\mathbb{R}} h(u)\xi(t-u)\,du\right]$$

236 Weakly stationary random signals

and then, by putting $t - u = v$ and introducing the autocorrelation function of the r.s. ξ,

$$R_{\eta\xi}(\tau) = R_{\xi\xi}(\tau) * h(\tau) \qquad (5.73)$$

The cross-correlation function $R_{\eta\xi}$ therefore does not depend on t. It is the convolution of the autocorrelation function of the r.s. to be filtered and the impulse response of the filter. The cross-correlation function $R_{\xi\eta}$ is obtained by applying Eqn (5.5). The r.s.s ξ and η are therefore mutually weakly stationary.

1.7.5. Cross-correlation function of two filtered random signals

Let ξ_1 and ξ_2 be two mutually weakly stationary r.s.s and consider two filters having impulse responses h_1 and h_2 respectively. Let the filtered r.s. be

$$\begin{aligned} \eta_1(t) &= \xi_1(t) * h_1(t) \\ \eta_2(t) &= \xi_2(t) * h_2(t) \end{aligned} \qquad (5.74)$$

This is illustrated by Fig. 5.14.

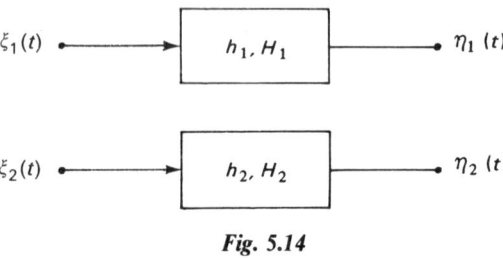

Fig. 5.14

The cross-correlation function of the r.s.s ξ_1 and ξ_2 may be obtained from Eqn (4.73). We get, reasoning as in Section 1.7.3,

$$R_{\eta_1\eta_2}(\tau) = R_{\xi_1\xi_2}(\tau) * R_{h_1h_2}(\tau) \qquad (5.75)$$

The cross-correlation function of the filtered r.s. is the convolution of the cross-correlation function of the r.s. to be filtered and the cross-correlation function of the filters. The filtered r.s.s are thus mutually weakly stationary.

It is obvious that the relation (5.75) applies in the case when the r.s.s ξ_1 and ξ_2 are both identical with one and the same r.s. ξ. The cross-correlation function $R_{\eta_1\eta_2}$ is then the convolution of the autocorrelation function $R_{\xi\xi}$ with the cross-correlation function $R_{h_1h_2}$.

Example 1.9 Complementary filters

Consider the situation illustrated by Fig. 5.14. The cross-correlation spectrum of the r.s.s η_1 and η_2 is obtained by taking the Fourier transforms of

Random signals in continuous time 237

the two sides of Eqn (5.75). It follows (cf. Eqn (3.24)) that

$$G_{\eta_1\eta_1}(f) = G_{\xi_1\xi_2}(f)H_1(f)H_2^*(f)$$

Suppose that the transfer functions of the two filters satisfy

$$\forall f \in \mathbb{R} \quad H_1(f)H_2(f) = 0 \quad (5.76)$$

which expresses the fact that the supports of the transfer functions are disjoint. The filters are then called **complementary**. It is easy to verify that the cross-correlation coefficient of the r.s.s η_1 and η_2 is zero and consequently that the filtered r.s.s are uncorrelated. When, in addition, the r.s.s ξ_1 and ξ_2 are mutually gaussian, the filtered r.s.s are also mutually gaussian. They are thus independent since they are uncorrelated.

Example 1.10 Filtering of a gaussian white noise and whitening of a spectrum

(a) Let $\xi(t)$, $t \in \mathbb{R}$, be an r.s. in continuous time, weakly stationary, centred and gaussian, whose spectrum $G_{\xi\xi}$ is a function. Consider a filter having a real impulse response and transfer function H such that (cf. Eqn (3.72))

$$|H(f)|^2 = H(f)\check{H}(f) = \frac{G_{\xi\xi}(f)}{N_0/2} \quad (5.77)$$

Finally, let $x(t)$, $t \in \mathbb{R}$, be the r.s. generated by the filtering by the filter H of a gaussian white noise $n(t)$, $t \in \mathbb{R}$, with spectral density $N_0/2$. We obtain from Eqns (5.70) and (5.74)

$$E[x] = E[n]H(0) = 0$$

$$G_{xx}(f) = \frac{N_0}{2}|H(f)|^2 = G_{\xi\xi}(f)$$

The r.s.s x and ξ are weakly equivalent (cf. Chapter 4, Section 1.4). They are in fact gaussian and centred and they have the same autocorrelation function. Their probability laws are thus identical. This enables us to say that the r.s. ξ can be generated by the filtering of a gaussian white noise with spectral density $N_0/2$ by a filter whose transfer function is defined by Eqn (5.77).

(b) Conversely and by the same reasoning, the r.s. ξ can be filtered so as to generate gaussian white noise with spectral density $N_0/2$. The transfer function K of the appropriate filter is defined as a function of the spectrum $G_{\xi\xi}$ by

$$|K(f)|^2 = K(f)\check{K}(f) = \frac{N_0/2}{G_{\xi\xi}(f)} \quad (5.78)$$

We say that the spectrum $G_{\xi\xi}$ is 'whitened' by the filter K.

(c) The situation is represented in Fig. 5.15. The filters H and K are not necessarily realizable. For them to be realizable, it is necessary and sufficient

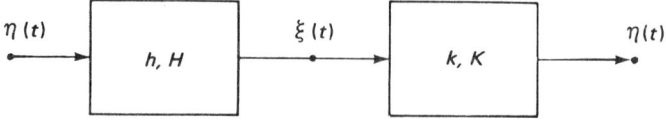

Fig. 5.15

that the impulse responses h and k corresponding to the transfer functions defined by Eqns (5.77) and (5.78) be of the form of Eqn (3.71).

Consider the particular case in which we are trying to achieve this in practice, when the spectrum $G_{\xi\xi}$ is a rational fraction. It is then necessary, for the filter H to be realizable, that the degree of the numerator be less than or equal to that of the denominator, whereas the reverse is true for the filter K (cf. Chapter 3, Example 4.1). It is impossible in these conditions that the filters H and K both be realizable. Being gaussian, the r.s. ξ has finite mean power, which implies that the degree of the numerator of the rational fraction $G_{\xi\xi}$ is strictly smaller than that of its denominator. The whitening filter K is therefore not realizable. In contrast, the filter H is realizable, for the stated necessary condition is also sufficient. It is in fact easy to verify, $G_{\xi\xi}$ being a real and even function, that it is possible to find a rational fraction H satisfying Eqn (5.77) whose numerator is of smaller degree than its denominator and all of whose poles are in the lower half-plane (cf. Chapter 3, Example 4.1).

Example 1.11 Prediction of a gaussian random signal

Let $\xi(t), t \in \mathbb{R}$, be an r.s. generated by linear filtering of a gaussian white noise $n(t), t \in \mathbb{R}$, by a filter whose impulse response h is a square integrable function. This is a weakly stationary r.s., gaussian and centred, which can be written

$$\xi(t) = \int_{-\infty}^{t} n(u)h(t-u)\,du \qquad (5.79)$$

Let t_1 and t_2 be two instants such that $t_1 \leq t_2$ and suppose that we wish to predict the value taken by the r.v. $\xi_2 = \xi(t_2)$ as a function of the values taken by the noise n at times prior to the instant t_1. The r.v. ξ_2 is centred and of integrable square. We know that in these conditions the best estimate, in m.s., of the r.v. ξ_2 is the conditional expectation r.v.

$$\hat{\xi}_2 = E[\xi_2 | \mathscr{B}_n(-\infty, t_1)]$$

where we denote by $\mathscr{B}_n(-\infty, t_1)$ the sub-Borel field generated by the noise n prior to the instant t_1 (cf. Chapter 4, Section 1.5). We can then write

$$\xi_2 = \int_{-\infty}^{t_1} n(u)h(t_2 - u)\,du + \int_{t_1}^{t_2} n(u)h(t_2 - u)\,du = v_1 + v_2$$

It follows from Eqn (1.100) that

$$E[\xi_2 | \mathcal{B}_n(-\infty, t_1)] = E[v_1 | \mathcal{B}_n(-\infty, t_1)] + E[v_2 | \mathcal{B}_n(-\infty, t_2)]$$
$$= v_1 + E[v_2]$$
$$= v_1$$

On the one hand, the r.v. v_1 is in fact a function of the r.v.s of the family $\{n(u) | u \leq t_1\}$. It is therefore measurable with respect to the sub-Borel field $\mathcal{B}_n(-\infty, t_1)$ and Eqn (1.101) can be applied to it. On the other hand, the r.v. v_2 is a function of the r.v.s of the family $\{n(u) | t_1 < u \leq t_2\}$ which is independent of the sub-Borel field $\mathcal{B}_n(-\infty, t_1)$ since n is a gaussian white noise. The relation (2.53) may be applied and we get $E[v_2] = 0$. Finally,

$$\hat{\xi}_2 = \int_{-\infty}^{t_1} n(u) h(t_2 - u) \, du \tag{5.80}$$

Example 1.12 Analytic random signal
(a) Let $\xi(t)$, $t \in \mathbb{R}$, be a weakly stationary centred r.s. and let $\eta(t)$, $t \in \mathbb{R}$, be the r.s. defined by

$$\eta(t) = -\xi(t) * \frac{1}{\pi t} = \mathcal{H}\xi(t) \tag{5.81}$$

It is possible to consider formally that the r.s. η is generated by filtering of the r.s. ξ by the non-realizable filter which has as impulse response and transfer function respectively (cf. Chapter 3, Example 2.2)

$$h(t) = -1/\pi t \qquad H(f) = i \operatorname{sgn} f$$

The r.s. η is the **Hilbert conjugate** of the r.s. ξ.
It follows by applying Eqns (5.70) and (5.72) that

$$E[\eta] = 0 \qquad G_{\eta\eta} = G_{\xi\xi} \tag{5.82}$$

These relations have meaning if and only if the spectrum $G_{\xi\xi}$ has no line at frequency 0. When it exists, the r.s. η is then centred and has the same spectrum as the r.s. ξ. It follows from Eqns (5.73) and (5.5) that

$$R_{\eta\xi}(\tau) = -R_{\xi\eta}(\tau) = -R_{\xi\xi}(\tau) * \frac{1}{\pi\tau} = \mathcal{H} R_{\xi\xi}(\tau) \tag{5.83}$$

(b) Let $\sigma(t)$, $t \in \mathbb{R}$, be the r.s. defined by

$$\sigma(t) = \xi(t) - i\mathcal{H}\xi(t) \tag{5.84}$$

It is weakly stationary and centred and its autocorrelation function can be written, by applying Eqns (5.82) and (5.83),

$$R_{\sigma\sigma}(\tau) = 2\{R_{\xi\xi}(\tau) - i\mathcal{H} R_{\xi\xi}(\tau)\}$$

It follows from this (cf. Eqn (3.14)) that

$$G_{\sigma\sigma}(f) = 4G_{\xi\xi}(f)Y(f) \tag{5.85}$$

This relation has meaning when the r.s. $\mathcal{H}\xi$ exists, i.e. when the spectrum of the r.s. ξ has no line at frequency 0. The r.s. σ is then the analytic r.s. associated with the r.s. ξ. Its spectrum extends only over the space of positive frequencies. The notion of an analytic r.s. is thus comparable with that of an analytic deterministic signal (d.s.) (cf. Chapter 3, Example 2.2).

(c) **Analytic white noise:** The above applies to white noise, of spectral density $N_0/2$. The analytic r.s. which is associated with it has spectral density equal to $2N_0$, for positive frequencies.

Example 1.13 **Distortion-free filtering of a random signal**
(a) Let $\xi(t)$, $t \in \mathbb{R}$, be a weakly stationary r.s. and let $\eta(t)$, $t \in \mathbb{R}$, be the r.s. obtained by filtering of the r.s. ξ by a filter with impulse response h. The mean value and the spectrum of the filtered r.s. collectively characterize the set of its realizations. It is often necessary, however, to characterize the latter individually. In particular, it is important to define the conditions under which the realizations of the filtered r.s. result from distortion-free transmission of realizations of the r.s. to be filtered.

For this let us consider the r.s. $\varepsilon(t)$, $t \in \mathbb{R}$, defined by

$$\varepsilon(t) = A\xi(t-\gamma) - \eta(t) \qquad A \in \mathbb{C} \qquad \gamma \in \mathbb{R}_+^*$$

The autocorrelation function of the r.s. ε is obtained by applying Eqns (5.71) and (5.73). We get

$$R_{\varepsilon\varepsilon}(\tau) = R_{\xi\xi}(\tau) * \{|A|^2\delta(\tau) + R_{hh}(\tau) - A\tilde{h}^*(\tau) * \delta(\tau-\gamma) - A^*h(\tau) * \delta(\tau+\gamma)\}$$

The spectrum $G_{\varepsilon\varepsilon}$ is therefore written

$$G_{\varepsilon\varepsilon}(f) = G_{\xi\xi}(f)|A - H(f)\exp(-i2\pi\gamma f)|^2$$

Suppose that the transfer function H is such that

$$\forall f \in \text{supp } G_{\xi\xi} \qquad H(f) = A \exp(-i2\pi\gamma f) \tag{5.86}$$

It follows from this that the mean power of the r.s. ε is zero, which enables us to write (cf. Eqn (1.39))

$$\forall t \in \mathbb{R} \qquad E[|\varepsilon(t)|^2] = 0 \Leftrightarrow P\{\xi(t) \neq 0\} = 0$$

The r.s.s $A\xi(t-\gamma)$ and $\eta(t)$ are then strongly equivalent, which does not in general imply that their realizations are a.s. identical. In order that they should be a.s. identical, it is sufficient that the realizations of the r.s. ξ be a.s. continuous to the right (or the left) (cf. Chapter 4, Section 1.4.2). This case is sufficiently general for us to state that in practice Eqn (5.86) gives the necessary and sufficient condition for a.s. **distortion-free transmission** of an r.s. This is identical with that which ensures distortion-free transmission of a d.s. (cf. Chapter 3, Example 4.4).

(b) Suppose that the r.s. ξ is band limited with a cut-off frequency f_c such that

$$\text{supp } G_{\xi\xi} = [-f_c, +f_c]$$

Suppose moreover that the spectrum $G_{\xi\xi}$ does not have lines at the frequencies $\pm f_c$ and consider the ideal filter with transfer function

$$H(f) = A\mathring{I}_{2f_c}(f)\exp(-i2\pi\gamma f)$$

It satisfies the condition (5.86) and therefore transmits a.s. without distortion the realizations of the r.s. ξ when the latter are a.s. continuous to the right (or to the left). In this case we can conclude from this that the supports of the complex spectra of the realizations $\zeta(\omega, t)$ are a.s. contained within the interval $[-f_c, +f_c]$.

Exercises

Exercise 1.1 Almost surely periodic realizations
Let $\xi(t)$, $t \in \mathbb{R}$, be a weakly stationary r.s. with autocorrelation function such that

$$\exists T \in \mathbb{R}_+^* \qquad R_{\xi\xi}(0) = R_{\xi\xi}(T)$$

(1) Prove that the function $R_{\xi\xi}$ is periodic with period T. (Use the Schwarz inequality to establish that the quantity $|R_{\xi\xi}(\tau) - R_{\xi\xi}(\tau + T)|$ is bounded by 0.)

(2) Show that the r.s.s $\xi(t)$ and $\xi(t + T)$, $t \in \mathbb{R}$, are strongly equivalent (cf. Chapter 4, Section 1.4.2). Hence deduce that, if the realizations of the r.s. ξ are a.s. continuous to the right, they are a.s. periodic with period T (cf. Chapter 4, Exercise 1.11).

Exercise 1.2 Quadratic detection of a gaussian random signal
The block diagram of a **quadratic detector** is represented in Fig. 5.16. It is constructed by placing in series the non-linear function $(.)^2$ and a linear low-pass filtering function.

Let $\xi(t)$, $t \in \mathbb{R}$, be a weakly stationary gaussian r.s., centred and with variance σ^2, and let $\eta(t)$, $t \in \mathbb{R}$, be the r.s. defined by

$$\forall t \in \mathbb{R} \qquad \eta(t) = \xi^2(t)$$

Let $\hat{\eta}(t)$, $t \in \mathbb{R}$, be the filtered r.s. at the output of the detector.

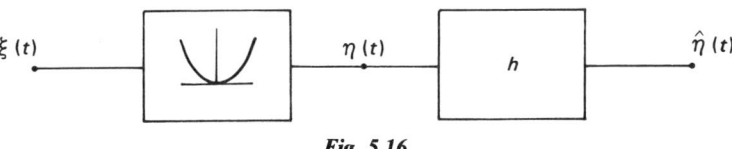

Fig. 5.16

242 Weakly stationary random signals

(1) Prove that the mean value and autocorrelation function of the r.s. η are respectively

$$E[\eta] = \sigma^2 \qquad R_{\eta\eta}(\tau) = \sigma^4 + 2R_{\xi\xi}{}^2(\tau) \qquad (5.87)$$

(Apply Eqns (5.25), (5.26) and (5.27) and use the classical result

$$\int_\mathbb{R} x^{2n} \exp(-ax^2)\,dx = \frac{(2n)!}{2^{2n}a^n n!}\left(\frac{\pi}{a}\right)^{1/2} \qquad n \in \mathbb{N} \qquad a \in \mathbb{R}_+^*$$

for this purpose.)

(2) Suppose now that the spectrum of the r.s. ξ can be written

$$G_{\xi\xi}(f) = \frac{N_0}{2}\mathring{I}_{2B}(f) \qquad B \in \mathbb{R}_+^*$$

Prove that the spectrum of the r.s. η is

$$G_{\eta\eta}(f) = N_0{}^2 B^2 \delta(f) + N_0{}^2 B\left(1 - \frac{|f|}{2B}\right)\mathring{I}_{4B}(f)$$

It is represented in Fig. 5.17. (Use the table in Appendix 2 and Eqn (3.5).)

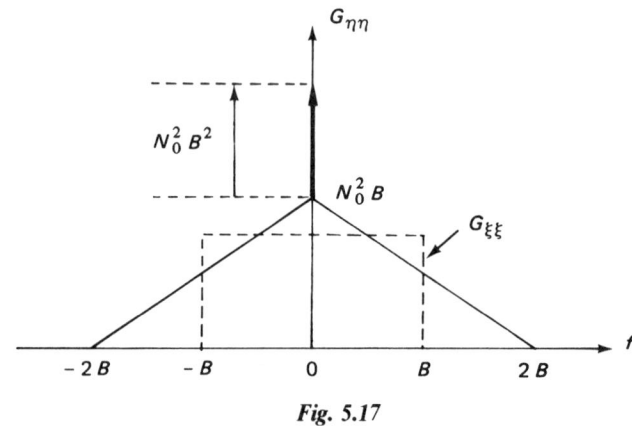

Fig. 5.17

(3) Suppose that the transfer function H of the low-pass filter is such that

$$|H(f)|^2 = \mathring{I}_{2b}(f) \qquad b \ll B$$

Hence deduce that

$$\forall t \in \mathbb{R} \qquad E[\hat{\eta}(t)] = N_0 B = P_\xi \qquad V[\hat{\eta}(t)] \approx 2N_0{}^2 Bb$$

Let us write

$$\hat{\eta}(t) = P_\xi + \hat{\eta}(t) - E[\hat{\eta}(t)] = P_\xi + \varepsilon(t)$$

The r.v. $\hat{\eta}(t)$ is equal to the mean power P_ξ of the r.s. ξ to within an error of $\varepsilon(t)$.

It follows from the above that the quadratic detector enables us, by measuring the value taken by the r.v. $\hat{\eta}(t)$, to obtain the value of the mean power P_ξ with an error whose relative standard deviation is equal to $(2b/B)^{1/2}$. The larger the ratio B/b is, the smaller is this standard deviation.

Exercise 1.3   Random signal function of a gaussian random signal

Let $\xi(t)$, $t \in \mathbb{R}$, be a weakly stationary centred gaussian r.s. and let $\eta(t)$, $t \in \mathbb{R}$, be the r.s. defined by

$$\forall t \in \mathbb{R} \qquad \eta(t) = \exp\{ia\xi(t)\} \qquad a \in \mathbb{R}$$

Prove that the autocorrelation function of the r.s. η may be written

$$R_{\eta\eta}(\tau) = \exp[-a^2\{R_{\xi\xi}(0) - R_{\xi\xi}(\tau)\}] \tag{5.88}$$

(Use Eqn (5.25) and the table in Appendix 2.)

Exercise 1.4   Representation in (x, y) of a gaussian random signal

This exercise complements Section 1.3. Consider a real, weakly stationary, centred r.s. $\xi(t)$, $t \in \mathbb{R}$, and the r.s. $\eta(t)$, $t \in \mathbb{R}$, defined by

$$\eta(t) = x(t)\cos(\hat{\omega}_0 t + \varphi_0 + \varphi) + g(t)\sin(\hat{\omega}_0 t + \varphi_0 + \varphi)$$

where $x(t)$ and $y(t)$, $t \in \mathbb{R}$, are mutually weakly stationary r.s.s, gaussian, real and centred, such that (cf. Eqn (5.35))

$$\begin{aligned} R_{xx} &= R_{yy} = R_{\xi\xi}{}^+ + R_{\xi\xi}{}^- \\ R_{xy} &= -R_{yx} = -i(R_{\xi\xi}{}^+ - R_{\xi\xi}{}^-) \end{aligned} \tag{5.89}$$

φ_0 is an r.v. uniformly distributed over the interval $[-\pi, \pi]$, independent of the r.s.s x and y, φ is an arbitrary phase angle and $\hat{f}_0$ is an arbitrary positive frequency allowing definition of the spectra $G_{\xi\xi}{}^+$ and $G_{\xi\xi}{}^-$ (cf. Eqn (5.30)).

(1) Prove that the r.s. η is conditionally gaussian with respect to the r.v. φ_0. (Fix φ_0 and show that every r.v. $(\eta(t_1), \ldots, \eta(t_n))$ is gaussian by applying the results proved in Chapter 2, Example 1.7.)

(2) Verify that the autocorrelation function of the r.s. η, conditional with respect to the r.v. φ_0, does not depend on φ_0 and may be written

$$R_{\eta\eta}(\tau) = R_{\xi\xi}(x) = R_{xx}(\tau)\cos(\hat{\omega}_0\tau) - R_{xy}(\tau)\sin(\hat{\omega}_0\tau) \tag{5.90}$$

(Fix φ_0, calculate $E[\eta(t)\eta(t-\tau)|\varphi_0]$ and apply Eqns (5.33) and (5.34).)

(3) Deduce from the above that the r.s. η is (unconditionally) gaussian and that the gaussian r.s.s ξ and η are weakly equivalent; this enables us to write the following representation in (x, y) (cf. Eqn (5.43)):

$$\xi(t) = x(t)\cos(\hat{\omega}_0 t + \varphi_0 + \varphi) + y(t)\sin(\hat{\omega}_0 t + \varphi_0 + \varphi) \tag{5.91}$$

(Note that the covariance matrix of the r.v. $(\eta_1, \ldots, \eta_n)$ does not depend on φ_0.)

Exercise 1.5 Telegraph signal (Parzen, 1962, p. 36)
Let $\xi(t)$, $t \in \mathbb{R}_+$, be the real causal r.s. defined by

$$\xi(t) = \xi_0(-1)^{N(t)}$$

where ξ_0 is a binary r.v. taking the values $+A$ and $-A$ with respective probabilities

$$P\{\xi_0 = +A\} = p_+ \qquad P\{\xi_0 = -A\} = p_-$$

and $N(t)$, $t \in \mathbb{R}_+$, is a Poisson r.s. with parameter ϑ (cf. Chapter 4, Example 1.3).

The r.s. ξ takes the values $+A$ or $-A$, the instants of transition between the two states being the point events of the Poisson process. A realization of the r.s. ξ is represented in Fig. 5.18.

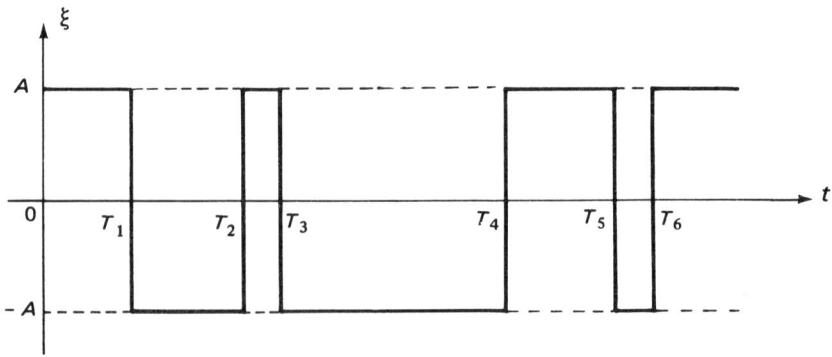

Fig. 5.18

(1) Show that we have, for all $t \in \mathbb{R}_+$,

$$P\{\xi(t) = A\} = \tfrac{1}{2} + \tfrac{1}{2}(p_+ - p_-)\exp(-\vartheta t)$$
$$P\{\xi(t) = -A\} = \tfrac{1}{2} - \tfrac{1}{2}(p_+ - p_-)\exp(-\vartheta t)$$

(2) Prove that the autocorrelation function of the r.s. ξ is

$$R_{\xi\xi}(t_1, t_2) = A^2 \exp\{-2\vartheta(t_2 - t_1)\} \qquad 0 < t_1 < t_2$$

(Use Eqn (4.48).)

(3) Deduce from the above that the r.s. ξ becomes weakly stationary when t tends to infinity.

This allows us to define a weakly stationary r.s. $\xi(t)$, $t \in \mathbb{R}$, by supposing that the Poisson process has lasted since $t = -\infty$. This is a binary r.s., taking at every instant the value $+A$ or $-A$ with probability $\tfrac{1}{2}$. The number of transitions in a time interval $\tau > 0$ is a Poisson r.v. with parameter $\vartheta\tau$ (cf. Chapter 1, Example 2.5). The r.s. ξ thus defined is the mathematical model of the telegraph signal.

(3.1) Prove that
$$\forall t \in \mathbb{R} \quad E[\xi(t)] = 0$$
The telegraph signal is centred.

(3.2) Prove that

$$R_{\xi\xi}(\tau) = A^2 \exp(-2\vartheta|\tau|) \quad G_{\xi\xi}(f) = \frac{\vartheta}{\vartheta^2 + \pi^2 f^2} \quad (5.92)$$

The autocorrelation function and the spectrum are represented in Fig. 5.19. The cut-off frequency at mid-power of the spectrum is equal to ϑ/π. It is larger, the larger the mean number ϑ of transitions per second. In the limit, the spectrum becomes white as ϑ tends to infinity and the telegraph signal becomes white noise.

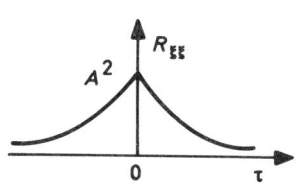
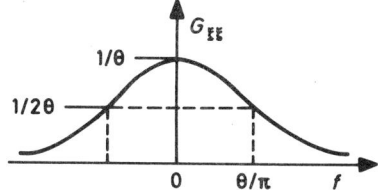

Fig. 5.19

Exercise 1.6   Binary non-return to zero random signal

Let $(\xi_n, n \in \mathbb{Z})$ be an independent sequence of binary r.v.s, such that
$$\forall n \in \mathbb{Z} \quad P\{\xi_n = A\} = P\{\xi_n = -A\} = \tfrac{1}{2}$$
Consider the r.s. $\eta(t)$, $t \in \mathbb{R}$, such that
$$\forall t \in [nT, (n+1)T[\quad \forall n \in \mathbb{Z} \quad \xi(t) = \xi_n$$
Finally let $\xi(t)$, $t \in \mathbb{R}$, be the r.s. defined by
$$\xi(t) = \eta(t - \vartheta)$$
ϑ being an r.v. uniformly distributed over the interval $[0, T]$ and independent of the sequence (ξ_n). A realization of the r.s. ξ is represented in Fig. 5.20. It

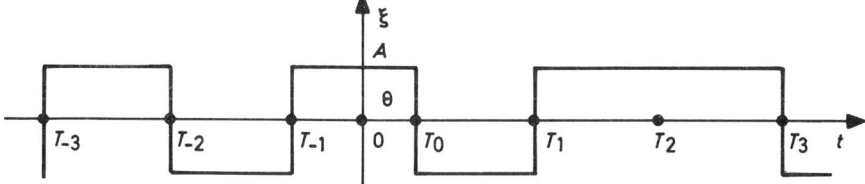

Fig. 5.20

shows the random instants $T_n = nT + \vartheta, n \in \mathbb{Z}$. The r.s. ξ is a binary non-return to zero (NRZ) signal.

(1) Prove that
$$\forall t \quad E[\xi(t)] = 0$$

The r.s. ξ is centred.

(2) Prove that
$$R_{\xi\xi}(\tau) = 0 \qquad |\tau| > T$$
$$R_{\xi\xi}(\tau) = A^2\left(1 - \frac{|\tau|}{T}\right) \qquad |\tau| < T \qquad (5.93)$$

(Verify that $R_{\xi\xi}(\tau)$ is not zero if and only if the interval $[t - \tau, t[$ contains no point of the sequence (T_n), the probability of such an event being equal to $1 - |\tau|/T$.)

(3) Deduce from the above that
$$G_{\xi\xi}(f) = A^2 T \left\{\frac{\sin(\pi Tf)}{\pi Tf}\right\}^2 \qquad (5.94)$$

(Use the table in Appendix 2.)

The autocorrelation function and the spectrum are represented in Fig. 5.21.

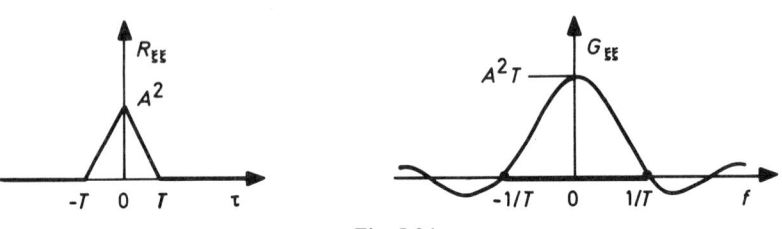

Fig. 5.21

The cut-off frequency of the spectrum is of the order of $1/T$. It is larger, the smaller the duration T. In the limit, the spectrum becomes white when T tends to zero and the binary NRZ r.s. becomes white noise.

Exercise 1.7 Multiplexing in quadrature of two independent random signals

Let $a(t)$ and $b(t)$, $t \in \mathbb{R}$, be two r.s.s, weakly stationary and independent. Consider the r.s. $\xi(t)$, $t \in \mathbb{R}$, defined by
$$\xi(t) = a(t)\cos(2\pi f_0 t + \varphi_0) + b(t)\sin(2\pi f_0 t + \varphi_0) \qquad (5.95)$$

φ_0 being an r.v. uniformly distributed over the interval $[-\pi, \pi]$, independent of the r.s.s a and b. We say that the r.s.s a and b are **multiplexed in quadrature** to construct the r.s. ξ.

(1) Prove that
$$\forall t \in \mathbb{R} \quad E[\xi(t)] = 0$$

(2) Prove that the spectrum $G_{\xi\xi}$ of the r.s. ξ may be written

$$G_{\xi\xi}(f) = \tfrac{1}{2}\{G_{aa}(f) + G_{bb}(f)\} * \{\delta(f - f_0) + \delta(f + f_0)\} \quad (5.96)$$

It follows from this that, when the spectra G_{aa} and G_{bb} have the same support, the r.s. ξ occupies the same spectral band as if one of the two r.s.s a and b were absent. Thus the multiplexing conserves the spectral width. (Use the results given in Chapter 4, Example 2.3.)

(3) Consider the r.s. $\eta(t)$, $t \in \mathbb{R}$, defined by

$$\eta(t) = \xi(t) 2 \cos(2\pi f_0 t + \varphi_0)$$

Show that its spectrum may be written

$$G_{\eta\eta}(f) = G_{aa}(f) + \tfrac{1}{4}\{G_{aa}(f) + G_{bb}(f)\} * \{\delta(f + 2f_0) + \delta(f - 2f_0)\}$$

Hence deduce that it is possible, under certain conditions, to recover the r.s. a by suitable low-pass filtering of the r.s. η.

Exercise 1.8 Integration in mean square of a random signal

Let $n(t)$, $t \in \mathbb{R}$, be a complex r.s., weakly stationary, and let $\xi(t)$, $t \in \mathbb{R}$, be a complex r.s., weakly stationary, of finite mean power P_ξ independent of the r.s. n. Consider the r.v. X defined by the integral in m.s.

$$X = \int_0^T n(t)\xi(t)\,dt \quad (5.97)$$

(1) Prove that

$$E[X] = E[n]E[\xi]T \quad (5.98)$$

(2) Prove that

$$E[|X|^2] = T\int_{\mathbb{R}} R_{nn}(\tau) R_{\xi\xi}(\tau)\left(1 - \frac{|\tau|}{T}\right) \mathring{I}_{2T}(\tau)\,d\tau$$

$$= T^2 \int_{\mathbb{R}} G_{nn}(f)\left[\check{G}_{\xi\xi}(f) * \left\{\frac{\sin(\pi Tf)}{\pi Tf}\right\}^2\right] df \quad (5.99)$$

(Reason as in Section 1.2.1.2, introduce the spectrum and apply the duality relation

$$\int_{\mathbb{R}} R_{nn}(\tau) g(\tau)\,d\tau = \int_{\mathbb{R}} \bar{\mathcal{F}} G_{nn}(\tau) g(\tau)\,d\tau$$

$$= \int_{\mathbb{R}} G_{nn}(f) \bar{\mathcal{F}} g(\tau)\,d\tau$$

to achieve the result.)

(3) Suppose that the r.s. n is white noise with spectral density $N_0/2$. Deduce

from this that the r.v. X is centred and has as variance

$$V[X] = \frac{N_0}{2} TR_{\xi\xi}(0) = \frac{N_0}{2} TP_\xi \qquad (5.100)$$

(Apply Eqn (3.3) to the first line of Eqn (5.99).)

(4) Now suppose that ξ is a band-limited r.s. with cut-off frequency f_c and that n is band-limited white noise with the same cut-off frequency:

$$\operatorname{supp} G_{\xi\xi} = [-f_c, f_c] \qquad G_{nn}(f) = \frac{N_0}{2} \hat{1}_{2f_c}(f)$$

Verify that

$$V[X] \leq \frac{N_0}{2} TP_\xi$$

(Use the second line of Eqn (5.99) and reason on the respective supports of G_{nn} and the convolution product in square brackets.) Note that the equality holds, with good approximation, if and only if the range T of integration is large enough for us to be able to write $1/T \ll f_c$.

Exercise 1.9 Smoothing of a random signal

Let $\xi(t)$, $t \in \mathbb{R}$, be a real weakly stationary r.s. of finite mean power and let $X(t)$, $t \in \mathbb{R}$, be the r.s. defined by the integral

$$\forall t \in \mathbb{R} \qquad X(t) = \frac{1}{T} \int_{t-T}^{t} \xi(u)\, du \qquad T \in \mathbb{R}_+^* \qquad (5.101)$$

(1) Prove that the r.s. X is generated by linear filtering of the r.s. ξ, the impulse response of the filter being the function

$$h(t) = \frac{1}{T}\{Y(t) - Y(t - T)\}$$

(2) Hence deduce that

$$E[X] = E[\xi] \qquad G_{XX}(f) = G_{\xi\xi}(f)\left\{\frac{\sin(\pi T f)}{\pi T f}\right\}^2 \qquad (5.102)$$

(3) Prove that

$$\forall t \in \mathbb{R} \qquad V[X(t)] \leq V[\xi(t)]$$

This justifies us in saying that the r.s. X is obtained by 'smoothing' of the r.s. ξ since the dispersion of the values taken by the r.s. X is smaller.

(4) Prove that

$$\forall t \in \mathbb{R} \qquad \lim_{T \to \infty} V[X(t)] = G_{\xi\xi}(\{0\}) - (E[\xi])^2$$

(Apply the theorem of dominated convergence: Chapter 6, Section 4.1.3.5.)

Exercises 249

Exercise 1.10 Poisson process and shot noise

Let $(T_n, n \in \mathbb{Z})$ be the sequence defining the instants of realization of the point events of the renewal process defined in Chapter 4, Exercise 1.10, and let $\xi(t)$, $t \in \mathbb{R}$, be the r.s. defined by

$$\xi(t) = \sum_{n \in \mathbb{Z}} qh(t - T_n) = h(t) * \sum_{n \in \mathbb{Z}} q\delta(t - T_n) \qquad (5.103)$$

h being the impulse response of a filter. A realization of the r.s. ξ is represented in Fig. 5.22 for the case where the filter is of low-pass RC type.

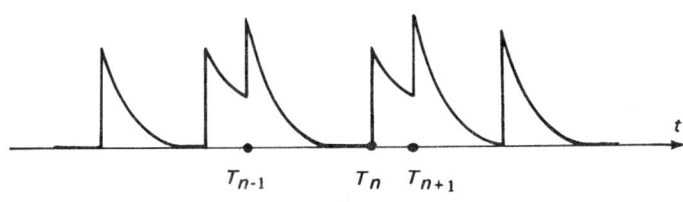

Fig. 5.22

(1) Prove that the mean value m_ξ of the r.s. ξ is the function

$$m_\xi(t) = q \int_\mathbb{R} h(t - u)\lambda(u)\, du$$

$$= qh(t) * \lambda(t) \qquad (5.104)$$

(Apply the results of Chapter 4, Exercise 1.10, more particularly Eqn (4.64), noting that $\lambda(u) = \sum_{n \in \mathbb{Z}} p_{T_n}(u)$.)

(2) Prove that the autocorrelation function $R_{\xi\xi}$ is written

$$R_{\xi\xi}(t, t - \tau) = q^2 \int_\mathbb{R} h(t - u)h^*(t - \tau - u)\lambda(u)\, du + m_\xi(t) m_\xi^*(t - \tau) \qquad (5.105)$$

(Apply the results of Chapter 4, Exercise 1.10, and reason as in the preceding question.)

(3) Deduce from the above the expression for the variance:

$$\sigma_\xi^2(t) = q^2 \int_\mathbb{R} |h(t - u)|^2 \lambda(u)\, du$$

$$= q^2 |h(t)|^2 * \lambda(t) \qquad (5.106)$$

(4) Suppose now that the renewal process under consideration is a Poisson process with parameter ϑ (cf. Chapter 4, Example 1.3). Verify that the r.s. ξ is weakly stationary and that its mean value, its variance and its spectrum are

250 Weakly stationary random signals

written respectively

$$m_\xi = q\vartheta H(0)$$

$$\sigma_\xi^2 = q^2\vartheta \int_{\mathbb{R}} |H(f)|^2 \, df \qquad (5.107)$$

$$G_{\xi\xi}(f) = q^2\vartheta|H(f)|^2 + q^2\vartheta^2|H(0)|^2\delta(f)$$

(Note that the function λ is constant and equal to ϑ.)

Comment: The r.s. ξ is the mathematical model of the electric current at the output of a filter into which is applied a current of variable intensity $q\lambda(t)$. In particular, it is the model of the current passing through a transistor and of the current generated by a photodetector.

The current applied has constant intensity when the renewal process is poissonian. The centred r.s. $\xi - m_\xi$ is then called shot noise (transistor) or quantum noise (photodetector). It follows from Eqn (5.107) that this noise results from the filtering of a white noise of spectral density $q^2\vartheta$ (cf. Eqn (5.72)).

Exercise 1.11   RLC band-pass filter

Consider the RLC band-pass filter studied in Chapter 3, Exercise 4.6. Suppose that the input current $i(t)$ is gaussian white noise with spectral density $N_0/2$ (N_0 being in ampères per hertz). Suppose in addition that the Q factor is larger than unity.

(1) Show that the spectrum of the output voltage is written

$$G_{vv}(f) \approx \frac{N_0}{2} \frac{R^2}{1 + 4Q^2 f^2/f_0^2} * \{\delta(f - f_0) + \delta(f + f_0)\}$$

This is the approximation of the narrow band. (Apply Eqns (3.102) and (5.72).)

(2) Hence deduce

$$R_{vv}(\tau) = \frac{N_0}{2} \frac{\pi f_0^2}{Q} R^2 \exp\left(-\frac{\pi f_0}{Q}|\tau|\right) \cos(\omega_0 \tau) \qquad (5.108)$$

(Use the table in Appendix 2.)

(3) Show that we can write the representation in (x, y) (cf. Exercise 1.4)

$$v(t) = x(t) \cos(\omega_0 t + \varphi_0 + \varphi) + y(t) \sin(\omega_0 t + \varphi_0 + \varphi)$$

x and y being independent gaussian centred r.s.s having spectra

$$G_{xx}(f) = G_{yy}(f) = N_0 \frac{R^2}{1 + (2Q/f_0)^2 f^2} \qquad (5.109)$$

(Apply Eqn (5.89).)

(4) Hence deduce that the r.s.s x and y are markovian. (Apply Eqn (5.20).)

Exercise 1.12 Ideal band-pass filter

Consider an ideal band-pass filter defined by the transfer function

$$H(f) = \mathring{I}_B(f) * \{\exp(i\psi_0)\,\delta(f - f_0) + \exp(-i\psi_0)\,\delta(f + f_0)\} \qquad B < 2f_0 \tag{5.110}$$

It is used to filter a gaussian white noise $n(t)$, $t \in \mathbb{R}$, with spectral density $N_0/2$. Let $\hat{n}(t)$, $t \in \mathbb{R}$, be the filtered noise.

(1) Show that the spectrum of the filtered noise may be written

$$G_{\hat{n}\hat{n}}(f) = \frac{N_0}{2}\, \mathring{I}_B(f) * \{\delta(f - f_0) + \delta(f + f_0)\} \tag{5.111}$$

(Apply Eqn (5.72).)

(2) We wish to write the weakly equivalent representation

$$\hat{n}(t) = x(t)\cos(2\pi f_0 t + \varphi_0 + \varphi) + y(t)\sin(2\pi f_0 t + \varphi_0 + \varphi) \tag{5.112}$$

where the r.s.s x and y are mutually gaussian and centred, φ_0 being an r.v. taken uniformly over the interval $[-\pi, \pi]$ and φ an arbitrary phase angle (cf. Exercise 1.4).

Show that the r.s.s x and y are defined by the spectra

$$\begin{aligned} G_{xx}(f) &= G_{yy}(f) = N_0 \mathring{I}_B(f) \\ G_{xy}(f) &= G_{yx}(f) = 0 \end{aligned} \tag{5.113}$$

Hence deduce that the r.s.s x and y are independent. (Apply Eqn (5.89).) The spectra are represented in Fig. 5.23. The representation (5.112) is widely used in the theory of modulations (Dupraz, 1973).

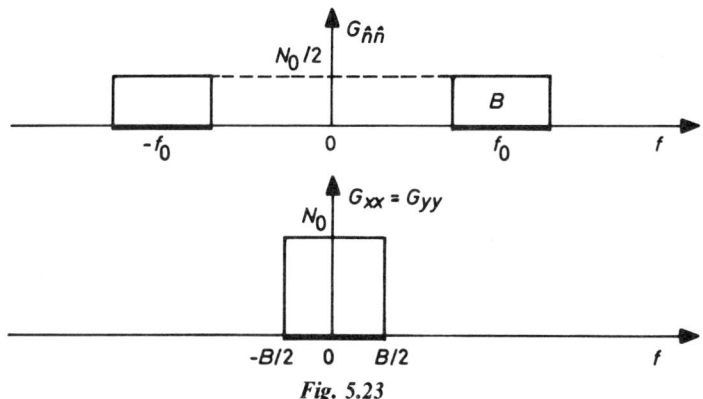

Fig. 5.23

(3) We wish to write the weakly equivalent representation

$$\hat{n}(t) = x(t)\cos(2\pi f_1 t + \varphi_0 + \varphi) + y(t)\sin(2\pi f_1 t + \varphi_0 + \varphi)$$

$$f_1 = f_0 - B/2$$

the other notations remaining unchanged.

(3.1) Show that the r.s.s x and y are defined by the spectra

$$G_{xy}(f) = -G_{yx}(f) = i\frac{N_0}{2} \operatorname{sgn} f \, \mathring{I}_{2B}(f)$$

$$G_{xy}(f) = -G_{yx}(f) = i\frac{N_0}{2} \operatorname{sgn} f \, \mathring{I}_{2B}(f)$$

This is illustrated in Fig. 5.24.

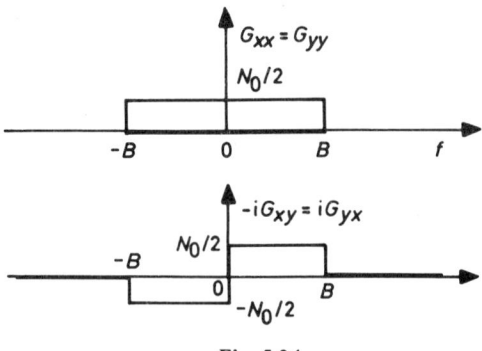

Fig. 5.24

(3.2) Deduce from the above that the r.s.s x and y are not independent and have cross-correlation coefficient

$$r_{xy}(\tau) = -\frac{\sin^2(\pi B\tau)}{\pi B\tau}$$

(Use the table in Appendix 2.)

Exercise 1.13 Random signal with single side-band

Let $a(t)$, $t \in \mathbb{R}$, be a weakly stationary r.s. whose spectrum does not contain a line at zero frequency, and let $\mathscr{H}a(t)$, $t \in \mathbb{R}$, be the Hilbert conjugate (cf. Example 1.12). Consider the r.s. $b(t)$, $t \in \mathbb{R}$, defined by

$$b(t) = a(t)\cos(2\pi f_0 t + \varphi_0) \pm \mathscr{H}a(t)\sin(2\pi f_0 t + \varphi_0) \quad (5.114)$$

where f_0 is a positive frequency and φ_0 is an r.v. uniformly distributed over the interval $[-\pi, \pi]$, independent of the r.s. a.

Show that the spectrum of the r.s. b may be written

$$G_{bb}(f) = Y(\pm f)G_{aa}(f) * \delta(f - f_0) + Y(\mp f)G_{aa}(f) * \delta(f + f_0)$$

(5.115)

Y being the Heaviside function (cf. Eqn (3.46)). (Calculate the autocorrelation function R_{bb} by applying the results given in Example 1.12. Note that the

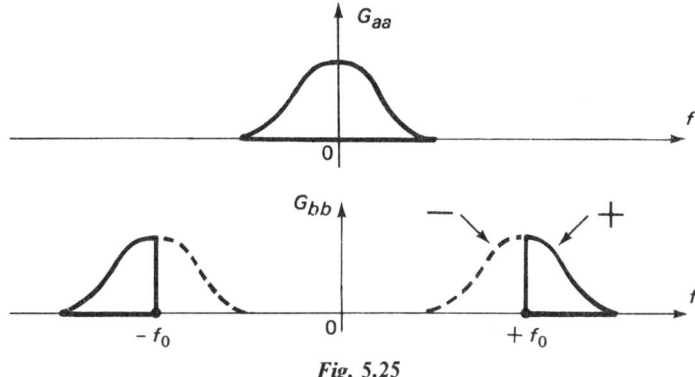

Fig. 5.25

supposition made about the spectrum G_{aa} implies that the products $Y(\pm f)G_{aa}(f)$ are well defined.)

This is illustrated in Fig. 5.25. The r.s. b is said to be with **single side-band**. The positive sign corresponds to the upper band, the negative sign to the lower band.

Exercise 1.14 Ergodicity in the mean

Let $\xi(t)$, $t \in \mathbb{R}$, be an ergodic r.s. and η an r.v. independent of the r.s. ξ. This means that, for all $t \in \mathbb{R}$, the r.v.s $\xi(t)$ and η are independent. Let $\zeta(t)$, $t \in \mathbb{R}$, be the r.s. such that $\zeta(t) = \xi(t) + \eta$. Show that the r.s. ζ is ergodic in the mean if and only if the r.v. η is a.s. constant.

Exercise 1.15 Spectrum of a centred random signal

(1) Let $\xi(t)$, $t \in \mathbb{R}$, be a weakly stationary r.s. with a non-zero mean value m_ξ. Verify that the spectrum $G_{\xi\xi}$ has a line of magnitude greater than m_ξ at frequency 0. (Write $\xi(t) = \overset{\circ}{\xi}(t) + m_\xi$, calculate the spectra of the two sides and take into account the fact that they are positive measures.)

(2) The converse of the above property is not true. Consider, in fact, the constant r.s. whose realizations can be written

$$\forall t \in \mathbb{R} \quad \xi(\omega, t) = \eta(\omega)$$

η being a centred r.v. of variance σ^2. Prove that

$$m_\xi = 0 \quad G_{\xi\xi}(f) = \sigma^2 \delta(f)$$

The r.s. ξ is centred and its spectrum contains a line at zero frequency.

Exercise 1.16 Random signal integrable in mean square

Let $\xi(t)$, $t \in \mathbb{R}$, be a weakly stationary r.s.
(1) Verify that

$$\int_\mathbb{R} \frac{dG_{\xi\xi}(f)}{f^2} df < +\infty \Rightarrow G_{\xi\xi}(\{0\}) = 0$$

254 Weakly stationary random signals

This means that the function $1/f^2$ is integrable with respect to the measure $G_{\xi\xi}$ and that, consequently, the singleton $\{0\}$ has zero measure with respect to $G_{\xi\xi}$ (cf. Chapter 6, Exercise 4.1).

(2) Deduce from the above that

$$\int_{\mathbb{R}} \frac{dG_{\xi\xi}(f)}{f^2} df < +\infty \Rightarrow E[\xi] = 0$$

It follows from this that an r.s. integrable in m.s. is centred. (Show that the hypothesis $E[\xi] \neq 0$ is absurd using the result of Exercise 1.15, Question (1).)

Exercise 1.17   Product of two mutually gaussian random signals

Let $\xi(t)$ and $\eta(t)$, $t \in \mathbb{R}$, be two r.s.s which are mutually weakly stationary, gaussian and centred, and let ζ be the weakly stationary r.s. $\zeta = \xi\eta$.

(1) Prove that the autocorrelation function of the r.s. ζ can be written

$$R_{\zeta\zeta}(\tau) = R_{\xi\eta}^2(0) + R_{\xi\xi}(\tau)R_{\eta\eta}(\tau) + R_{\xi\eta}(\tau)R_{\eta\xi}(\tau)$$

(Apply Eqn (2.81).)

(2) Put $\eta(t) = \xi(t - T)$. Deduce from the above that

$$R_{\zeta\zeta}(\tau) = R_{\xi\xi}^2(T) + R_{\xi\xi}^2(\tau) + R_{\xi\xi}(\tau + T)R_{\xi\xi}(\tau - T)$$

2. Random signals in discrete time

2.1. PROPERTIES OF THE CORRELATION FUNCTIONS

The correlation functions of weakly stationary r.s.s in discrete time have the properties of the correlation functions of r.s.s in continuous time which do not depend on the topology of $\mathbb{R}$. They are repeated here without further justification than that given in Section 1.1; we adopt notation appropriate to r.s.s in discrete time.

2.1.1. Autocorrelation function

Let $\xi(t)$, $t \in T\mathbb{Z}$, $T \in \mathbb{R}_+^*$, be a complex r.s. in discrete time, weakly stationary and with finite mean power. We shall write systematically in what follows

$$\forall n \in \mathbb{Z} \qquad \xi(nT) = \xi_n \qquad (5.116)$$

An r.s. in discrete time is thus in fact a sequence of r.v.s.

The autocorrelation function $R_{\xi\xi}$ is a complex function defined over the set $T\mathbb{Z}$ by Eqn (4.69). We have

$$\forall k \in \mathbb{Z} \qquad R_{\xi\xi}(kT) = E[\xi_n \xi_{n-k}^*] \qquad (5.117)$$

2.1.1.1. Symmetry and parity
The function $R_{\xi\xi}$ has hermitian symmetry:
$$\forall k \in \mathbb{Z} \qquad R_{\xi\xi}(kT) = R_{\xi\xi}{}^*(-kT) \qquad (5.118)$$
It is even when the r.s. ξ is real.

2.1.1.2. Value at the origin
When the r.s. ξ has finite mean power P_ξ, the function $R_{\xi\xi}$ is bounded and
$$\forall k \in \mathbb{Z} \qquad |R_{\xi\xi}(kT)| \leq R_{\xi\xi}(0) = P_\xi \qquad (5.119)$$

2.1.1.3. Non-negative definiteness
The function $R_{\xi\xi}$ is non-negative definite, which may be written (cf. Eqn (5.4))
$$\forall (c_1, \ldots, c_n) \in \mathbb{C}^n \quad \forall (k_1, \ldots, k_n) \in \mathbb{Z}^n \qquad \sum_{i=1}^n \sum_{j=1}^n c_i c_j{}^* R_{\xi\xi}(k_i T - k_j T) \geq 0$$

2.1.2. Cross-correlation function

Let $\xi(t)$ and $\eta(t)$, $t \in T\mathbb{Z}$, $T \in \mathbb{R}_+^*$, be two complex r.s.s in discrete time, mutually weakly stationary and of finite mean power. The cross-correlation function $R_{\xi\eta}$ is a complex function defined over the set $T\mathbb{Z}$ by Eqn (4.73). This gives
$$\forall k \in \mathbb{Z} \qquad R_{\xi\eta}(kT) = E[\xi_n \eta_{n-k}{}^*] \qquad (5.120)$$
The functions $R_{\xi\eta}$ and $R_{\eta\xi}$ satisfy
$$\forall k \in \mathbb{Z} \qquad R_{\xi\eta}(kT) = R_{\eta\xi}{}^*(-kT) \qquad |R_{\xi\eta}(kT)| \leq (P_\xi P_\eta)^{1/2} \qquad (5.121)$$

2.2. SPECTRAL PROPERTIES

Let $\xi(t)$, $t \in T\mathbb{Z}$, $T \in \mathbb{R}_+^*$, be an r.s. in discrete time, weakly stationary and of finite mean power. Its autocorrelation function is defined by the numerical sequence (5.117). It has a z transform which we shall write (cf. Eqn (3.116))
$$R_{\xi\xi}(z) = \sum_{k \in \mathbb{Z}} R_{\xi\xi}(kT) z^{-k} \qquad (5.122)$$
where we write in the same way, to avoid accumulation of notation, the autocorrelation function and its z transform. The risk of confusion is minimal if we take care to indicate precisely what we are dealing with.

The sequence $(R_{\xi\xi}(kT), k \in \mathbb{Z})$ is bounded (cf. Eqn (5.119)). It is therefore possible to associate with it, by reasoning exactly as in Chapter 3, Section 5.2.1, the tempered distribution (cf. Eqn (3.114))
$$R_{\xi\xi}\{\exp(i2\pi Tf)\} = \sum_{k \in \mathbb{Z}} R_{\xi\xi}(kT) \exp(-i2\pi kTf) \qquad (5.123)$$

256 Weakly stationary random signals

This is a periodic distribution over the frequency space, of period $1/T$. It can be expanded in a Fourier series in the space $\mathscr{S}'$ and the coefficient c_n is the value $R_{\xi\xi}(-nT)$ taken by the autocorrelation function for $\tau = -nT$.

Put

$$\forall f \in [-1/2T, 1/2T[\qquad G_{\xi\xi}(f) = TR_{\xi\xi}\{\exp(i2\pi Tf)\} \qquad (5.124)$$

We get then from Eqn (3.115) the fundamental relation

$$P_\xi = R_{\xi\xi}(0) = T \int_{-1/2T}^{1/2T} R_{\xi\xi}\{\exp(i2\pi Tf)\}\,df \qquad (5.125)$$

It can be shown, by applying Bochner's theorem, that $G_{\xi\xi}$ is a positive bounded measure over the interval $[-1/2T, 1/2T[$ (together with its Borel sets). It defines the spectral distribution of the mean power P_ξ. This is the **spectrum** of the r.s. ξ in discrete time.

Integral (5.125) is most frequently obtained through a contour integral, by applying the theorem of residues. It follows then, returning to the complex variable z, that

$$P_\xi = \frac{1}{i2\pi} \int_\Gamma R_{\xi\xi}(z) \frac{dz}{z} \qquad (5.126)$$

Γ being the circle of unit radius traversed in the positive direction.

The spectrum of an r.s. in continuous time is the Fourier transform of its autocorrelation function. Its support is a set in $\mathbb{R}$. The spectrum of an r.s. in discrete time is the z transform of its autocorrelation function taken for $z = \exp(i2\pi Tf)$. Its support is contained in the interval $[-1/2T, 1/2T[$.

Example 2.1 Sinusoidal random signal in discrete time

Consider the complex sinusoidal r.s., in discrete time, defined by the sequence

$$\forall n \in \mathbb{Z} \qquad \xi_n = \exp\{i(2\pi n Tf_0 + \varphi_0)\} \qquad (5.127)$$

φ_0 being an r.v. uniformly distributed over the interval $[-\pi, \pi]$ and f_0 an arbitrary positive frequency.

We obtain from Eqn (5.117)

$$\forall k \in \mathbb{Z} \qquad R_{\xi\xi}(kT) = \exp(i2\pi k Tf_0)$$

Hence it follows that (cf. Eqn (5.123))

$$R_{\xi\xi}\{\exp(i2\pi Tf)\} = \sum_{k\in\mathbb{Z}} \exp\{-i2\pi kT(f-f_0)\}$$

Then, using the expansion (3.61),

$$R_{\xi\xi}\{\exp(i2\pi Tf)\} = \frac{1}{T} \sum_{k\in\mathbb{Z}} \delta\left(f - f_0 - \frac{k}{T}\right)$$

Random signals in discrete time 257

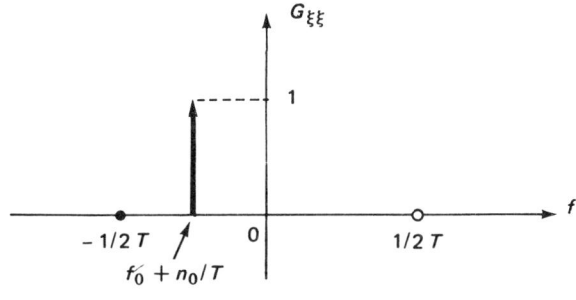

Fig. 5.26

The spectrum of the r.s. ξ may be written finally

$$G_{\xi\xi}(f) = \delta\left(f - f_0 - \frac{n_0}{T}\right) \quad -\frac{1}{2T} \leqslant f_0 + \frac{n_0}{T} < \frac{1}{2T} \quad (5.128)$$

n_0 being the integer such that the frequency $f_0 + n_0/T$ belongs to the interval $[-1/2T, 1/2T[$. It contains a single line and is represented in Fig. 5.26.

Example 2.2 White noise in discrete time
White noise in discrete time is an r.s. $w(t)$, $t \in T\mathbb{Z}$, $T \in \mathbb{R}_+^*$, weakly stationary and centred, such that

$$R_{ww}(0) = \sigma^2 \qquad \forall k \in \mathbb{Z}^* \quad R_{ww}(kT) = 0 \quad (5.129)$$

White noise in discrete time is thus in fact a sequence $(w_n, n \in \mathbb{Z})$ of centred r.v.s with the same variance σ^2, pairwise uncorrelated.

It is immediately obvious that the z transform of its autocorrelation function is the constant

$$R_{ww}(z) = \sigma^2 \quad (5.130)$$

and consequently that its spectrum is the function

$$G_{ww}(f) = T\sigma^2 \qquad -\frac{1}{2T} \leqslant f \leqslant \frac{1}{2T} \quad (5.131)$$

The spectral density is constant and equal to $T\sigma^2$, whence the name of white noise given to the r.s. w. It is represented in Fig. 5.27.

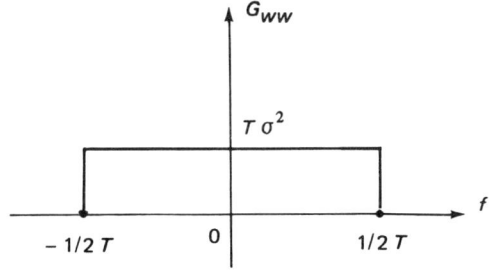

Fig. 5.27

The white noise defined by Eqn (5.129) is gaussian when the r.v.s w_n are gaussian.

2.3. NUMERICAL LINEAR FILTERING (Oppenheim and Schafer, 1975; Bellanger, 1981)

2.3.1. Generalities

Let $\xi(t), t \in T\mathbb{Z}, T \in \mathbb{R}_+^*$, be an r.s. in discrete time which is weakly stationary. Consider the block diagram in Fig. 5.28. The r.s. ξ is filtered by a numerical

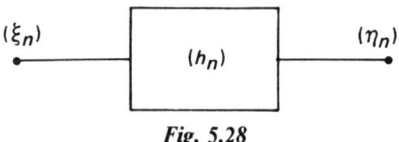

Fig. 5.28

filter having as impulse response the causal sequence $(h_n, n \in \mathbb{N})$ and as z transform the power series (cf. Chapter 3, Section 5.3.2)

$$H(z) = \sum_{n \in \mathbb{N}} h_n z^{-n}$$

In what follows the sequence (h_n) can be complex.

Let $\eta(t)$, $t \in T\mathbb{Z}$, $T \in \mathbb{R}_+^*$, be the filtered r.s. in discrete time. Each of its realizations is obtained by filtering a realization of the r.s. ξ. This enables us to write (cf. Eqn (3.130))

$$\forall n \in \mathbb{Z} \quad \eta_n = \sum_{m \in \mathbb{N}} h_m \xi_{n-m} \quad (5.132)$$

The same remarks as were made in Section 1.7.1, with reference to the filtering of r.s.s in continuous time, apply here.

2.3.2. Mean value of the filtered random signal

It follows from Eqn (5.132), through the linearity of the operator E, that

$$\forall n \in \mathbb{Z} \quad E[\eta_n] = \sum_{m \in \mathbb{N}} h_m E[\xi_{n-m}]$$

and then, the mean value of the r.s. ξ being constant,

$$E[\eta] = E[\xi]\{H(z)\}_{z=1} = E[\xi][H\{\exp(i2\pi Tf)\}]_{f=0} \quad (5.133)$$

The mean value of the r.s. η is therefore constant and equal to the product of

Random signals in discrete time 259

the mean value of the r.s. ξ and the value at $f = 0$ of the transfer function of the filter (cf. Eqn (3.127)).

This result is formally the same as that expressed by formula (5.70) with reference to filtering of r.s.s in continuous time. The z transform of the numerical filter plays the same part as the transfer function (of f) of the analogue filter.

2.3.3. Autocorrelation function and spectrum of the filtered random signal

The autocorrelation function of the r.s. η is obtained from Eqn (5.117). We get

$$E[\eta_n \eta_{n-k}^*] = \sum_{n \in \mathbb{N}} \sum_{l \in \mathbb{N}} h_m h_l^* E[\xi_{n-m} \xi_{n-k-l}^*]$$

Then, putting $m - l = n$ and introducing the autocorrelation function of the r.s. ξ and the autocorrelation function of the numerical filter (cf. Chapter 3, Exercise 5.1),

$$\forall k \in \mathbb{Z} \qquad R_{\eta\eta}(kT) = \sum_{n \in \mathbb{Z}} R_{hh}(nT) R_{\xi\xi}(kT - nT) \qquad (5.134)$$

The sequence $(R_{\eta\eta}(kT))$ is therefore the convolution of the sequences $(R_{\xi\xi}(kT))$ and $(R_{hh}(kT))$ (cf. Chapter 3, Section 5.3.1). It follows from this that the z transform of the autocorrelation function of the filtered r.s. is the product of the z transform of the filter and the z transform of the autocorrelation function of the r.s. to be filtered. This may be written (cf. Eqn (3.148))

$$R_{\eta\eta}(z) = R_{\xi\xi}(z) R_{hh}(z)$$
$$= R_{\xi\xi}(z) H(z) H_*(z^{-1}) \qquad (5.135)$$

We recall that $H(z)$ is the z transform of the sequence (h_n) and $H_*(z)$ is that of the sequence (h_n^*).

The spectrum of the r.s. η is obtained from Eqns (5.124) and (3.149). We get

$$\forall f \in [-1/2T, 1/2T[\qquad G_{\eta\eta}(f) = G_{\xi\xi}(f) |H\{\exp(i 2\pi Tf)\}|^2 \qquad (5.136)$$

This formula is analogous to Eqn (5.72). The z transform of the numerical filter thus plays the same part as the transfer function (of f) of the analogue filter.

The mean power P_η of the filtered r.s. is obtained from Eqn (5.125) or (5.126). It follows from (5.126), taking (5.135) into account, that

$$P_\eta = \frac{1}{i 2\pi} \int_\Gamma R_{\xi\xi}(z) H(z) H_*(z^{-1}) \frac{dz}{z} \qquad (5.137)$$

Γ being the circle of unit radius run in the positive direction (counterclockwise).

Example 2.3 Filtering of white noise in discrete time

Consider the white noise $w(t)$, $t \in T\mathbb{Z}$, in discrete time, defined in Example 2.2, and suppose that it is filtered by a numerical filter whose z transform is $H(z)$. Let $\hat{w}(t)$, $t \in T\mathbb{Z}$, be the filtered noise, in discrete time. Its spectrum and its mean power can be obtained from Eqns (5.136) and (5.137) respectively. It follows, taking account of Eqn (5.130), that

$$P_{\hat{w}} = \frac{1}{i2\pi} \sigma^2 \int_{\Gamma} H(z) H_*(z^{-1}) \frac{dz}{z} \tag{5.138}$$

Γ being the circle of unit radius run in the positive direction.

The above integral is usually calculated by the method of residues. It follows that

$$P_{\hat{w}} = \sigma^2 \sum \left\{ \text{residues of } H(z) H_*(z^{-1}) \frac{1}{z} \right\} \tag{5.139}$$

the sum being taken over the poles situated within the circle of unit radius. We recall that when the filter is stable the poles of $H(z)$ are within the circle of unit radius.

2.3.4. Comparison of analogue and numerical filtering

Linear filtering plays an extremely important part in signal theory. Table 5.1 presents the essential results, establishing the parallel between analogue filtering (cf. Section 1.7) and numerical filtering.

Table 5.1

Analogue filtering	Numerical filtering
$\xi(t) \to \boxed{h(t)} \to \eta(t)$	$(\xi_n) \to \boxed{(h_n)} \to (\eta_n)$
$h(t)$, impulse response $H(f)$, transfer function	(h_n), impulse response $H(z)$, z transform of (h_n) $H_*(z)$, z transform of (h_n^*)
Mean value $E[\eta] = E[\xi]H(0)$	Mean value $E[\eta] = E[\xi][H\{\exp(i2\pi Tf)\}]_{f=0}$
Autocorrelation function, $\tau \in \mathbb{R}$ $R_{\eta\eta}(\tau) = R_{\xi\xi}(\tau) * R_{hh}(\tau)$ $= R_{\xi\xi}(\tau) * h(\tau) * \tilde{h}^*(\tau)$	Autocorrelation function, $k \in \mathbb{Z}$ $R_{\eta\eta}(kT) = R_{\xi\xi}(kT) * R_{hh}(kT)$ $= R_{\xi\xi}(kT) * (h_k) * (h_{-k}^*)$
Spectrum, $f \in \mathbb{R}$ $G_{\eta\eta}(f) = G_{\xi\xi}(f)H(f)H^*(f)$	z transform, $z \in \mathbb{C}$ $R_{\eta\eta}(z) = R_{\xi\xi}(z)H(z)H_*(z^{-1})$
Spectrum, $f \in \mathbb{R}$ $G_{\eta\eta}(f) = G_{\xi\xi}(f)\|H(f)\|^2$	Spectrum, $f \in [-1/2T, 1/2T[$ $G_{\eta\eta}(f) = G_{\xi\xi}(f)\|H\{\exp(i2\pi Tf)\}\|^2$

2.4. SAMPLING OF A RANDOM SIGNAL

2.4.1. Ideal sampled random signal

(a) Let $\xi(t), t \in \mathbb{R}$, be a complex r.s. in continuous time, weakly stationary and of finite mean power. Consider the ideal sampled r.s. $\tilde{\xi}(t), t \in \mathbb{R}$, obtained by sampling, at frequency $f_e = 1/T$ with random phase ϑ, the realizations of the r.s. ξ (cf. Chapter 3, Section 5.1). It can be written formally

$$\tilde{\xi}(t) = \xi(t) \sum_{n \in \mathbb{Z}} \delta(t - nT - \vartheta) = \sum_{n \in \mathbb{Z}} \xi_n \delta(t - nT - \vartheta) \qquad (5.140)$$

where $\xi_n = \xi(nT + \vartheta)$ is a complex square integrable r.v. representing the sample drawn at the random instant $nT + \vartheta$, $n \in \mathbb{Z}$, and ϑ is an r.v. uniformly distributed over the interval $[-T/2, T/2]$, independent of the sequence (ξ_n).

The r.s. $\tilde{\xi}$ is the product of two independent r.s.s. Its autocorrelation function can therefore be written (cf. Eqns (4.37) and (4.86))

$$R_{\tilde{\xi}\tilde{\xi}}(\tau) = R_{\xi\xi}(\tau) \frac{1}{T^2} \sum_{n \in \mathbb{Z}} \exp\left(\frac{in2\pi\tau}{T}\right)$$

It can also be written (cf. Eqns (3.61) and (3.3))

$$R_{\tilde{\xi}\tilde{\xi}}(\tau) = \frac{1}{T} \sum_{n \in \mathbb{Z}} R_{\xi\xi}(nT) \delta(\tau - nT)$$

The spectrum of the r.s. $\tilde{\xi}$ can therefore be expressed in the following equivalent forms:

$$G_{\tilde{\xi}\tilde{\xi}}(f) = \frac{1}{T^2} G_{\xi\xi}(f) * \sum_{n \in \mathbb{Z}} \delta\left(f - \frac{n}{T}\right) \qquad (5.141)$$

$$= \frac{1}{T} \sum_{n \in \mathbb{Z}} R_{\xi\xi}(nT) \exp(-in2\pi Tf) \qquad (5.142)$$

This is an infinite sum of terms obtained by multiplying the spectrum $G_{\xi\xi}$ by $1/T^2$ and translating it onto the frequencies n/T, $n \in \mathbb{Z}$.

(b) Let $\bar{\xi}(t), t \in T\mathbb{Z}$, be the complex r.s., in discrete time, defined from the sequence (ξ_n) by

$$\bar{\xi}(nT) = \xi(nT + \vartheta) = \xi_n \qquad (5.143)$$

It is immediately obvious that its autocorrelation function is the complex function $R_{\bar{\xi}\bar{\xi}}$, defined over the set $T\mathbb{Z}$ by

$$R_{\bar{\xi}\bar{\xi}}(nT) = R_{\xi\xi}(nT)$$

Comparison of Eqns (5.124) and (5.142) shows that the spectrum of the r.s. $\bar{\xi}$ can be expressed as a function of the spectrum of the r.s. $\tilde{\xi}$ by

$$G_{\bar{\xi}\bar{\xi}}(f) = T^2 G_{\tilde{\xi}\tilde{\xi}}(f) \qquad f \in [-1/2T, 1/2T[\qquad (5.144)$$

262 Weakly stationary random signals

It is therefore possible, starting from an r.s. ξ in continuous time, to define either the ideal sampled r.s. $\tilde{\xi}$ or the r.s. in discrete time $\bar{\xi}$. Their spectra can be obtained from the spectrum $G_{\xi\xi}$ by applying Eqns (5.141) and (5.144) respectively.

2.4.2. Regularized sampled random signal

The r.s. $\tilde{\xi}$ does not actually exist but it can be regularized by filtering to generate an r.s. which can exist. Let $\hat{\xi}(t)$, $t \in \mathbb{R}$, be the complex r.s., in continuous time, obtained by filtering of the r.s. $\tilde{\xi}$ by an analogue filter having impulse response h. We get

$$\hat{\xi}(t) = \tilde{\xi}(t) * h(t) = \left\{ \xi(t) \sum_{n \in \mathbb{Z}} \delta(t - nT - \vartheta) \right\} * h(t) \qquad (5.145)$$

The operation of sampling and of filtering is represented by the block diagram in Fig. 5.29 (cf. Fig. 3.23).

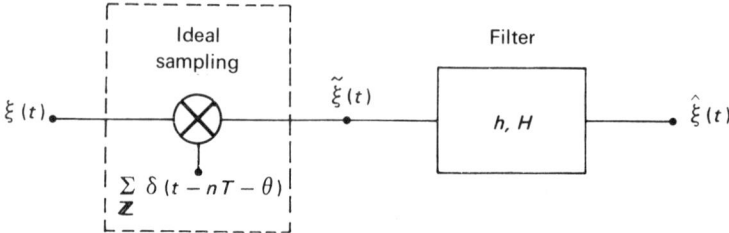

Fig. 5.29

The filtering regularizes the r.s. $\tilde{\xi}$ by interpolation between the sample values ξ_n. Figure 5.30 represents a realization of the r.s. $\hat{\xi}$ in the case when the impulse response of the filter is the causal rectangular impulse

$$h(t) = Y(t) - Y(t - T)$$

There is then sampling and holding (cf. Chapter 3, Example 5.3).

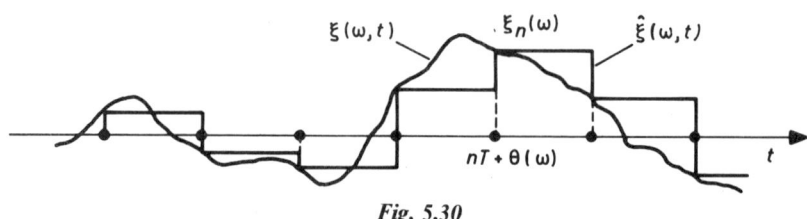

Fig. 5.30

Random signals in discrete time   263

The mean value and the spectrum of the r.s. $\hat{\xi}$ are obtained using Eqns (5.70) and (5.73), bearing in mind that the r.s. $\tilde{\xi}$ is the product of two independent r.s.s. It follows that

$$E[\hat{\xi}] = E[\xi] \frac{1}{T} H(0) \qquad (5.146)$$

$$G_{\hat{\xi}\hat{\xi}}(f) = \left\{ G_{\xi\xi}(f) * \sum_{n \in \mathbb{Z}} \delta\left(f - \frac{n}{T}\right) \right\} \frac{1}{T^2} |H(f)|^2 \qquad (5.147)$$

The spectrum $G_{\hat{\xi}\hat{\xi}}$ is the product of the function $|H|^2/T^2$ and the sum of the spectra obtained from the spectrum $G_{\xi\xi}$ by translations of amplitude $n/T, n \in \mathbb{Z}$. This result should be compared with that obtained in Chapter 3, Example 5.3.

2.4.3. Sampling error

The sampling error introduced by replacing the r.s. ξ by the r.s. $\hat{\xi}$ is the r.s. $\varepsilon(t)$, $t \in \mathbb{R}$, defined by

$$\varepsilon(t) = \hat{\xi}(t) - \xi(t)$$

Its mean value may be written (cf. Eqn (5.146))

$$E[\varepsilon] = E[\xi]\left\{ \frac{1}{T} H(0) - 1 \right\} \qquad (5.148)$$

To calculate the spectrum $G_{\varepsilon\varepsilon}$, we write $\varepsilon = \alpha + \beta$, applying Eqn (3.61) and defining the r.s.s α and β by

$$\alpha(t) = \left\{ \xi(t) \frac{1}{T} \sum_{n \in \mathbb{Z}^*} \exp\left(in \frac{2\pi}{T} t\right) \exp\left(-in \frac{2\pi}{T} \vartheta\right) \right\} * h(t)$$

$$\beta(t) = \frac{1}{T} \xi(t) * h(t) - \xi(t)$$

It is easy to prove, taking account of the probability law of the r.v. ϑ, that the cross-correlation function $R_{\alpha\beta}$ is zero. It follows from this (cf. Eqn (4.45)) that

$$G_{\varepsilon\varepsilon} = G_{\alpha\alpha} + G_{\beta\beta}$$

Application of Eqn (5.73) enables us to write

$$G_{\alpha\alpha}(f) = |H(f)|^2 \left\{ G_{\xi\xi}(f) * \frac{1}{T^2} \sum_{n \in \mathbb{Z}^*} \delta\left(f - \frac{n}{T}\right) \right\}$$

$$G_{\beta\beta}(f) = \left| \frac{1}{T} H(f) - 1 \right|^2 G_{\xi\xi}(f)$$

264 Weakly stationary random signals

Finally, the spectrum $G_{\varepsilon\varepsilon}$ may be written

$$G_{\varepsilon\varepsilon}(f) = |H(f)|^2 \left\{ \frac{1}{T^2} \sum_{n \in \mathbb{Z}} G_{\xi\xi}\left(f - \frac{n}{T}\right) \right\} + G_{\xi\xi}(f) \left[1 - \frac{1}{T} \{H(f) + H^*(f)\} \right]$$

(5.149)

The mean square value of the sampling error is then given by the integral

$$E[|\varepsilon|^2] = \int_{\mathbb{R}} dG_{\varepsilon\varepsilon}(f)$$

Let us write

$$\int_{\mathbb{R}} |H(f)|^2 G_{\xi\xi}\left(f - \frac{n}{T}\right) df = \int_{\mathbb{R}} \left|H\left(f - \frac{n}{T}\right)\right|^2 dG_{\xi\xi}(f)$$

It follows that

$$E[|\varepsilon|^2] = \int_{\mathbb{R}} \left[\frac{1}{T^2} \sum_{n \in \mathbb{Z}} \left|H\left(f - \frac{n}{T}\right)\right|^2 - \frac{1}{T} \{H(f) + H^*(f)\} + 1 \right] dG_{\xi\xi}(f)$$

(5.150)

This fundamental formula enables us to calculate the mean square value of the error introduced by replacing the r.s. ξ by the sampled and filtered r.s. $\hat{\xi}$.

Example 2.4 Optimal interpolation

Suppose that the measure $G_{\xi\xi}$ is defined by a density with respect to Lebesgue measure (which we shall denote by the same symbol for simplicity), and let us consider the real positive function H_0 defined by

$$\frac{H_0(f)}{T} = G_{\xi\xi}(f) \bigg/ \sum_{n \in \mathbb{Z}} G_{\xi\xi}\left(f - \frac{n}{T}\right) \leqslant 1$$

(5.151)

We get then from Eqn (5.149)

$$E[|\varepsilon|^2] = \int_{\mathbb{R}} G_{\xi\xi}(f) \left\{ 1 - \frac{H_0(f)}{T} \right\} df + \frac{1}{T} \int_{\mathbb{R}} \frac{G_{\xi\xi}(f)}{H_0(f)} |H(f) - H_0(f)|^2 df$$

From this form it is clear that the mean square error is minimal and equal to

$$E[|\varepsilon|^2]_{\min} = \int_{\mathbb{R}} G_{\xi\xi}(f) \left\{ 1 - G_{\xi\xi}(f) \bigg/ \sum_{n \in \mathbb{Z}} G_{\xi\xi}\left(f - \frac{n}{T}\right) \right\} df \quad (5.152)$$

when the transfer function of the interpolation filter is the function H_0. It must be noted that the filter which corresponds to this is not necessarily realizable.

2.4.4. The sampling theorem

(a) Suppose that the filter used to regularize the r.s. ξ has as impulse response

Random signals in discrete time

and transfer function respectively the functions

$$h(t) = \frac{\sin(\pi t/T)}{\pi t/T} \tag{5.153}$$

$$H(f) = T\mathring{I}_{1/T}(f)$$

It follows then, for every value of the frequency except $1/2T + k/T, k \in \mathbb{Z}$, that

$$\frac{1}{T^2} \sum_{n \in \mathbb{Z}} \left| H\left(f - \frac{n}{T}\right) \right|^2 - \frac{1}{T}\{H(f) + H^*(f)\} + 1 = 2\{1 - \mathring{I}_{1/T}(f)\}$$

If we suppose that the spectrum $G_{\xi\xi}$ does not contain lines at frequencies $1/2T + k/T, k \in \mathbb{Z}$, it follows from Eqn (5.150) that

$$E[|\varepsilon|^2] = 2 \int_{\mathbb{R}} \{1 - \mathring{I}_{1/T}(f)\} \, dG_{\xi\xi}(f)$$

(b) Suppose in addition that the r.s. $G_{\xi\xi}$ is band limited and that (cf. Eqn (5.12))

$$\operatorname{supp} G_{\xi\xi} = [-f_c, f_c] \qquad 2f_c \leq f_e = \frac{1}{T} \tag{5.154}$$

It follows from this that the m.s. error is zero

$$\forall t \in \mathbb{R} \qquad E[|\varepsilon(t)|^2] = 0$$

The r.s. ξ and $\hat{\xi}$ are then strongly equivalent. As the r.s. ξ is band limited, its realizations are a.s. continuous and consequently the realizations of the r.s.s ξ and $\hat{\xi}$ are a.s. identical (cf. Chapter 4, Section 1.4.2).

(c) It follows from the above that the r.s. ξ is a.s. completely defined by the sequence (ξ_n) of its samples when the sampling frequency satisfies Eqn (5.154). This is the sampling theorem for weakly stationary band-limited r.s.s in continuous time. It is stated, subject to an a.s. condition, in the same way as for band-limited d.s.s.

Example 2.5 **Band-limited white noise in continuous time and white noise in discrete time**

Band-limited white noise in continuous time is a weakly stationary centred r.s. $v(t)$, $t \in \mathbb{R}$, having as spectrum the function

$$G_{vv}(f) = \frac{N_0}{2} \mathring{I}_{2b}(f) \qquad b \in \mathbb{R}_+^* \tag{5.155}$$

The spectrum G_{vv} is represented in Fig. 5.31.

The r.s. v can be generated by filtering a white noise of spectral density $N_0/2$ by an ideal low-pass filter with cut-off frequency b (cf. Chapter 3, Example 4.4). It is immediately obvious that its mean power and its autocorrelation

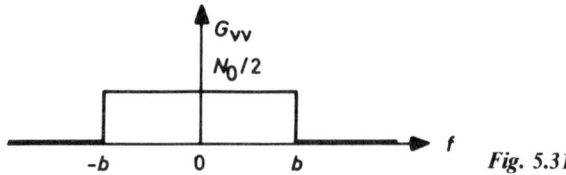

Fig. 5.31

coefficient can be written respectively

$$P_v = N_0 b \qquad r_{vv}(\tau) = \frac{\sin(2\pi b\tau)}{2\pi b\tau} \tag{5.156}$$

The sampling theorem can be applied to the r.s. v, the minimal sampling frequency being $f_e = 2b$. The r.s. v is thus a.s. completely represented by the sequence (v_k) of r.v.s defined by

$$\forall k \in \mathbb{Z} \qquad v_k = v\left(\frac{k}{2b} + \vartheta\right) \tag{5.157}$$

It follows from Eqn (5.156) that these r.v.s are pairwise uncorrelated.

The sequence (v_k) is therefore a white noise in discrete time having mean power $N_0 b$ (cf. Example 2.2). When the r.s. is gaussian, the sequence (v_k) is independent and it defines a gaussian white noise in discrete time.

Exercises

Exercise 2.1 Numerical filtering of white noise in discrete time

Consider the white noise $w(t), t \in T\mathbb{Z}$, in discrete time defined in Example 2.2.

(1) The noise w is filtered by a recursive filter of order 1 (cf. Chapter 3, Example 5.6) and by a moving average transversal filter (cf. Chapter 3, Exercise 5.3) having z transforms respectively

$$H_1(z) = \frac{\mu z}{z - \lambda}$$

$$H_2(z) = \frac{1}{N} \frac{1 - z^{-N}}{1 - z} \tag{5.158}$$

Prove, supposing that the recursive filter is stable, that the mean powers of the filtered noises $\hat{w}_1$ and $\hat{w}_2$ are respectively

$$P_{\hat{w}_1} = \frac{\sigma^2 |\mu|^2}{1 - |\lambda|^2} \qquad P_{\hat{w}_2} = \frac{\sigma^2}{N} \tag{5.159}$$

(Apply Eqn (5.139).)

(2) The noise w is filtered by a recursive filter of order 2 (cf. Chapter 3,

Exercise 5.4) which has as z transform

$$H(z) = \frac{\mu z^2}{z^2 - \lambda_1 z - \lambda_2} \qquad \lambda_1, \lambda_2, \mu \in \mathbb{R} \qquad (5.160)$$

Show that, when the filter is stable, the mean power of the filtered noise $\hat{w}$ is

$$P_{\hat{w}} = \sigma^2 \frac{\mu^2(1 - \lambda_2)}{(1 + \lambda_2)\{(1 - \lambda_2)^2 - \lambda_1^2\}} \qquad (5.161)$$

(Apply Eqn (5.139).)

Exercise 2.2 Estimation of a constant by filtering

A physical quantity a is measured by an instrument which introduces a measuring error or noise whose effect it is desirable to reduce. For this purpose we carry out a sequence of measurements of which the model is studied here.

Let $(w_n, n \in \mathbb{N})$ be an independent sequence of centred gaussian r.v.s, with the same variance σ^2, defining a gaussian white noise in discrete time (cf. Example 2.2).

Let $(x_n, n \in \mathbb{N})$ be the sequence of r.v.s $x_n = a + w_n$, a being the unknown constant which it is required to evaluate. The r.v. x_n is the result of the nth measurement and w_n is the noise affecting this measurement.

Let $(y_n, n \in \mathbb{N})$ be the sequence obtained by numerical filtering of the sequence (x_n) by a recursive filter of order 1 having real parameters λ and μ (cf. Chapter 3, Example 5.6). The sequence (y_n) is defined by the recurrence relation (cf. Eqn (3.136))

$$y_0 = \mu x_0 \qquad \forall n \in \mathbb{N} \quad y_{n+1} = \lambda y_n + \mu x_{n+1}$$

The r.v. y_n is the estimate of the constant a after the nth measurement, the error of estimation being the r.v. $\varepsilon_n = a - y_n$.

(1) Establish the relation

$$\forall n \in \mathbb{N} \qquad y_n = \mu x_n + \ldots + \mu \lambda^{n-1} x_1 + \mu \lambda^n x_0$$

(2) Let $m_n = E[\varepsilon_n]$ be the mean value of the estimation error. Show that for the sequence (m_n) to converge to zero it is necessary and sufficient that

$$\mu = 1 - \lambda \qquad |\lambda| < 1$$

It is therefore necessary that the recursive filter be stable. Verify that the convergence is exponential (with respect to n).

(3) Let $v_n = V[\varepsilon_n]$ be the variance of the estimation error. Show that when the above conditions are satisfied the sequence (v_n) converges to the limit $v = \sigma^2(1 - \lambda)/(1 + \lambda)$. One could start by establishing that the sequence (v_n) satisfies the recurrence relation

$$v_{n+1} = \lambda^2 v_n + (1 - \lambda)^2 \sigma^2$$

(4) Deduce from the above that the interesting values of λ are situated in the interval $[0, 1[$ and that, the smaller the value of the limit of the variance of the

error, the less rapid is the convergence to zero of the mean of the error. The choice of λ therefore follows from the classic compromise between precision and speed.

Exercise 2.3 Correlation of samples

Let $\xi(t), t \in \mathbb{R}$, be a weakly stationary random signal and let $\xi_n = \xi(nT), n \in \mathbb{Z}$, be the sequence of samples drawn at frequency $f_e = 1/T$. Let $r_{\xi\xi}$ be the autocorrelation coefficient of the r.s. ξ and let $g = \mathscr{F} r_{\xi\xi}$ be its Fourier transform.

(1) Prove that, for the r.v.s ξ_n to be pairwise uncorrelated, it is necessary and sufficient that the function g satisfies

$$\sum_{n \in \mathbb{Z}} g\left(f - \frac{n}{T}\right) = T \qquad (5.162)$$

(Argue exactly as in Chapter 3, Exercise 5.5.)

(2) Suppose now that the r.s. ξ is band limited and such that supp $G_{\xi\xi} = [-1/2T, 1/2T]$. Deduce from Eqn (5.162) that the function g is written

$$g = T\mathring{I}_{1/T}$$

Next deduce that the spectrum of the r.s. ξ can be written

$$G_{\xi\xi}(f) = TV[\xi]\mathring{I}_{1/T}(f) + (E[\xi])^2 \delta(f) \qquad (5.163)$$

($\mathring{I}_{1/T}$ is the indicator function of the interval $[-1/2T, 1/2T]$. Use Eqn (4.71).)

In conclusion, the samples obtained by sampling, at frequency $1/T$, a weakly stationary r.s. ξ having cut-off frequency $f_c = 1/2T$ are pairwise uncorrelated if and only if the spectrum $G_{\xi\xi}$ is of the form of Eqn (5.163). It should be noted that this is never the case for an r.s. having concrete existence.

Exercise 2.4 Compression of a sequence of random variables

Let $(\xi_n, n \in \mathbb{Z})$ be an arbitrary sequence of square integrable real r.v.s. It can obtained by sampling an r.s. of finite mean power.

Consider the sequence $(\hat{\xi}_n, n \in \mathbb{Z})$ of conditional expectation r.v.s defined by

$$\forall n \in \mathbb{Z} \qquad \hat{\xi}_n = E[\xi_n | \mathscr{B}_{n-1}] \qquad (5.164)$$

$\mathscr{B}_{n-1}$ being the sub-Borel field generated by the family $\{\xi_k | k \in \mathbb{Z}, k < n\}$ of r.v.s. The r.v. $\hat{\xi}_n$ is the best approximation in quadratic mean of the r.v. ξ_n as a function of the r.v.s of the family defined above (cf. Chapter 1, Section 2.6.2.5).

Finally let $(x_n, n \in \mathbb{Z})$ be the sequence of square integrable r.v.s defined by

$$\forall n \in \mathbb{Z} \qquad x_n = \hat{\xi}_n - \xi_n \qquad (5.165)$$

The r.v. x_n is the prediction error of the r.v. ξ_n. It is the part of the r.v. ξ_n called residue which cannot be expressed as a function of the values taken by the preceding r.v.s $\xi_k, k < n$, and therefore cannot be predicted.

(1) Prove that
$$\forall n \in \mathbb{Z} \quad E[x_n] = 0 \tag{5.166}$$
(Apply Eqn (1.103).)

(2) Prove that
$$\forall n \in \mathbb{Z} \quad \forall m \in \mathbb{N}^* \quad E[x_n x_{n+m}] = 0 \tag{5.167}$$
(Note that $\mathscr{B}_{n-1} \subset \mathscr{B}_n \subset \mathscr{B}_{n+m-1}$ and that the r.v. $\hat{\xi}_n$ is measurable with respect to the sub-Borel field $\mathscr{B}_{n+m-1}$, and apply Eqn (1.121).)

(3) Prove that
$$\forall n \in \mathbb{Z} \quad V[x_n] = V[\xi_n] - V[\hat{\xi}_n] \leqslant V[\xi_n] \tag{5.168}$$
(Apply Eqns (1.121) and (1.105).)

Comment: The above results mean that it is always possible, at least in theory, starting from an arbitrary random sequence (ξ_n), to generate a random sequence (x_n) of r.v.s which are centred and pairwise uncorrelated and for which the variances are smaller than or equal to those of the r.v.s with the same index as the sequence (ξ_n).

In so far as the 'quantity of information' carried by an r.v. is greater the greater its variance is, we can say that the 'information' carried by the sequence (ξ_n) is compressed to generate the sequence (x_n) of residues. This result is very important in signal theory.

References

BELLANGER M. (1981): *Traitement Numérique du Signal*, Masson, Paris.
BLANC-LAPIERRE A. and PICINBONO B. (1981): *Fonctions Aléatoires*, pp. 143, 157, Masson, Paris.
COX D. R. and MILLER H. D. (1970): *The Theory of Stochastic Processes*, p. 279, Methuen, London.
DUPRAZ J. (1973): *Théorie de la Communication*, pp. 175, 195, Eyrolles, Paris.
DUPRAZ J. (1977): *La Théorie des Distributions*, Cepadues, Toulouse.
GAGLIARDI R. M. and KARP S. (1976): *Optical Communications*, p. 37, Wiley, New York.
OPPENHEIM A. W. and SCHAFER R. W. (1975): *Digital Data Processing*, p. 377, Prentice-Hall, Englewood Cliffs, NJ.
PARZEN E. (1962): *Stochastic Processes*, Holden-Day, San Francisco, CA.

Chapter 6
Elements of measure theory

The elements of measure theory on which the preceding chapters rest are assembled in this chapter. So as to spare the reader purely mathematical proofs which would be outside the scope of this book, the main theorems are merely cited. We endeavour to give only material which directly supports the other chapters and which is necessary for understanding them. The proposed exercises were chosen in the same spirit.

Four parts present successively measurable spaces, measurable functions, positive measures, and integration. The positive measures used in practice are limited in number: Dirac measure, Lebesgue measure, and absolutely continuous measures with respect to Lebesgue measure which are defined by measurable functions.

1. Measurable spaces

1.1. SIMPLE MEASURABLE SPACES

1.1.1. Borel field and algebra of subsets of a set

1.1.1.1. Definition

A family $\mathscr{A}$ of subsets of a set X is a **Borel field** or a **Boolean σ algebra** when it satisfies the following three axioms.

Axiom 1   The set X belongs to the family $\mathscr{A}$.
Axiom 2   The family $\mathscr{A}$ is closed under complementation:

$$A \in \mathscr{A} \Rightarrow \bar{A} \in \mathscr{A}$$

where $\bar{A}$ denotes the complement of A in the set X. (6.1)

Axiom 3   The family $\mathscr{A}$ is closed under denumerable union:

$$(\forall n \in \mathbb{N} \quad A_n \in \mathscr{A}) \Rightarrow \bigcup_{n \in \mathbb{N}} A_n \in \mathscr{A}$$

The pair $(X, \mathscr{A})$ is a **measurable space** and the elements of the Borel field $\mathscr{A}$ are the **measurable sets** of the set X.

Remark: It is easy to verify that axioms 2 and 3 imply that a Borel field is also closed under denumerable intersection. It is also easy to verify that axiom 3 is equivalent to the axiom of closure under increasing denumerable union (cf. Example 1.2).

1.1.1.2. Definition

A family $\mathscr{C}$ of subsets of a set X is an **algebra** or a **Boolean algebra** when it satisfies the following three axioms.

Axiom 1   The set X belongs to the family $\mathscr{C}$.
Axiom 2   The family $\mathscr{C}$ is closed under complementation.
Axiom 3   The family $\mathscr{C}$ is closed under finite union.

Axioms 2 and 3 obviously imply closure under finite intersection. It is immediate that a Borel field is an algebra.

Example 1.1 Borel field of subsets

It is immediate that the set $\mathscr{P}(X)$ of subsets of a set X is a Borel field which contains the singletons. This is the **Borel field of subsets** of the set X. It is also immediately obvious that every Borel field is contained in the Borel field of subsets which is therefore the largest Borel field of a set.

When the set X is finite and contains n elements, the Borel field $\mathscr{P}(X)$ contains 2^n elements. To see this, it is sufficient to note that every subset of X either does or does not contain each of the n elements of X. The number of different subsets is then equal to the number of binary sequences of n digits 0 or 1, a 0 in the ith position ($1 \leqslant i \leqslant n$) meaning that element number i of X is not included in the subset under consideration.

When the set X is infinite, whether denumerable or not, the Borel field $\mathscr{P}(X)$ is non-denumerable since the order of infinity of the set $\{0, 1\}^\mathbb{N}$ is larger than that of the set $\mathbb{N}$. The Borel field of subsets is thus either finite or non-denumerable.

Example 1.2 Smallest Borel field

The family $\{\varnothing, X\}$ is a Borel field of subsets of the set X. It is obviously contained in all the Borel fields, so that it is the smallest Borel field of subsets of the set X.

1.1.2. Borel field generated by a mapping

(a) Let g be a mapping from a set X into a measurable space $(Y, \mathscr{B})$. The Borel field $\mathscr{A}_g$ **generated** by the mapping g in the set X is the family

$$\mathscr{A}_g = g^{-1}(\mathscr{B}) = \{g^{-1}(B) \,|\, B \in \mathscr{B}\} \tag{6.2}$$

(b) The family $\mathscr{A}_g$ is composed of the inverse images through g of the measurable sets of the set Y. It is the inverse image of the Borel field $\mathscr{B}$. Let us verify that it is a Borel field. For this we must use the properties of inverse images (cf. Eqn (6.17)). We get successively, for the three axioms (6.1),

1 $$Y \in \mathscr{B} \Rightarrow X = g^{-1}(Y) \in g^{-1}(\mathscr{B}) = \mathscr{A}_g$$

2 $$A \in \mathscr{A} \Rightarrow \exists B \in \mathscr{B} \quad A = g^{-1}(B)$$

$$\bar{B} \in \mathscr{B} \Rightarrow g^{-1}(\bar{B}) = \overline{g^{-1}(B)} = \bar{A} \in g^{-1}(\mathscr{B}) = \mathscr{A}_g$$

3 $$(\forall n \in \mathbb{N} \quad A_n \in \mathscr{A}) \Rightarrow (\forall n \in \mathbb{N} \quad \exists B_n \in \mathscr{B} \quad A_n = g^{-1}(B_n))$$

$$\bigcup_{n \in \mathbb{N}} A_n = \bigcup_{n \in \mathbb{N}} g^{-1}(B_n) = g^{-1}\left(\bigcup_{n \in \mathbb{N}} B_n\right) \in g^{-1}(\mathscr{B}) = \mathscr{A}_g$$

1.1.3. Borel field induced by a mapping

(a) Let g be a mapping from a measurable space $(X, \mathscr{A})$ into a set Y. The Borel field $\mathscr{B}_g$ **induced** by the mapping g in the set Y is the family

$$\mathscr{B}_g = \{B \subset Y \mid \exists A \in \mathscr{A} \quad A = g^{-1}(B)\} \tag{6.3}$$

(b) The family $\mathscr{B}_g$ is composed of the sets whose inverse images are the measurable sets of the set X. We can verify without difficulty, as in the preceding section, that the family $\mathscr{B}_g$ is a Borel field.

It follows from the definition that the Borel field $g^{-1}(\mathscr{B}_g)$ is contained in the Borel field $\mathscr{A}$. We say that it is a **sub-Borel field** of $\mathscr{A}$. It must be noted, in contrast, that the image $g(\mathscr{A})$ of the Borel field $\mathscr{A}$ is not in general a Borel field.

1.1.4. Borel field generated by a family of subsets

1.1.4.1. Definition

Let X be a set and F a family of subsets of X. The Borel field generated by the family F is the Borel field $\tilde{F}$ such that

(1) the Borel field $\tilde{F}$ contains the family F
(2) every Borel field containing the family F contains the Borel field $\tilde{F}$ (6.4)

This is equivalent to saying that the Borel field $\tilde{F}$ is the smallest Borel field containing the family F.

1.1.4.2. Existence and uniqueness

We must verify that the Borel field $\tilde{F}$ defined by the axioms (6.4) exists and is unique.

Let $\mathscr{A}_1$ and $\mathscr{A}_2$ be two Borel fields satisfying the axioms (6.4). It is immediately obvious that $\mathscr{A}_1 \subset \mathscr{A}_2$ and $\mathscr{A}_2 \subset \mathscr{A}_1$ which implies that $\mathscr{A}_1 = \mathscr{A}_2$. Uniqueness is therefore established.

Next let $(\mathscr{A}_i, i \in I)$ be the family of Borel fields containing the family F. It is not empty since it contains the Borel field of subsets $\mathscr{P}(X)$. Consider the family

$$\bigcap_{i \in I} \mathscr{A}_i$$

It is easy to show that it is a Borel field of subsets of the set X which satisfies the axioms (6.4) (cf. Exercise 1.4). It is therefore the Borel field $\tilde{F}$.

The Borel field $\tilde{F}$ is thus the intersection of the Borel fields containing the family F. It is in fact the smallest of the Borel fields containing the family F.

Example 1.3 Borel field generated by a subset

Let X be a set and A a subset of X, $A \subset X$. It follows immediately that the smallest Borel field containing the subset A is the family $\{A, \bar{A}, X, \emptyset\}$. This is the Borel field generated by the subset A. It is also the Borel field generated by the complementary subset $\bar{A}$.

1.1.4.3. Property

Let X be a set and F_1 and F_2 be two families of subsets of X. Then

$$F_1 \subset F_2 \Rightarrow \tilde{F}_1 \subset \tilde{F}_2 \tag{6.5}$$

The verification of this property is immediate.

1.1.4.4. Definition

A Borel field is **separable** when it can be generated by a denumerable family.

1.1.4.5. Property

Let $g: X \to Y$ be a mapping, F_Y a family of subsets of the set Y and $F_X = g^{-1}(F_Y)$ the inverse image of the family F_Y. The Borel fields generated by the families F_X and F_Y satisfy

$$\tilde{F}_X = \tilde{g}^{-1}(F_Y) = g^{-1}(\tilde{F}_Y) \tag{6.6}$$

This property is very useful, as we shall see later. Let us prove it. The family $g^{-1}(\tilde{F}_Y)$ is the Borel field generated in X by the mapping g starting from the Borel field $\tilde{F}_Y$ (cf. Eqn (6.2)). It then follows from the axioms (6.4) that

$$F_Y \subset \tilde{F}_Y \Rightarrow g^{-1}(F_Y) \subset g^{-1}(\tilde{F}_Y) \Rightarrow \tilde{F}_X \subset g^{-1}(\tilde{F}_Y)$$

Conversely, consider the Borel field $\mathscr{B}_g$ induced in Y by the mapping g starting from the Borel field $\tilde{F}_X$. Then $g^{-1}(\mathscr{B}_g) \subset \tilde{F}_X$ (cf. Section 1.1.3). Next, Eqn (6.3) enables us to write

$$g^{-1}(F_Y) = F_X \subset \tilde{F}_X \Rightarrow F_Y \subset \mathscr{B}_g \Rightarrow \tilde{F}_Y \subset \mathscr{B}_g \Rightarrow g^{-1}(\tilde{F}_Y) \subset g^{-1}(\mathscr{B}_g)$$

Hence $g^{-1}(\tilde{F}_Y) \subset \tilde{F}_X$. The equality (6.6) is thus proved.

1.1.4.6. Generated Borel field

Let X be a topological space with a family of open sets $\mathcal{O}$. The family $\mathcal{O}$ is not a Borel field since the complement of an open set is not an open set and thus axiom 2 is not satisfied.

The Borel field $\tilde{\mathcal{O}}$ generated by the family $\mathcal{O}$ is the **generated Borel field** of the set X and its elements are the **Borel sets** of the set X.

Example 1.4 Generated Borel field of $\mathbb{R}$

Consider the set $\mathbb{R}$ with its usual topology $\mathcal{O}$. The generated Borel field associated with this topology is denoted by $\mathcal{R}$. The family of Borel sets is extremely rich. In practice, we generally take as an example of a Borel set an arbitrary interval, open, closed or half-open. It must be noted that the singletons are Borel sets. They are the atoms of the Borel field $\mathcal{R}$ (cf. Exercise 1.10).

Consider the family $\mathcal{D}$ of open half-lines $]-\infty, a[, a \in \mathbb{R}$; let us show that it generates the Borel field $\mathcal{R}$. It follows immediately from Eqn (6.5) that

$$\mathcal{D} \subset \mathcal{O} \Rightarrow \tilde{\mathcal{D}} \subset \tilde{\mathcal{O}} = \mathcal{R}$$

We then write

$$\left]-\infty, a+\frac{1}{n}\right[\in \tilde{\mathcal{D}} \Rightarrow]-\infty, a] = \bigcap_{n \in \mathbb{N}} \left(\left]-\infty, a+\frac{1}{n}\right[\right) \in \tilde{\mathcal{D}}$$

It follows from this that

$$]a, b[= (]-\infty, b[) \cap \overline{(]-\infty, a])} \in \tilde{\mathcal{D}}$$

Every open set of $\mathcal{O}$, being a denumerable union of open intervals, thus belongs to the Borel field $\tilde{\mathcal{D}}$. This enables us to write

$$\mathcal{O} \subset \tilde{\mathcal{D}} \Rightarrow \tilde{\mathcal{O}} = \mathcal{R} \subset \tilde{\mathcal{D}}$$

The result stated is thus proved: $\mathcal{R} = \tilde{\mathcal{D}}$. There are other ways of generating the Borel field $\mathcal{R}$ (cf. Exercise 1.6).

Finally, let $g: X \to (\mathbb{R}, \mathcal{R})$ be a real function. It follows from Eqn (6.6) that

$$g^{-1}(\mathcal{R}) = g^{-1}(\tilde{\mathcal{D}}) = \tilde{g}^{-1}(\mathcal{D}) \tag{6.7}$$

The Borel field $\mathcal{A}_g$ generated by the mapping g is thus the Borel field generated by the family $g^{-1}(\mathcal{D})$.

1.2. MEASURABLE PRODUCT SPACE

1.2.1. Borel field generated by projections

Let $(X_i, \mathcal{A}_i)$ be a family of measurable spaces, indexed by an arbitrary set I, and consider the product set

$$\prod_{i \in I} X_i$$

When $X_i = X$ for every $i \in I$, the product set is denoted by X^I. This is the function set composed of all the functions defined over the indexation space I, with values in X.

Let φ_i be the projection

$$\prod_{i \in I} X_i \to X_i$$

This is the mapping defined by

$$(\ldots, x_h, x_i, x_j, \ldots) \mapsto x_i$$

The inverse image $\varphi_i^{-1}(\mathscr{A}_i)$ of the Borel field $\mathscr{A}_i$ is the Borel field generated in the product space by the mapping φ_i. It is composed of the sets

$$\varphi_i^{-1}(A_i) = \ldots \times X_h \times A_i \times X_j \times \ldots \qquad A_i \in \mathscr{A}_i$$

The family

$$F = \bigcup_{i \in I} \varphi_i^{-1}(\mathscr{A}_i)$$

is not a Borel field of the product set (cf. Exercise 1.4), but it generates a Borel field $\tilde{F}$ which is called the product Borel field.

1.2.2. Definition

The **product Borel field** of the product set

$$\prod_{i \in I} X_i$$

is the Borel field generated by the family F. It is denoted by $\bigotimes_I \mathscr{A}_i$:

$$F = \bigcup_{i \in I} \varphi_i^{-1}(\mathscr{A}_i) \qquad \bigotimes_I \mathscr{A}_i = \tilde{F} \tag{6.8}$$

It is the smallest Borel field containing the Borel fields $\varphi_i^{-1}(\mathscr{A}_i)$ (cf. Section 1.1.4). The pair

$$\prod_{i \in I} X_i, \bigotimes_I \mathscr{A}_i$$

is a measurable product space.

It should be noted that the product set

$$\prod_{i \in I} X_i$$

can be provided with Borel fields other than the product Borel field.

1.2.3. Property

A set
$$C \subset \prod_{i \in I} X_i$$
is a **measurable block** if it is of the form
$$C = \prod_{i \in I} A_i \qquad A_i \in \mathcal{A}_i$$
with $A_i = X_i$ for all the indices not belonging to a finite set $J \subset I$.

The family $\mathcal{M}$ of measurable blocks is not an algebra since axiom 2 (cf. axioms (6.1)) is not satisfied. In fact, the complement of a measurable block is not a measurable block, since we have, for example,
$$\overline{(\ldots X_h \times A_i \times A_j \times X_k \ldots)} = (\ldots X_h \times \bar{A}_i \times X_j \ldots) \cup (\ldots X_i \times \bar{A}_j \times X_h \ldots)$$
However, the set $\mathscr{C}$ of finite unions of disjoint measurable blocks is an algebra. It can be shown that the family $\mathcal{M}$ and the algebra $\mathscr{C}$ both generate the product Borel field
$$\bigotimes_I \mathcal{A}_i$$
(Neveu, 1964). It follows from this that
$$\bigotimes_I \mathcal{A}_i = \tilde{F} = \tilde{\mathcal{M}} = \tilde{\mathscr{C}} \qquad (6.9)$$

This property enables us to characterize the product Borel field in a way which is more useful than by using the family F. In particular it enables us to define the product measure by applying the extension theorem (cf. Section 3.2.1).

1.2.4. Property

Let
$$\left(\prod_{i \in I} X_i, \bigotimes_I \mathcal{A}_i\right) \quad \text{and} \quad \left(\prod_{j \in J} X_j, \bigotimes_J \mathcal{A}_j\right)$$
be measurable product spaces. Their product is the measurable space
$$\left(\prod_{k \in K} X_k, \bigotimes_K \mathcal{A}_k\right)$$
with $K = I \cup J$.

This follows from the proposition proved in Section 2.2.2 (cf. Example 2.2).

Exercises

Exercise 1.1 Upper limit and lower limit of a sequence of sets

Let X be an arbitrary set. Remember that inclusion defines a partial order relation on the set $\mathscr{P}(X)$ of subsets of X. Now let $(A_n, n \in \mathbb{N})$ be a sequence of subsets of X.

The **upper limit** of the sequence, when n tends to infinity, is defined by

$$\limsup A_n = \bigcap_{n \in \mathbb{N}} \bigcup_{k \geq n} A_k = \inf_{n \in \mathbb{N}} \sup_{k \geq n} A_k \qquad (6.10)$$

while the **lower limit**, when n tends to infinity, is defined by

$$\liminf A_n = \bigcup_{n \in \mathbb{N}} \bigcap_{k \geq n} A_k = \sup_{n \in \mathbb{N}} \inf_{k \geq n} A_k \qquad (6.11)$$

The sequence (A_n) has by definition a limit A when

$$\limsup A_n = \liminf A_n = A \qquad (6.12)$$

(1) Verify that the set $\limsup A_n$ contains the points of X which belong to an infinity of sets A_n and that the set $\liminf A_n$ contains the points of X which belong to all the sets A_n except a finite number of them. Hence deduce that

$$\liminf A_n \subset \limsup A_n \qquad (6.13)$$

(2) Prove that for a sequence (A_n) to have a limit it is necessary and sufficient that there exists no point belonging simultaneously to an infinity of sets A_n and an infinity of sets $\bar{A}_n$.

(3) Prove that

$$\overline{\liminf A_n} \subset \limsup \bar{A}_n \qquad (6.14)$$

(4) Suppose now that the sequence (A_n) is increasing (with respect to inclusion). Show that it has as limit

$$\lim_{n \to \infty} A_n = \bigcup_{n \in \mathbb{N}} A_n = \sup_{n \in \mathbb{N}} A_n$$

Then show that a decreasing sequence (A_n) has as limit

$$\lim_{n \to \infty} A_n = \bigcap_{n \in \mathbb{N}} A_n = \inf_{n \in \mathbb{N}} A_n$$

(Apply Eqn (6.12).)

Exercise 1.2 Closure under increasing denumerable union

Show that closure under increasing denumerable union is equivalent to closure under denumerable union (cf. axiom 3 of Section 1.1.1.1). This implies that the limit of an increasing sequence of measurable sets is a measurable set (cf. Exercise 1.1).

Exercise 1.3 Trace Borel field
Let $(X, \mathscr{A})$ be a measurable space and B be a non-empty subset of X. Show that the family

$$\mathscr{A}_B = \{A \cap B \mid A \in \mathscr{A}\} \tag{6.15}$$

is a Borel field of subsets of the set B. This is the **trace** of the Borel field $\mathscr{A}$ on the set B.

Exercise 1.4 Borel field generated by a family
Let X be an arbitrary set and $\mathscr{A}_1$ and $\mathscr{A}_2$ be two Borel fields of subsets of X.
(1) Verify that the family $\mathscr{A}_1 \cup \mathscr{A}_2 = \{A \mid A \in \mathscr{A}_1 \vee A \in \mathscr{A}_2\}$ is not necessarily either a Borel field or an algebra.
(2) Verify that the family $\mathscr{A}_1 \cap \mathscr{A}_2 = \{A \mid A \in \mathscr{A}_1 \wedge A \in \mathscr{A}_2\}$ is a Borel field and, more generally, that the intersection of a denumerable family of Borel fields is a Borel field.
(3) Prove that the Borel field generated by the family $\mathscr{A}_1 \cup \mathscr{A}_2$ is identical with that generated by the family

$$\{A_1 \cap A_2 \mid A_1 \in \mathscr{A}_1, A_2 \in \mathscr{A}_2\}$$

as well as that generated by the family

$$\{A_1 \cup A_2 \mid A_1 \in \mathscr{A}_1, A_2 \in \mathscr{A}_2\}$$

(Apply Eqn (6.5).)

Exercise 1.5 Borel field generated by a partition
Let X be an arbitrary set and $F = (A_i, i \in I)$ a family of subsets of X.
(1) Consider the family B defined by

$$B = \left\{ \bigcap_{i \in I} \hat{A}_i \,\Big|\, \hat{A}_i = A_i \text{ or } \bar{A}_i \right\}$$

Prove that it is a partition of X. Put $B = (B_k, k \in K)$, K being the indexation set. Verify that the partition is finite if I is finite and is non-denumerable in the opposite case.
(2) Let $\tilde{F}$ and $\tilde{B}$ be the Borel fields generated by the families F and B respectively. Prove that we always have $\tilde{B} \subset \tilde{F}$ and that we have $\tilde{B} = \tilde{F}$ if the set I is finite.

Exercise 1.6 Generated Borel field of $\mathbb{R}$
Let the set $X = \mathbb{R}$ and let F be the family of subsets of $\mathbb{R}$ consisting of the half-lines $]-\infty, a[$, the half-lines $[b, +\infty[$ and the intervals $[a, b[$ with $a < b$.
(1) Verify that the family F is not an algebra.
(2) Let $\mathscr{C}$ be the family consisting of finite unions of pairwise disjoint elements of F. Verify that the family $\mathscr{C}$ is an algebra.
(3) Prove that the families F and $\mathscr{C}$ generate the generated Borel field $\mathscr{R}$, which is written $\tilde{F} = \tilde{\mathscr{C}} = \mathscr{R}$. (Use Example 1.4 and Eqn (6.5).)

Exercise 1.7 Borel field generated by a denumerable partition

Let X be a set and $F = (X_n, n \in \mathbb{N})$ be a denumerable family of subsets constituting a denumerable partition of the set X. Consider also the family

$$\mathscr{A} = \left\{ \bigcup_{i \in I} X_i \,\middle|\, I \subset \mathbb{N} \right\}$$

composed of the denumerable unions of elements of the family F.

(1) Verify that the family $\mathscr{A}$ is a Borel field of subsets of the set X. (Verify the axioms (6.1) by applying Eqns (6.17).)

(2) Prove that the Borel field $\mathscr{A}$ is identical with the Borel field $\tilde{F}$ generated by the family F. Every element of the Borel field $\tilde{F}$ is thus a denumerable union of elements of the family F. (Apply axioms (6.1) and (6.4) to verify that $\mathscr{A} \subset \tilde{F}$ and $\tilde{F} \subset \mathscr{A}$.)

Exercise 1.8 Borel field generated by a discrete mapping

Let g be a discrete mapping $g: X \to (Y, \mathscr{B})$ taking a denumerable set of values $y_i, i \in \mathbb{N}$, and let X_i be the sets $X_i = g^{-1}(\{y_i\})$.

(1) Verify that the family $F = (X_i, i \in \mathbb{N})$ is a partition of the set X.

(2) Suppose that the singletons belong to the Borel field $\mathscr{B}$. Prove that the Borel field $\mathscr{A}_g$ generated by the mapping g is the family

$$\mathscr{A}_g = \left\{ \bigcup_{i \in I} X_i \,\middle|\, I \subset \mathbb{N} \right\}$$

The Borel field $\mathscr{A}_g$ is therefore composed of the denumerable unions of elements of the partition F. It is identical with the Borel field $\tilde{F}$ generated by the family F. (Use the results of Exercise 1.7 and apply Eqn (6.2).)

Exercise 1.9 Section in a product space

Let $(X_1 \times X_2, \mathscr{A}_1 \otimes \mathscr{A}_2)$ be a measurable product space. The section by $x_1 \in X_1$ of a subset $A \subset X_1 \times X_2$ is the set

$$A_{x_1} = \{x_2 \,|\, x_2 \in X_2, (x_1, x_2) \in A\}$$

Show that the set $\mathscr{B}_{x_1}$ of subsets of $X_1 \times X_2$, whose section by x_1 is a measurable set of X_2, is a Borel field. Hence deduce the inclusion $\mathscr{A}_1 \otimes \mathscr{A}_2 \subset \mathscr{B}_{x_1}$. It follows from this that the section of a measurable subset is measurable, but the converse is false. (Apply axioms (6.4).)

Exercise 1.10 Atoms of a Borel field

Let $\mathscr{A}$ be a Borel field of subsets of a set X. A non-empty set $A \in \mathscr{A}$ is an **atom** of the Borel field $\mathscr{A}$ if it is such that

$$\forall B \in \mathscr{A} \qquad A \cap B = \emptyset \text{ or } A$$

(1) Suppose that the Borel field $\mathscr{A}$ is arbitrary, not necessarily denumerable.

(1.1) Show that two distinct atoms are disjoint.

(1.2) Let A be an atom. Show that, if B and C are two non-empty measurable sets such that $A = B \cup C$, then $A = B = C$, which means that an atom cannot be divided into two measurable parts (it is indivisible).

(2) Suppose now that the Borel field $\mathscr{A}$ is denumerable and such that $\mathscr{A} = (A_n, n \in \mathbb{N})$.

(2.1) Let $x \in X$ be an arbitrary point and consider the sets $I = \{n \in \mathbb{N} \mid x \in A_n\}$ and $J = \{n \in \mathbb{N} \mid x \in \bar{A}_n\}$. Show that the set

$$A_x = \left(\bigcap_{n \in I} A_n\right) \cap \left(\bigcap_{n \in J} \bar{A}_n\right)$$

is an atom of the Borel field $\mathscr{A}$.

(2.2) Let $\mathscr{A}_0$ be the family of atoms of the Borel field $\mathscr{A}$. Prove that it is a partition of the set X. Hence deduce that every element $A \in \mathscr{A}$ is a union of atoms and that the family $\mathscr{A}_0$ generates the Borel field $\mathscr{A}$: $\tilde{\mathscr{A}}_0 = \mathscr{A}$.

Exercise 1.11 Borel field generated by a mapping into a product space
Let X be a set,

$$\left(\prod_{i \in I} Y_i, \bigotimes_I \mathscr{B}_i\right)$$

a measurable product space, and g a mapping from the set X into the product space. Let φ_i be the projection

$$\varphi_i : \prod_{i \in I} Y_i \to Y_i$$

Finally, let $\mathscr{B}_g$ be the Borel field generated in X by the mapping g and $\mathscr{B}_{g_i}$ the Borel field generated in X by the mapping $g_i = \varphi_i \circ g$. Prove that

$$\mathscr{B}_g = \widetilde{\bigcup_{i \in I} \mathscr{B}_{g_i}}$$

The Borel field generated by the mapping g is thus the smallest Borel field containing the Borel fields generated by the mappings g_i. (Use Eqns (6.8) and (6.6).)

2. Measurable functions

2.1. DEFINITION

Let $(X, \mathscr{A})$ and $(Y, \mathscr{B})$ be two measurable spaces and g be a mapping $g: X \to Y$. The mapping g is **measurable** when the Borel field $\mathscr{A}_g$ which it generates is a sub-Borel field of $\mathscr{A}$:

$$\mathscr{A}_g = g^{-1}(\mathscr{B}) \subset \mathscr{A} \tag{6.16}$$

It is useful to recall the following properties of inverse images:

$$g\{g^{-1}(B)\} \subset B \qquad g^{-1}\{g(A)\} \supset A \qquad (6.17)$$

$$g^{-1}\left(\bigcup_{i \in I} B_i\right) = \bigcup_{i \in I} g^{-1}(B_i) \qquad g^{-1}\left(\bigcap_{i \in I} B_i\right) = \bigcap_{i \in I} g^{-1}(B_i)$$

It must be noted that the mapping g is always measurable when $\mathscr{A}$ is the Borel field of subsets $\mathscr{P}(X)$ since the latter is the largest Borel field of the set X (cf. Example 1.1).

2.2. PROPERTIES

2.2.1. *Composition of measurable mappings*

2.2.1.1. Statement
Let g and h be measurable mappings:

$$(X, \mathscr{A}) \xrightarrow{g} (Y, \mathscr{B}) \xrightarrow{h} (Z, \mathscr{C})$$

The composite mapping $h \circ g : (X, \mathscr{A}) \to (Z, \mathscr{C})$ is measurable.

2.2.1.2. Proof
The proof follows immediately from the fact that

$$\forall C \in \mathscr{C} \qquad (h \circ g)^{-1}(C) = g^{-1}\{h^{-1}(C)\} \in \mathscr{A}$$

2.2.2. *Measurable mapping into a product space*

2.2.2.1. Statement
Let g be a mapping

$$g : (X, \mathscr{A}) \to \left(\prod_{i \in I} X_i, \bigotimes_I \mathscr{A}_i\right)$$

It is measurable if and only if the composite mappings $\varphi_i \circ g$ are measurable, φ_i being the projection

$$\prod_{i \in I} X_i \to X_i$$

2.2.2.2. Proof
Suppose that the mapping g is measurable. By definition

$$g^{-1}\left(\bigotimes_I \mathscr{A}_i\right) \subset \mathscr{A}$$

Now, the inverse image through φ_i of a measurable set $A_i \in \mathscr{A}_i$ is the measurable block

$$\varphi_i^{-1}(A_i) = \ldots X_h \times A_i \times X_j \ldots$$

It belongs to the product Borel field $\underset{I}{\otimes} \mathscr{A}_i$ and consequently

$$\forall A_i \in \mathscr{A}_i \quad (\varphi_i \circ g)^{-1}(A_i) = g^{-1}\{\varphi_i^{-1}(A_i)\} \in \mathscr{A}$$

The mapping $\varphi_i \circ g$ is thus measurable for all $i \in I$.

Conversely, suppose that the mappings $\varphi_i \circ g$ are measurable for all $i \in I$ and consider the set $\mathscr{M}$ of measurable blocks. It generates the product Borel field (cf. Eqn (6.9)) and it is always possible to write for an arbitrary block C, J being a finite subset of I,

$$C = \bigcap_{i \in J}(\ldots X_h \times A_i \times X_j \ldots) = \bigcap_{i \in J} \varphi_i^{-1}(A_i)$$

Next,

$$g^{-1}(C) = g^{-1}\left\{\bigcap_{i \in J} \varphi_i^{-1}(A_i)\right\} = \bigcap_{i \in J}(\varphi_i \circ g)^{-1}(A_i) \in \mathscr{A}$$

which implies that $g^{-1}(\mathscr{M}) \subset \mathscr{A}$. We get from this, since

$$\tilde{\mathscr{M}} = \underset{I}{\otimes} \mathscr{A}_i$$

and by using Eqn (6.6), that

$$\tilde{g}^{-1}(\mathscr{M}) = g^{-1}(\tilde{\mathscr{M}}) = g^{-1}\left(\underset{I}{\otimes} \mathscr{A}_i\right) \subset \mathscr{A}$$

The mapping g is therefore measurable.

Example 2.1 Projection mappings
The identity mapping I_d from the space

$$\left(\prod_{i \in I} X_i, \underset{I}{\otimes} \mathscr{A}_i\right)$$

onto itself is obviously measurable. It follows from the above that the projections

$$\varphi_i: \prod_{i \in I} X_i \to X_i$$

are measurable since they are such that $\varphi_i = \varphi_i \circ I_d$.

Example 2.2 Measurable product spaces
Consider the measurable product spaces

$$\left(\prod_{i \in I} X_i, \underset{I}{\otimes} \mathscr{A}_i\right) \quad \text{and} \quad \left(\prod_{j \in J} X_j, \underset{J}{\otimes} \mathscr{A}_j\right)$$

Their product is the measurable space

$$\left(\prod_{k \in K} X_k, \underset{K}{\otimes} \mathscr{A}_k\right)$$

with $K = I \cup J$.

In fact, the Borel field of the product space

$$\prod_{k \in K} X_k$$

is the smallest Borel field with respect to which the projections

$$g: \prod_{k \in K} X_k \to \prod_{i \in I} X_i \quad \text{and} \quad h: \prod_{k \in K} X_k \to \prod_{j \in J} X_j$$

are measurable. It is also the smallest Borel field which makes all the mappings $\varphi_i \circ g$ and $\varphi_j \circ h$ measurable, i.e. all the projections φ_k when $K = I \cup J$.

2.2.3. Mapping measurable with respect to a Borel field $\tilde{F}$

2.2.3.1. Statement
Let g be a mapping $g:(X, \mathscr{A}) \to (Y, \tilde{F})$ (cf. Section 1.1.4). It is measurable if and only if $g^{-1}(F) \subset \mathscr{A}$.

2.2.3.2. Proof
Suppose that the mapping g is measurable. By definition $g^{-1}(\tilde{F}) \subset \mathscr{A}$, which implies that $g^{-1}(F) \subset \mathscr{A}$ since $F \subset \tilde{F}$. Conversely, if $g^{-1}(F) \subset \mathscr{A}$, the family F is contained in the Borel field induced by g in Y (cf. Eqn (6.3)). This is also the case for the Borel field $\tilde{F}$ which is the smallest Borel field containing the family F. It follows from this that $g^{-1}(\tilde{F}) \subset \mathscr{A}$, which implies that the mapping g is measurable.

Example 2.3 Borel function
Let X and Y be two topological spaces, with their generated Borel fields $\mathcal{O}_X$ and $\mathcal{O}_Y$. A mapping $g: X \to Y$ is by definition continuous when $g^{-1}(\mathcal{O}_Y) \subset \mathcal{O}_X$. It follows from this, by application of Eqns (6.6) and (6.5), that

$$\tilde{g}^{-1}(\mathcal{O}_Y) = g^{-1}(\tilde{\mathcal{O}}_Y) \subset \tilde{\mathcal{O}}_X$$

The mapping g is thus measurable and consequently every continuous mapping is measurable with respect to the generated Borel fields. It is called **borelian** for this reason.

The mapping $y \mapsto y^n$ from the space $(Y, \mathcal{O}_Y)$ into the space $(Y, \mathcal{O}_Y)$ is measurable since it is continuous. It follows from this that, if the mapping $g:(X, \mathcal{A}) \to (Y, \mathcal{O}_Y)$ is measurable, the mapping $g^n:(X, \mathcal{A}) \to (Y, \mathcal{O}_Y)$ is also measurable.

Example 2.4 Measurable indicator function

Let $(X, \mathcal{A})$ and $(\mathbb{R}, \mathcal{R})$ be measurable sets and A be a set $A \subset X$. The indicator function I_A of the set A is the mapping from X into $\mathbb{R}$ defined by

$$I_A(a) = 1 \qquad a \in A$$
$$I_A(a) = 0 \qquad a \in \bar{A}$$

It follows from this definition that

$$I_A^{-1}(]-\infty, x[) = \varnothing \qquad x \leqslant 0$$
$$I_A^{-1}(]-\infty, x[) = \bar{A} \qquad 0 < x \leqslant 1$$
$$I_A^{-1}(]-\infty, x[) = X \qquad 1 < x$$

The Borel field generated by the function I_A is therefore the family (cf. Eqn (6.7))

$$I_A^{-1}(\mathcal{R}) = I_A^{-1}(\tilde{\mathcal{D}}) = \{\varnothing, X, A, \bar{A}\} \tag{6.18}$$

It is thus the Borel field generated by the set A. It follows from this that the indicator function I_A is measurable if and only if the set A is measurable.

2.2.4. Sum and product of measurable mappings

2.2.4.1. Statement

Consider two measurable mappings g_1 and $g_2 : (X, \mathcal{A}) \to (\mathbb{R}, \mathcal{R})$. The mappings $g_1 + g_2$ and $g_1 g_2 : (X, \mathcal{A}) \to (\mathbb{R}, \mathcal{R})$ are measurable.

2.2.4.2. Proof

In fact let the mapping $g:(X, \mathcal{A}) \to (\mathbb{R}^2, \mathcal{R}^2)$ be such that $g(x) = (g_1(x), g_2(x))$. It is measurable, according to property 2.2.2, since $\varphi_i \circ g = g_i$. In addition, the mappings $(x_1, x_2) \mapsto x_1 + x_2, x_1 x_2$, from $(\mathbb{R}^2, \mathcal{R}^2)$ into $(\mathbb{R}, \mathcal{R})$, are measurable because they are continuous (cf. Example 2.3). The mappings $g_1 + g_2$ and $g_1 g_2$ are thus measurable since they are compositions of measurable mappings (cf. Section 2.2.1).

It follows immediately, in addition, that the mapping $ag_1, a \in \mathbb{R}$, is measurable.

2.2.5. Limit of a sequence of measurable functions

Theorem: Let $(g_n, n \in \mathbb{N})$ be a sequence of real measurable functions, converging simply to a real function g. The function g is measurable.

We shall accept this theorem (Bertrandias, 1969, p. 91).

Exercises

Exercise 2.1 Measurable functions
Let $(X, \mathscr{A})$ and $(Y, \mathscr{B})$ be two measurable spaces.
(1) Show that a constant mapping $(X, \mathscr{A}) \to (Y, \mathscr{B})$ and the identity mapping $I_d : (X, \mathscr{A}) \to (X, \mathscr{A})$ are both measurable.
(2) Hence deduce, by using property 2.2.2, that the mapping $h_{x_1} : (X_2, \mathscr{A}_2) \to (X_1 \times X_2, \mathscr{A}_1 \otimes \mathscr{A}_2)$ defined by
$$\forall x_1 \in X_1 \quad \forall x_2 \in X_2 \quad h_{x_1}(x_2) = (x_1, x_2)$$
is measurable.
(3) Hence deduce that for every measurable set $A \in \mathscr{A}_1 \otimes \mathscr{A}_2$ the set $A_{x_1} \subset X_2$ defined by $A_{x_1} = \{x_2 \mid (x_1, x_2) \in A\}$ is measurable.
(Note that $I_{A_{x_1}} = I_A \circ h_{x_1}$.)

Exercise 2.2 Measurable functions
Let g_1 and g_2 be two real measurable functions defined over the space $(X, \mathscr{A})$. Prove that the real functions $\sup(g_1, g_2)$ and $\inf(g_1, g_2)$ are measurable. Hence deduce that, if the function g is measurable, so is the function $|g|$ (cf. Section 4.1.3.1). Note that the converse is false.

Exercise 2.3 Borel field generated by a mapping
Let $(X, \mathscr{A})$ be a measurable space,
$$\left(\prod_{i \in I} X_i, \bigotimes_I \mathscr{A}_i \right)$$
the product space, and g a mapping from X into the product space such that
$$g(x) = \{g_1(x), \ldots, g_n(x)\}$$
the mappings $g_i : X \to X_i$ being measurable. Prove that the Borel field generated by g in X is identical with that generated by the family F:
$$F = \left\{ \bigcap_{i=1}^n B_i \mid B_i \in \mathscr{B}_i \right\}$$
where $\mathscr{B}_i$ is the Borel field generated in X by the mapping g_i. (Use Eqn (6.6).)

Exercise 2.4 Composite mappings
Consider the measurable mappings
$$(X, \mathscr{A}) \xrightarrow{g} (\mathbb{R}, \mathscr{R}) \xrightarrow{h} (\mathbb{R}, \mathscr{R})$$

(1) Prove that the Borel fields $\mathscr{A}_g$ and $\mathscr{A}_{h \circ g}$ generated by the mappings g and $h \circ g$ respectively satisfy
$$\mathscr{A}_{h \circ g} \subset \mathscr{A}_g \subset \mathscr{A} \tag{6.19}$$

(2) Prove that

$$\widetilde{\bigcup_h} \mathscr{A}_{h \circ g} = \mathscr{A}_g$$

the union being taken over the set h of all borelian mappings. (Note that the identity mapping is borelian.)

Exercise 2.5 Measurable product set
Let

$$\left(\prod_{i \in I} X_i, \bigotimes_I \mathscr{A}_i \right)$$

be a measurable product set, $J \subset I$ a subset, and $\mathscr{B}$ the set

$$\mathscr{B} = \left\{ A \times \prod_{i \in I - J} X_i \, \Big| \, A \in \bigotimes_J \mathscr{A}_i \right\}$$

Show that the family $\mathscr{B}$ is a sub-Borel field of the measurable product space

$$\left(\prod_{i \in I} X_i, \bigotimes_I \mathscr{A}_i \right)$$

(Use proposition 2.2.2.)

Exercise 2.6 Measurable mappings and atoms
Let $(X, \mathscr{A})$ and $(Y, \mathscr{B})$ be two measurable spaces and g be a mapping $g: X \to Y$.
(1) Suppose that the mapping g is arbitrary, not necessarily measurable.

(1.1) Prove that, when it is not empty, the inverse image of an atom of the Borel field $\mathscr{B}$ is an atom of the Borel field $\mathscr{A}_g = g^{-1}(\mathscr{B})$ generated by the mapping g. (Use Exercise 1.10.)
(1.2) Show that, if the Borel field $\mathscr{B}$ is denumerable, every atom of the Borel field $g^{-1}(\mathscr{B})$ is the non-empty inverse image of an atom of the Borel field $\mathscr{B}$. (Use Exercise 1.10, Question (2.2), and Eqns (6.17).)

(2) Suppose that the Borel field $\mathscr{B}$ contains the singletons of the set Y. Prove that, if the mapping g is measurable, it takes a constant value over each atom of the Borel field $\mathscr{A}$. (Take an atom $a \in \mathscr{A}$, a point $y_a \in g(a)$, and show that $g(a) = \{y_a\}$.)
Prove that the converse is true if we suppose in addition that the Borel field $\mathscr{A}$ is denumerable.

3. Positive measure

3.1. SIMPLE MEASURE

3.1.1. Definitions

A **positive measure** μ defined over a measurable space $(X, \mathscr{A})$ is a **denumerably additive** or a **σ additive** mapping $\mu: \mathscr{A} \to \bar{\mathbb{R}}_+$.

The mapping μ is denumerably additive when it satisfies the following relation for every sequence $(A_n, n \in \mathbb{N})$ of measurable, pairwise disjoint, sets:

$$\mu\left(\bigcup_{n \in \mathbb{N}} A_n\right) = \sum_{n \in \mathbb{N}} \mu(A_n) \tag{6.20}$$

The positive measure μ associates with every measurable set A a positive number $\mu(A)$ which is its measure through μ. This number can be equal to $+\infty$.

A positive measure μ is **bounded** if $\mu(X) < +\infty$. It is **normed** if $\mu(X) = 1$. The measure μ is borelian if X is a topological space and if $\mathscr{A}$ is its generated Borel field.

The triplet $(X, \mathscr{A}, \mu)$ is a **measured space**.

Example 3.1 Dirac measure

Let $(X, \mathscr{A})$ be a measurable space and a a point $a \in X$. The mapping δ_a from $\mathscr{A}$ into $\bar{\mathbb{R}}_+$ such that

$$\begin{aligned}\delta_a(A) &= 1 \quad a \in A \\ \delta_a(A) &= 0 \quad a \notin A\end{aligned} \tag{6.21}$$

is a positive normed measure. It is the **Dirac measure** at the point a. The measure δ_a is usually denoted by the improper expression $\delta(x - a)$ when it is used to integrate a function (cf. Eqn (6.41)). It plays an extremely important part, as we shall have occasion to see later.

Let $(\mathbb{R}, \mathscr{R})$ be a measurable space, a a point $a \in \mathbb{R}$, and $]-\infty, x[$, $x \in \mathbb{R}$, a Borel set. We get from Eqn (6.21)

$$\delta_a(]-\infty, x[) = Y(x - a)$$

Y being the **Heaviside function** defined by

$$\begin{aligned}Y(x) &= 1 \quad x > 0 \\ Y(x) &= 0 \quad x \leqslant 0\end{aligned} \tag{6.22}$$

It is the indicator function of the interval $]0, +\infty[$.

Example 3.2 Counting measure

Let $(X, \mathscr{A})$ be a measurable set. The mapping $A \mapsto \text{card } A$ from $\mathscr{A}$ into $\bar{\mathbb{R}}_+$ is a positive measure called **counting measure**. It is bounded only if the set X is finite.

3.1.2. *Extension of a measure* (Marle, 1974, p. 34)

We have defined directly a positive measure over a Borel field. This implies that we can in practice attach a measure to all the members of the Borel field. Now, in general, the elements of a Borel field are not known explicitly and it is difficult to attach a measure to them. It is easier to construct a measure over an algebra whose elements are known.

A positive measure over an algebra $\mathscr{C}$ is a σ additive mapping from $\mathscr{C}$ into $\bar{\mathbb{R}}_+$. The σ additivity must be understood to mean here that Eqn (6.20) applies to every sequence $(A_n, n \in \mathbb{N})$ of pairwise disjoint sets of $\mathscr{C}$ whose union also belongs to the algebra $\mathscr{C}$.

The following fundamental theorem enables us to extend a measure defined over an algebra to a measure defined over the Borel field generated by the algebra. We shall confine ourselves to stating the extension theorem (Marle, 1974, p. 40).

Theorem: Let X be a set, $\mathscr{C}$ an algebra of subsets of X and μ a positive measure defined over the algebra $\mathscr{C}$. There exists a unique measure $\tilde{\mu}$ defined over the Borel field $\tilde{\mathscr{C}}$ generated by $\mathscr{C}$ such that

$$\forall C \in \mathscr{C} \qquad \tilde{\mu}(C) = \mu(C)$$

The measure $\tilde{\mu}$ is the extension of the measure μ. We shall always treat the two measures as identical, denoting them by μ.

Example 3.3 Lebesgue measure over the space $(\mathbb{R}, \mathscr{R})$ (Bertrandias, 1969, p. 99)

Consider the set $\mathbb{R}$ and the algebra $\mathscr{C}$ composed of finite unions of pairwise disjoint sets of the form $]-\infty, a[, [b, +\infty[$ or $[a, b[$. It is easy to verify that the algebra $\mathscr{C}$ generates the Borel field $\mathscr{R}$ (cf. Exercise 1.6).

The **Lebesgue measure** μ_0 over the measurable set $(\mathbb{R}, \mathscr{R})$ is the extension of the measure μ_0 defined over the algebra $\mathscr{C}$ by

$$\mu_0(]-\infty, a[) = \mu_0([b, +\infty[) = +\infty$$
$$\mu_0([a, b[) = b - a \qquad b \geqslant a \tag{6.23}$$

We shall accept the σ additivity of the measure μ_0 over the algebra $\mathscr{C}$. It follows from the Borel–Lebesgue theorem (Dieudonné, 1969, p. 63).

3.1.3. *Properties of positive measures*

The object of this section is to review the principal properties of positive measures μ defined over a measurable space $(X, \mathscr{A})$. We shall suppose throughout that the measures considered are not necessarily bounded but that they are not uniformly equal to $+\infty$ over the Borel field $\mathscr{A}$.

3.1.3.1. Measure of the empty set
The measure of the empty set is zero.

In fact, according to the supposition that we have made, there exists a measurable set A such that $\mu(A) < +\infty$. It follows then from Eqn (6.20), since the sets A and $\emptyset$ are disjoint, that

$$\mu(A \cup \emptyset) = \mu(A) = \mu(A) + \mu(\emptyset)$$

Hence the result $\mu(\emptyset) = 0$.

3.1.3.2. Increasing property
Let A and B be two measurable sets. Then

$$A \subset B \Rightarrow \mu(A) \leqslant \mu(B) \qquad (6.24)$$

Consider, in fact, the measurable set $B - A = B \cap \bar{A}$. The sets A and $B - A$ are disjoint and $B = A \cup (B - A)$. It follows then from Eqn (6.20), since the measure μ is positive, that

$$\mu(B) = \mu(A) + \mu(B - A) \geqslant \mu(A)$$

3.1.3.3. Subadditivity
Let $(A_n, n \in \mathbb{N})$ be a sequence of measurable sets. Then

$$\mu\left(\bigcup_{n \in \mathbb{N}} A_n\right) \leqslant \sum_{n \in \mathbb{N}} \mu(A_n) \qquad (6.25)$$

the equality holding if and only if the sets of the sequence are pairwise disjoint. We say that the measure μ is **subadditive**.

Consider, in fact, the sequence $(B_n, n \in \mathbb{N})$ with

$$B_n = \bar{A}_0 \cap \ldots \cap \bar{A}_{n-1} \cap A_n$$

It is easy to verify that the sets B_n are pairwise disjoint and to show by recurrence that

$$\forall n \in \mathbb{N} \qquad \bigcup_{i=1}^{n} B_i = \bigcup_{i=1}^{n} A_i$$

It follows then from Eqns (6.20) and (6.24), since $B_n \subset A_n$ for every $n \in \mathbb{N}$, that

$$\mu\left(\bigcup_{n \in \mathbb{N}} A_n\right) = \mu\left(\bigcup_{n \in \mathbb{N}} B_n\right) = \sum_{n \in \mathbb{N}} \mu(B_n) \leqslant \sum_{n \in \mathbb{N}} \mu(A_n)$$

When the sets A_n are pairwise disjoint, $B_n = A_n$ for every $n \in \mathbb{N}$ and the equality follows from this.

3.1.3.4. Continuity under increasing union

Let $(A_n, n \in \mathbb{N})$ be an increasing sequence of measurable sets. Then

$$\mu\left(\bigcup_{n \in \mathbb{N}} A_n\right) = \mu\left(\lim_{n \to \infty} A_n\right) = \lim_{n \to \infty} \mu(A_n) \tag{6.26}$$

To prove this, consider the measurable sets defined by

$$B_0 = A_0 \qquad \forall n \in \mathbb{N}^* \qquad B_n = A_n - A_{n-1} = A_n \cap \bar{A}_{n-1}$$

This is illustrated in Fig. 6.1.

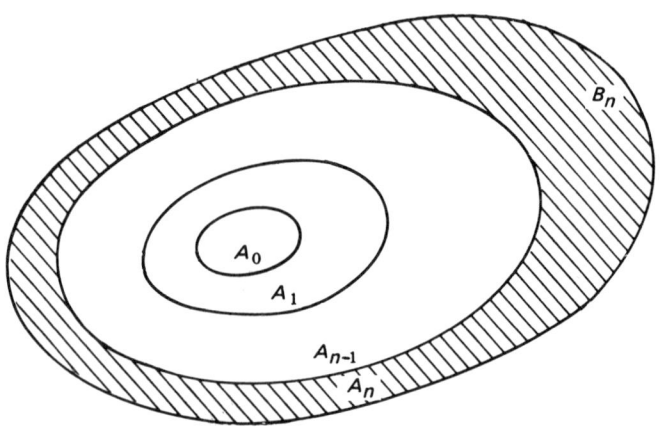

Fig. 6.1

It is immediately obvious that the sets B_n are pairwise disjoint and that

$$A_n = \bigcup_{i=0}^{n} B_i$$

The increasing sequence (A_n) has a limit (cf. Exercise 1.1) and this limit may be written

$$\lim_{n \to \infty} A_n = \bigcup_{n \in \mathbb{N}} A_n = \bigcup_{n \in \mathbb{N}} B_n$$

It follows then from Eqn (6.20) that

$$\mu\left(\bigcup_{i=0}^{n} B_i\right) = \mu(A_n) = \sum_{i=0}^{n} \mu(B_i)$$

Next we get from this, by making n tend to infinity, that

$$\mu\left(\lim_{n \to \infty} A_n\right) = \mu\left(\bigcup_{n \in \mathbb{N}} B_i\right) = \lim_{n \to \infty} \sum_{i=1}^{n} \mu(B_i) = \lim_{n \to \infty} \mu(A_n)$$

The property is thus proved.

Positive measure

3.1.3.5. Continuity under decreasing intersection

Let $(A_n, n \in \mathbb{N})$ be a decreasing sequence of measurable sets with $\mu(A_0) < +\infty$. Then

$$\mu\left(\bigcap_{n \in \mathbb{N}} A_n\right) = \mu\left(\lim_{n \to \infty} A_n\right) = \lim_{n \to \infty} \mu(A_n) \qquad (6.27)$$

To prove this, consider the measurable sets defined by

$$\forall n \in \mathbb{N}^* \qquad B_n = A_0 - A_n = A_0 \cap \bar{A}_n$$

This is illustrated in Fig. 6.2.

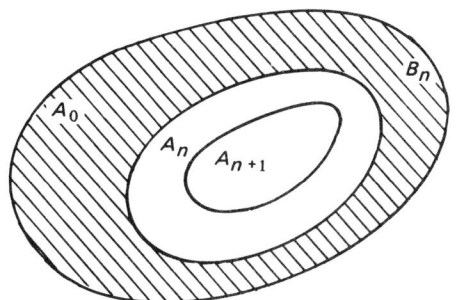

Fig. 6.2

The sequence (B_n) is obviously increasing and

$$\forall n \in \mathbb{N}^* \qquad A_0 = A_n \cup B_n = \left(\bigcap_{i=0}^{n} A_i\right) \cup \left(\bigcup_{i=1}^{n} B_i\right) \qquad A_n \cap B_n = \emptyset$$

It follows, by applying Eqn (6.20), that

$$\mu(A_0) = \mu(A_n) + \mu(B_n) = \mu\left(\bigcup_{i=1}^{n} B_i\right) + \mu\left(\bigcap_{i=1}^{n} A_i\right) < +\infty$$

The result stated is obtained by making n tend to infinity and applying Eqn (6.26).

3.1.4. Image of a measure

Let $(X, \mathscr{A}, \mu)$ be a measured space, $(Y, \mathscr{B})$ a measurable space and g a measurable mapping $g: (X, \mathscr{A}) \to (Y, \mathscr{B})$. The **image** of the measure μ through the mapping g is the measure μ_g defined over the measurable space $(Y, \mathscr{B})$ by

$$\forall B \in \mathscr{B} \qquad \mu_g(B) = \mu\{g^{-1}(B)\} \qquad (6.28)$$

It should be noted that it is necessary that the mapping g be measurable for the sets $g^{-1}(B)$ to be measurable.

Example 3.4 Translated measure

Consider a measure over the space $(\mathbb{R}, \mathcal{R})$ and the real function $g: x \mapsto x + a$, $a \in \mathbb{R}$. The function g is borelian (measurable) since it is continuous. Application of Eqn (6.28) enables us to write

$$\forall \beta \in \mathcal{R} \qquad \mu_g(\beta) = \mu(\{x \mid x + a \in \beta\}) = \mu(\beta - a) \qquad (6.29)$$

The measure through μ_g of a Borel set β is equal to the measure through μ of the Borel set $\beta - a$ obtained from β by a translation $-a$. The measure μ_g is thus obtained from the measure μ by a translation $+a$. This is illustrated in Fig. 6.3.

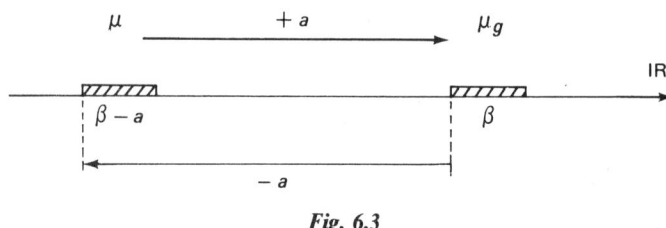

Fig. 6.3

Let μ_0 be the Lebesgue measure over the space $(\mathbb{R}, \mathcal{R})$ and ν_0 be the measure obtained from μ_0 by a translation a. The Borel field $\mathcal{R}$ is generated by the algebra $\mathcal{C}$ defined in Example 3.3. It follows immediately that

$$\forall C \in \mathcal{C} \qquad \nu_0(C) = \mu_0(C - a) = \mu_0(C)$$

It follows from this, according to the extension theorem (cf. Section 3.1.2), that the measures μ_0 and ν_0 are identical. Lebesgue measure is thus invariant under translation

3.1.5. Negligible sets

3.1.5.1. Definitions

Let $(X, \mathcal{A}, \mu)$ be a measured space. A subset N of X is **negligible** with respect to the measure μ (we say also that it is a μ-negligible set) when there exists a measurable set A such that $N \subset A$ and $\mu(A) = 0$.

A property relating to the points of the set X is **almost everywhere true** (a.e. true) when it is satisfied for all the points of X except those which belong to a negligible set.

When the measurable set $(X, \mathcal{A})$ has several measures $\mu, \nu, \ldots$, we say more precisely that a property is μ a.e. true when it is satisfied for all the points of the set X except those which belong to a μ-negligible set.

Remark: A negligible set is not necessarily measurable.

3.1.5.2. Property

A denumerable union of negligible sets is negligible.

In fact, let $(N_i, i \in I)$ be a family of negligible sets. There exists a family $(A_i, i \in I)$ of measurable sets such that $N_i \subset A_i$ and $\mu(A_i) = 0$ for every $i \in I$. We then get

$$\bigcup_{i \in I} N_i \subset \bigcup_{i \in I} A_i \in \mathscr{A}$$

It next follows from Eqn (6.25) that

$$\mu\left(\bigcup_{i \in I} A_i\right) \leq \sum_{i \in I} \mu(A_i) = 0$$

The property is thus proved.

3.1.5.3. Definition
Let $(X, \mathscr{A}, \mu)$ be a measured space. The measure μ is **complete** when the μ-negligible sets of the set X belong to the Borel field $\mathscr{A}$ and are therefore measurable.

3.1.5.4. Property
Let $(X, \mathscr{A}, \mu)$ be a measured space and $\mathscr{N}$ be the family of μ-negligible sets of X. Consider the triplet $(X, \bar{\mathscr{A}}, \bar{\mu})$ defined by

$$\bar{\mathscr{A}} = \{A \cup N \mid A \in \mathscr{A}, N \in \mathscr{N}\} \tag{6.30}$$

$$\bar{\mu}(A \cup N) = \mu(A)$$

This is a measured space such that the $\bar{\mu}$-negligible sets of X belong to the Borel field $\bar{\mathscr{A}}$. The measure $\bar{\mu}$ is therefore complete.

The family $\bar{\mathscr{A}}$ contains the Borel field $\mathscr{A}$. In fact, the set $\emptyset$ belongs to $\mathscr{N}$ and every set $A \in \mathscr{A}$ can also be written $A \cup \emptyset$. The family $\bar{\mathscr{A}}$ contains the family $\mathscr{N}$. In fact, the set $\emptyset$ belongs to $\mathscr{A}$ and every set $N \in \mathscr{N}$ can be written $N \cup \emptyset$. It is easy to verify that the family $\bar{\mathscr{A}}$ is a Borel field (cf. Exercise 3.1).

Formula (6.30) defines a positive measure over the measurable space $(X, \bar{\mathscr{A}})$. The property of σ additivity is easy to verify (cf. Exercise 3.1). We must still assure ourselves that the definition of the measure $\bar{\mu}$ is not ambiguous and that two identical sets of the Borel field $\bar{\mathscr{A}}$ have equal measures. So let $A_1 \cup N_1$ and $A_2 \cup N_2$ be two identical sets of the Borel field $\bar{\mathscr{A}}$ and let us prove that $\mu(A_1) = \mu(A_2)$. Put $N = N_1 \cup N_2$. Then $A_1 \cup N = A_2 \cup N$, which implies that

$$\bar{A}_1 \cap A_2 \subset \bar{A}_1 \cap N \subset N \qquad A_1 \cap \bar{A}_2 \subset \bar{A}_2 \cap N \subset N$$

and then $\mu(\bar{A}_1 \cap A_2) = \mu(A_1 \cap \bar{A}_2) = 0$. It follows from this that

$$\mu(A_1) = \mu(A_2) = \mu(A_1 \cap A_2)$$

It remains to verify that a $\bar{\mu}$-negligible set belongs to the Borel field $\bar{\mathscr{A}}$. Let M be such a set. By definition there exists a set $A \cup N \in \bar{\mathscr{A}}$ such that $M \subset A \cup N$ and $\bar{\mu}(A \cup N) = \mu(A) = 0$. Now, the set N being μ negligible, there exists a set

$B \in \mathscr{A}$ such that $N \subset B$ and $\mu(B) = 0$. Consequently,

$$M \subset A \cup B \in \mathscr{A} \qquad \mu(A \cup B) \leq \mu(A) + \mu(B) = 0$$

The set M is therefore μ negligible. It thus belongs to the family $\mathscr{N}$ and consequently to the Borel field $\bar{\mathscr{A}}$.

It is therefore always possible, starting from an arbitrary measured space, to construct a measured space whose measure is complete. This will enable us to suppose implicitly that the measured spaces considered are of this type. It follows immediately that, in such a space, a negligible set is measurable and of measure zero.

Example 3.5 Negligible sets in the space $(\mathbb{R}, \mathscr{R})$

Consider the measurable space $(\mathbb{R}, \mathscr{R})$ and a point $a \in \mathbb{R}$. The singleton $\{a\}$ belongs to the Borel field $\mathscr{R}$. It is negligible with respect to the Lebesgue measure μ_0. In fact

$$\{a\} = \bigcap_{n \in \mathbb{N}} \left(\left[a, a + \frac{1}{n} \right[\right)$$

and continuity under decreasing intersection enables us to write (cf. Eqn (6.27))

$$\mu_0(\{a\}) = \lim_{n \to \infty} \mu_0 \left(\left[a, a + \frac{1}{n} \right[\right) = \lim_{n \to \infty} \frac{1}{n} = 0 \qquad (6.31)$$

Every denumerable union of singletons is therefore negligible with respect to the measure μ_0.

The singleton $\{a\}$ is negligible with respect to the Dirac measure δ_b when $b \neq a$. It is not negligible with respect to the measure δ_a since $\delta_a(\{a\}) = 1$. This follows immediately from Eqn (6.21).

3.2. PRODUCT MEASURE

3.2.1. Definitions

Let $(X_i, \mathscr{A}_i, \mu_i, i \in I)$ be a family of measured spaces, the measures μ_i being normed. Consider the measurable product space

$$\left(\prod_{i \in I} X_i, \bigotimes_I \mathscr{A}_i \right)$$

The Borel field

$$\bigotimes_I \mathscr{A}_i$$

is generated by the algebra $\mathscr{C}$ of finite unions of disjoint measurable blocks (cf. Section 1.2.3).

The **tensor product**

$$\bigotimes_I \mu_i$$

of the measures μ_i is the unique positive measure over the product space such that

$$\forall C \in \mathscr{C} \quad \bigotimes_I \mu_i(C) = \prod_{i \in I} \mu_i(A_i) \tag{6.32}$$

for all measurable blocks $C \in \mathscr{M}$ (cf. Section 1.2.3). This is the **product measure**. Its uniqueness follows from the extension theorem (cf. Section 3.1.2).

When the set I is finite this definition may be written

$$\forall \left(\prod_{i \in I} A_i \right) \quad \bigotimes_I \mu_i \left(\prod_{i \in I} A_i \right) = \prod_{i \in I} \mu_i(A_i) \tag{6.33}$$

Remark: When the set I is finite, the measures μ_i need not be normed. When the set I is infinite, it is essential to assume that the measures μ_i are normed, failing which the definition would be ambiguous since the block

$$\prod_{i \in I} X_i$$

would not have a well-defined measure.

Example 3.6 Lebesgue measure over $(\mathbb{R}^n, \mathscr{R}^n)$

Lebesgue measure over the space $(\mathbb{R}^n, \mathscr{R}^n)$ is the unique measure μ_0 such that

$$\mu_0 \left\{ \prod_{i=1}^n (a_i, b_i) \right\} = \prod_{i=1}^n (b_i - a_i) \quad a_i < b_i \tag{6.34}$$

3.2.2. *Property*

3.2.2.1. Statement

Let $(\prod_{i \in I} X_i, \bigotimes_I \mathscr{A}_i, \bigotimes_I \mu_i)$ be a measured product space. Let $J \subset I$ be a finite subset and consider the measured product space $\prod_{i \in J} X_i, \bigotimes_J \mathscr{A}_i, \bigotimes_J \mu_i$. The product measure $\bigotimes_J \mu_i$ is the image of the product measure $\bigotimes_I \mu_i$ through the projection φ_{IJ} of the product set $\prod_{i \in I} X_i$ onto the product set $\prod_{i \in J} X_i$. We say more simply that the measure $\bigotimes_J \mu_i$ is the **projection** of the measure $\bigotimes_I \mu_i$.

296 *Elements of measure theory*

3.2.2.2. Proof
To prove this property it is sufficient to apply Eqn (6.28) to a measurable block $\prod_{i \in J} A_i$. Let μ_{IJ} be the measure which is the image of the measure $\underset{J}{\otimes} \mu_i$ through the projection φ_{IJ}. We get from Eqns (6.32) and (6.33)

$$\mu_{IJ}\left(\prod_{i \in J} A_i\right) = \underset{I}{\otimes} \mu_i \left\{ \varphi_{IJ}^{-1}\left(\prod_{i \in J} A_i\right) \right\}$$

$$= \underset{I}{\otimes} \mu_i \left(\prod_{i \in J} A_i \times \prod_{i \in I-J} X_i\right)$$

$$= \prod_{i \in J} \mu_i(A_i)$$

$$= \underset{I}{\otimes} \mu_i \left(\prod_{i \in J} A_i\right)$$

The extension theorem enables us to conclude that $\mu_{IJ} = \underset{J}{\otimes} \mu_i$.

3.2.3. *Kolmogoroff's theorem* (Guichardet, 1969, p. 216)

Let $(X_i, \mathscr{A}_i, \mu_i, i \in I)$ be an infinite family of measured spaces, the measures μ_i being normed. Consider the measurable product space

$$\left(\prod_{i \in I} X_i, \underset{I}{\otimes} \mathscr{A}_i\right)$$

There can exist over this space measures other than the product measure

$$\underset{I}{\otimes} \mu_i$$

These measures therefore do not satisfy Eqn (6.32).

Consider the family $\{J\}$ of finite subsets of the set I and suppose that we are given a family $\{\mu_J\}$ of measures defined respectively over the product spaces

$$\left(\prod_{i \in J} X_i, \underset{J}{\otimes} \mathscr{A}_i\right)$$

Suppose moreover that the measures μ_J possess the property of projectivity stated in the preceding section.

Kolmogoroff's theorem affirms that, under these conditions and with the addition of supplementary assumptions concerning the topology of the spaces $(X_i, \mathscr{A}_i)$, there exists a unique measure μ over the space

$$\left(\prod_{i \in I} X_i, \underset{I}{\otimes} \mathscr{A}_i\right)$$

which is projected in accordance with the measures μ_j. We shall content ourselves with this incomplete statement and it will be sufficient for us to know that the theorem applies to the measurable product space

$$\left(\mathbb{R}^{\mathbb{R}}, \bigotimes_{\mathbb{R}} \mathscr{R}\right)$$

3.2.4. Convolution of two measures

3.2.4.1. Definition
Let μ_1 and μ_2 be two bounded measures defined over the measurable space $(\mathbb{R}, \mathscr{R})$. Their **convolution** is the measure over $(\mathbb{R}, \mathscr{R})$ defined by

$$\forall \beta \in \mathscr{R} \quad \mu_1 * \mu_2(\beta) = \mu_1 \otimes \mu_2\{(x_1, x_2) \mid x_1 + x_2 \in \beta\} \tag{6.35}$$

3.2.4.2. Properties
It follows from these definitions that the convolution product is the image of the tensor product through the measurable mapping $(\mathbb{R}^2, \mathscr{R}^2)$ in the set $(\mathbb{R}, \mathscr{R})$ defined by $(x_1, x_2) \mapsto x_1 + x_2$.

The measure $\mu_1 * \mu_2$ is bounded since (cf. Eqn (6.33))

$$\mu_1 * \mu_2(\mathbb{R}) = \mu_1 \otimes \mu_2(\mathbb{R}^2) = \mu_1(\mathbb{R})\mu_2(\mathbb{R}) < +\infty$$

Finally, it follows immediately that the convolution is commutative since the sets $\{(x_1, x_2) \mid x_1 + x_2 \in \beta\}$ are symmetric with respect to the principal diagonal of $\mathbb{R}^2$.

Exercises

Exercise 3.1 Definition of a complete measure
Let $(X, \mathscr{A}, \mu)$ be a measured space, $\mathscr{N}$ the set of μ-negligible subsets of X, $\bar{\mathscr{A}}$ the family defined by Eqn (6.30), and $\bar{\mu}$ the measure defined by Eqn (6.30).
(1) Prove that the family $\bar{\mathscr{A}}$ is a Borel field of subsets of the set X.
(2) Verify that $\bar{\mu}$ is a positive measure over the measurable space $(X, \bar{\mathscr{A}})$.

Exercise 3.2 Measure of a reunion of two sets
Let $(X, \mathscr{A}, \mu)$ be a measured space. Prove that we have, for two measurable sets A and B,

$$\mu(A \cup B) + \mu(A \cap B) = \mu(A) + \mu(B) \tag{6.36}$$

Exercise 3.3 Upper limit and lower limit of a sequence of sets
Let $(X, \mathscr{A}, \mu)$ be a measured space and $(A_n, n \in \mathbb{N})$ be a sequence of measurable sets.

(1) Consider the sequence $(B_n, n \in \mathbb{N})$ with

$$B_n = \bigcup_{k \geq n} A_k$$

Verify that it is decreasing. Hence deduce successively

$$\forall n \in \mathbb{N} \qquad \mu(B_n) \geq \sup_{k \geq n} \mu(A_k)$$

$$\limsup \mu(A_n) \leq \mu(\limsup A_n)$$

(Apply Eqn (6.27) and use the convergence of the numeric decreasing and positive sequence $\mu(B_n)$.)

(2) Consider the sequence $(C_n, n \in \mathbb{N})$ with

$$C_n = \bigcap_{k \geq n} A_k$$

Verify that it is increasing. Hence deduce successively

$$\forall n \in \mathbb{N} \qquad \mu(C_n) \leq \inf_{k \geq n} \mu(A_k)$$

$$\mu(\liminf A_n) \leq \liminf \mu(A_n)$$

(Apply Eqn (6.26) and use the convergence of the numeric increasing and bounded sequence $\mu(C_n)$.)

(3) Deduce from the above that

$$\mu(\liminf A_n) \leq \liminf \mu(A_n) \leq \limsup \mu(A_n) \leq \mu(\limsup A_n)$$

(6.37)

It follows from this that, if the sequence (A_n) converges, the numerical sequence $\mu(A_n)$ converges also.

4. Integration

4.1. INTEGRATION OVER AN ARBITRARY SPACE

4.1.1. *Integration of positive measurable step functions*

4.1.1.1. Definition
A real **step** function is a function taking a finite number of values in $\mathbb{R}$.

4.1.1.2. Property
Let γ be a real step function defined over a measurable space $(X, \mathscr{A})$. Put

$$\gamma(X) = \{a_1, \ldots, a_n\} \qquad \forall i \in [1, n] \qquad A_i = \gamma^{-1}(\{a_i\})$$

It is easy to verify that the family (A_i) is a finite partition of the set X: this enables us to write

$$\gamma = \sum_{i=1}^{n} a_i I_{A_i} \qquad (6.38)$$

The function γ is measurable if and only if the sets A_i are measurable. It is in fact easy to verify that the Borel field generated by γ is that generated by the partition (A_i) (cf. Exercise 1.8).

4.1.1.3. Definitions
Let $(X, \mathcal{A}, \mu)$ be a measured space. The **integral with respect to the measure** μ of the indicator function I_A of a measurable set A is defined and denoted by

$$\mu(I_A) = \int_X I_A(x)\,d\mu(x) = \mu(A) \qquad (6.39)$$

Let γ be the positive measurable step function defined by Eqn (6.38). Its integral with respect to the measure μ is defined by

$$\mu(\gamma) = \int_X \gamma(x)\,d\mu(x) = \sum_{i=1}^{n} a_i \mu(A_i) \qquad (6.40)$$

This is a number belonging to $\bar{\mathbb{R}}_+$.

4.1.1.4. Properties
Let γ, γ_1 and γ_2 be positive measurable step functions. Their integrals have the following properties of linearity and of increase:
(1) the linearity property is

$$\forall a \in \mathbb{R} \qquad \mu(a\gamma) = a\mu(\gamma)$$
$$\mu(\gamma_1 + \gamma_2) = \mu(\gamma_1) + \mu(\gamma_2)$$

(2) the increasing property is

$$\gamma_1 \leqslant \gamma_2 \Rightarrow \mu(\gamma_1) \leqslant \mu(\gamma_2)$$

These properties follow without difficulty from the definition (6.40).

Example 4.1 Integral with respect to Dirac measure
Let $(\Omega, \mathcal{A}, \delta_a), a \in X$, be a measured space and γ the positive measurable step function defined by Eqn (6.38). Since the family (A_i) is a partition of the set X, let A_k be the unique set of the family containing the point a. It follows then by applying Eqns (6.40) and (6.21) that

$$\delta_a(\gamma) = \sum_{i=1}^{n} a_i \delta_a(A_i) = a_k = \gamma(a)$$

The value of the integral with respect to the measure δ_a is the value of the function at the point a.
Physicists are accustomed to treating the measure δ_a as a function $\delta(x - a)$,

writing

$$\delta_a(\gamma) = \int_X \gamma(x)\delta(x-a)\,dx = \gamma(a) \tag{6.41}$$

We shall use this notation, which is improper from the mathematical point of view but whose use is convenient for writing integrals. This amounts to accepting that the 'Dirac function' $\delta(x-a)$ is the density with respect to Lebesgue measure of the Dirac measure δ_a (cf. Section 4.1.5).

Example 4.2 Integration with respect to convolution

Let γ be the positive measurable step function defined by Eqn (6.38). Let us calculate its integral with respect to the convolution $\mu_1 * \mu_2$ of two bounded measures. We get from applying Eqns (6.40) and (6.35)

$$\mu_1 * \mu_2(\gamma) = \sum_{i=1}^n a_i \mu_1 * \mu_2(A_i)$$

$$= \sum_{i=1}^n a_i \mu_1 \otimes \mu_2(\{(x_1, x_2) \mid x_1 + x_2 \in A_i\})$$

$$= \sum_{i=1}^n a_i \mu_1 \otimes \mu_2\{I_{A_i}(x_1 + x_2)\}$$

This enables us to write

$$\mu_1 * \mu_2(\gamma) = \mu_1 \otimes \mu_2\{\gamma(x_1 + x_2)\} \tag{6.42}$$

The integral of the function γ with respect to the convolution $\mu_1 * \mu_2$ is thus the integral with respect to the tensor product $\mu_1 \otimes \mu_2$ of the function defined over $\mathbb{R}^2$ by

$$(x_1, x_2) \mapsto \gamma(x_1 + x_2)$$

4.1.2. Integration of positive measurable functions

4.1.2.1. Property

Let g be a real positive measurable function and γ_n, $n \in \mathbb{N}^*$, be the positive step function defined by

$$\gamma_n = \sum_{k=0}^{2^{2n}} \frac{k}{2^n} I_{A_{nk}} \qquad A_{nk} = g^{-1}\left(\left[\frac{k}{2^n}, \frac{k+1}{2^n}\right[\right) \tag{6.43}$$

The sets A_{nk} are measurable since the function g is measurable. The functions γ_n are therefore measurable. It is easy to verify that the sequence (γ_n) is increasing and that it converges simply to the function g.

A real positive measurable function can thus always be considered as the limit of an increasing (not unique) sequence of positive and measurable step functions. This property is much used in the following to extend the properties of step functions by passage to the limit.

Example 4.3 Composition of measurable functions

Let $(X, \mathscr{A}, \mu)$ be a measured space, g and h be two real measurable functions defined over X, and $\mathscr{B}_g$ and $\mathscr{B}_h$ be the sub-Borel fields which they generate in X; we suppose that $\mathscr{B}_h \subset \mathscr{B}_g$. We shall prove that there then exists a Borel function k such that $h = k \circ g$.

Assume, to begin with, that h is a positive measurable step function (cf. Eqn (6.38)):

$$h = \sum_{i=1}^{n} a_i I_{A_i}$$

We have then, since $\mathscr{B}_h \subset \mathscr{B}_g$

$$A_i = h^{-1}(\{a_i\}) \in \mathscr{B}_h \quad \Rightarrow \quad \exists \beta_i \in \mathscr{R} \quad A_i = g^{-1}(\beta_i)$$

It is easy to verify, moreover, that

$$I_{A_i} = I_{g^{-1}(\beta_i)} = I_{\beta_i} \circ g$$

This implies that we can write

$$h = \left(\sum_{i=1}^{n} a_i I_{\beta_i} \right) \circ g = k \circ g$$

The function k defined by this equality is borelian since the sets β_i are Borel sets.

Suppose now that the measurable function h is positive. It is then the limit of a sequence of positive measurable step functions, which can be written, by virtue of the above, $h_n = k_n \circ g$, the functions k_n being borelian. It follows from this that the set A of points for which the sequence (k_n) converges is such that $g(X) \subset A$. Let us put

$$k(x) = \lim_{n \to \infty} k_n(x) \qquad x \in A$$

$$k(x) = 0 \qquad x \notin A$$

The function k defined thus is borelian since it is the limit of a sequence of borelian functions. It follows immediately that we can write $h = k \circ g$.

Finally, when the measurable function h is not positive, we can argue in the same way for the positive measurable functions h_+ and h_- defined by Eqn (6.46).

4.1.2.2. Definitions

Let g be a real function, positive, measurable and defined over a measured space $(X, \mathscr{A}, \mu)$. Let (γ_n) be an increasing sequence of real positive measurable step functions, converging to the function g.

The integral of the function g with respect to the measure μ is the limit in $\overline{\mathbb{R}}_+$ of the increasing numerical sequence $\{\mu(\gamma_n)\}$:

$$\mu(g) = \int_X g(x)\,d\mu(x) = \mu\left(\lim_{n\to\infty} \gamma_n\right) = \lim_{n\to\infty} \mu(\gamma_n) \qquad (6.44)$$

It can be shown that the limit does not depend on the particular sequence (γ_n) chosen.

The function g is integrable with respect to the measure μ (we say also that it is μ integrable) when the above limit is finite.

4.1.2.3. Properties

The properties of linearity and of increase of the integral of real positive step functions extend to the integral of real positive functions by passage to the limit of increasing sequences (γ_n) of real positive measurable step functions.

4.1.2.4. Theorem of monotonic convergence

Let (g_n) be an increasing sequence of real positive measurable functions, converging to a real positive measurable function g. Then

$$\lim_{n\to\infty} \mu(g_n) = \mu(g) \qquad (6.45)$$

We shall accept this theorem (Bertrandias, 1969, p. 18). It must be noted that the equality (6.45) holds in $\overline{\mathbb{R}}_+$, which means that the two members can be equal to $+\infty$.

4.1.3. Integration of complex measurable functions

4.1.3.1. Property

Let g be an arbitrary real measurable function. Put

$$g^+ = \max(g, 0) \qquad g^- = \min(g, 0) \qquad (6.46)$$

It follows from this that

$$g = g^+ + g^- \qquad |g| = g^+ - g^-$$

The functions g^+ and $-g^-$ are positive and measurable (cf. Exercise 2.2). The positive function $|g|$ is thus measurable (cf. Section 2.2.4). It is integrable when

$$\mu(|g|) = \mu(g^+) - \mu(g^-) < +\infty$$

This implies that the positive functions g^+ and $-g^-$ are integrable and consequently that $\mu(g^+)$ and $\mu(g^-)$ are finite numbers.

4.1.3.2. Definitions

Let g be a real measurable function defined over a measured space $(X, \mathscr{A}, \mu)$. It is integrable with respect to the measure μ when the real positive measurable function $|g|$ is integrable with respect to the measure μ. Its integral is the finite number

$$\mu(g) = \int_X g(x)\,d\mu(x) = \mu(g^+) + \mu(g^-) \tag{6.47}$$

Let g be a complex measurable function defined over a measured space $(X, \mathscr{A}, \mu)$. It is integrable with respect to the measure μ when the real measurable functions Re g and Im g are integrable. Its integral is the complex number

$$\mu(g) = \int_X g(x)\,d\mu(x) = \mu(\text{Re } g) + i\mu(\text{Im } g) \tag{6.48}$$

4.1.3.3. Properties

The properties of linearity and of increase of the integral of measurable functions follow from the similar properties of the integral of real positive measurable functions:

(1) the linearity property is

$$\forall a \in \mathbb{C} \quad \mu(ag) = a\mu(g)$$
$$\mu(g_1 + g_2) = \mu(g_1) + \mu(g_2) \tag{6.49}$$

(2) the increasing property (real functions) is

$$g_1 \leqslant g_2 \Rightarrow \mu(g_1) \leqslant \mu(g_2) \tag{6.50}$$

4.1.3.4. Property

Let g be a complex measurable function and h be a real positive measurable integrable function such that $|g| \leqslant h$. Then the function g is integrable and satisfies the relation

$$|\mu(g)| \leqslant \mu(|g|) \leqslant \mu(h) \tag{6.51}$$

This property is much used for proving the integrability of a function. To show this, start by assuming that the function g is real and integrable. We then get immediately from the definitions that

$$|\mu(g)| \leqslant \mu(|g|)$$

Suppose now that the function g is complex and integrable. Let us put $\vartheta = \arg \mu(g)$ and write, taking account of the fact that $\mu(g)$ is a complex number and applying Eqn (6.49),

$$0 \leqslant |\mu(g)| = \mu(g)\exp(-i\vartheta) = \mu\{\exp(-i\vartheta)\,g\} = \mu[\text{Re}\{\exp(-i\vartheta)\,g\}]$$

The real function $\text{Re}\{\exp(-i\vartheta)\,g\}$ is integrable since the complex function g is integrable. It follows from this, by application of the above inequality, that

$$0 \leqslant \mu[\text{Re}\{\exp(-i\vartheta)\,g\}] \leqslant \mu(|\text{Re}\{\exp(-i\vartheta)\,g\}|) \leqslant \mu(|g|)$$

Suppose finally that the complex function g has its modulus bounded by a positive integrable function h. It follows immediately from Eqn (6.49) that the function g is integrable and satisfies Eqn (6.51).

4.1.3.5. Theorem of dominated convergence

Let $(g_n, n \in \mathbb{N})$ be a sequence of measurable functions, real or complex, defined a.e. over a measured space $(X, \mathscr{A}, \mu)$ and converging a.e. to a function g. If the functions have a.e. moduli bounded by a real positive measurable integrable function h, the measurable function g is integrable and

$$\mu(g) = \mu\left(\lim_{n \to \infty} g_n\right) = \lim_{n \to \infty} \mu(g_n) \tag{6.52}$$

We shall accept this very important theorem (Guichardet, 1969, p. 18).

Example 4.4 Integral with respect to Dirac measure

Let g be a real positive measurable function defined over a measured space $(X, \mathscr{A}, \delta_a)$, $a \in X$. There exists an increasing sequence (γ_n) of real positive measurable step functions, converging to the function g. It follows then from Eqns (6.44) and (6.41) that

$$\delta_a(g) = \lim_{n \to \infty} \delta_a(\gamma_n) = \lim_{n \to \infty} \gamma_n(a) = g(a)$$

This result can be extended, by arguing as we have done above, to a measurable function, real or complex. We get, using the improper notation introduced in Eqn (6.41),

$$\delta_a(g) = \int_X g(x)\delta(x-a)\,\mathrm{d}x = g(a) \tag{6.53}$$

The function g is integrable if and only if $|g(a)| < +\infty$.

Example 4.5 Integral with respect to convolution

Consider a real measurable function g, defined over a measurable space $(X, \mathscr{A})$, and two bounded measures μ_1 and μ_2 defined over the same space. Let $h: \mathbb{R}^2 \to \mathbb{R}$ be the function such that $h(x_1, x_2) = g(x_1 + x_2)$. It follows from Eqn (6.42), considering the functions g^+ and $-g^-$ as limits of increasing sequences of real positive and measurable step functions, that

$$\mu_1 * \mu_2(g) = \mu_1 \otimes \mu_2(h) \tag{6.54}$$

or

$$\int_\mathbb{R} g(x)\,\mathrm{d}[\mu_1 * \mu_2](x) = \iint_{\mathbb{R}^2} g(x_1 + x_2)\,\mathrm{d}[\mu_1 \otimes \mu_2](x_1, x_2)$$

Example 4.6 Integration with respect to an image measure

Let $(X, \mathscr{A}, \mu)$ be a measured space, $(Y, \mathscr{B})$ a measurable space and $k: (X, \mathscr{A}) \to (Y, \mathscr{B})$ a measurable mapping. Let μ_k be the image measure of the measure μ through the mapping k. We get from Eqns (6.28) and (6.38), taking into account that $I_{k^{-1}(B)} = I_B \circ k$,

$$\forall B \in \mathscr{B} \qquad \mu_k(I_B) = \mu(I_B \circ k)$$

Then let g be a complex measurable function, defined over the space $(Y, \mathscr{B})$. Its integral with respect to the image measure is obtained by considering passage to the limit of an increasing sequence of step functions. It follows that

$$\mu_k(g) = \mu(g \circ k) \qquad (6.55)$$

or

$$\int_Y g(y) \, d\mu_g(y) = \int_X g\{k(x)\} \, d\mu(x)$$

4.1.4. Negligible functions

4.1.4.1. Definition

A complex measurable function g, defined over a measured space $(X, \mathscr{A}, \mu)$, is **negligible** (we say also, more precisely, μ negligible) when it is a.e. zero; this may be written, where we denote by $\{g \neq 0\}$ the set of points for which the function g is not zero,

$$\mu(\{g \neq 0\}) = 0$$

It must be noted that the set $\{g \neq 0\}$ is measurable since it is the inverse image of the complement of the Borel set $\{0\}$.

4.1.4.2. Proposition

A real measurable function g, defined over a measured space $(X, \mathscr{A}, \mu)$, is negligible if and only if $\mu(|g|) = 0$.

This property is much used to prove that a measurable function is negligible. We start by proving that a negligible function g is such that $\mu(|g|) = 0$. Let us put for this

$$A_0 = \{x \mid g(x) = 0\} = \{g = 0\} \qquad \forall n \in \mathbb{N} \qquad g_n = \inf(|g|, n)$$

It is easy to verify that the sequence (g_n) is increasing and converges to the function $|g|$. The theorem of monotonic convergence then enables us to write (cf. Eqn (6.45))

$$\lim_{n \to \infty} \mu(g_n) = \mu(|g|)$$

It follows next, since the set $\bar{A}_0$ is negligible, that

$$\forall n \in \mathbb{N} \quad \mu(g_n) = \int_{A_0} g_n(x)\,d\mu(x) + \int_{\bar{A}_0} g_n(x)\,d\mu(x)$$

$$= \int_{\bar{A}_0} g_n(x)\,d\mu(x) \leq n \int_{\bar{A}_0} d\mu(x) = n\mu(\bar{A}_0) = 0$$

Hence we have $\mu(|g|) = 0$.

Conversely, let g be a real measurable integrable function such that $\mu(|g|) = 0$. Let us put

$$\forall n \subset \mathbb{N}^* \quad A_n = \{x \mid |g(x)| \geq 1/n\}$$

and write, applying Eqn (6.50),

$$0 \leq \frac{1}{n} I_{A_n} \leq |g| \Rightarrow \frac{1}{n}\mu(A_n) \leq \mu(|g|) = 0$$

It follows from this that, for all $n \in \mathbb{N}^*$, $\mu(A_n) = 0$. Now, the sequence (A_n) is obviously increasing and it has as limit (cf. Exercise 1.1)

$$\bigcup_{n \in \mathbb{N}^*} A_n = \bar{A}_0$$

We get on applying Eqn (6.25)

$$\mu(\bar{A}_0) \leq \sum_{n \in \mathbb{N}^*} \mu(A_n) = 0$$

The function g is thus negligible since $\bar{A}_0 = \{g \neq 0\}$.

A real measurable negligible function thus has a zero integral. The converse is true only if the function is positive.

4.1.4.3. Measurable functions, almost everywhere equal

Equality a.e. establishes an equivalence relation over the set of complex measurable functions defined over a measured space $(X, \mathcal{A}, \mu)$.

Reflexivity and symmetry are obvious. To establish transitivity, consider three measurable functions g, h and k. Denoting by $\{g = h\}$ the set $\{x \mid g(x) = h(x)\}$, we get

$$\{g = h\} \cap \{h = k\} \subset \{g = k\}$$

which implies that

$$\{g \neq h\} \cup \{h \neq k\} \supset \{g \neq k\}$$

It follows then from Eqn (6.25) that

$$\mu\{g \neq k\} \leq \mu\{g \neq h\} + \mu\{h \neq k\}$$

The functions g and k are therefore a.e. equal when the function g is a.e. equal to the function h and the function h is a.e. equal to the function k.

Then let g be a complex measurable integrable function and h be a measurable function, a.e. equal to the function g. The real functions $\text{Re}(g - h)$ and $\text{Im}(g - h)$ are then negligible and it follows from the proposition of the preceding section that their integrals are zero, which enables us to write

$$g \underset{\text{a.e.}}{=} h \Rightarrow \mu(g) = \mu(h) \tag{6.56}$$

It is thus sufficient that a measurable function g be integrable for measurable functions which are a.e. equal to it to be integrable and to have integrals equal to $\mu(g)$.

When the measured space in question has two measures μ and ν, we say that two functions a.e. equal with respect to the measure μ are μ a.e. equal, with a corresponding statement for ν. This is intended to avoid confusion.

4.1.5. Absolutely continuous measure

4.1.5.1. Measure defined by a class of almost everywhere equal functions

Let ψ be a real measurable positive function, defined over a measured space $(X, \mathscr{A}, \mu)$. The relation

$$\forall A \in \mathscr{A} \qquad \nu(A) = \mu(\psi I_A)$$

defines a positive measure over the measurable space $(X, \mathscr{A})$.

First, the number $\nu(A)$ belongs to $\bar{\mathbb{R}}_+$ since the function ψI_A is positive. Next, let $(A_n, n \in \mathbb{N})$ be a sequence of measurable pairwise disjoint sets and let A be the set

$$A = \bigcup_{n \in \mathbb{N}} A_n$$

Then

$$I_A = \sum_{n \in \mathbb{N}} I_{A_n}$$

and we get from Eqn (6.49)

$$\nu\left(\bigcup_{n \in A} A_n\right) = \sum_{n \in \mathbb{N}} \nu(A_n)$$

Thus σ additivity has been established.

Let $\dot{\psi}$ be a real positive measurable function, μ a.e. equal to the function ψ. The function $\psi - \dot{\psi}$ is then μ negligible and consequently the function $(\psi - \dot{\psi})I_A$ is also μ negligible and its integral with respect to the measure μ is zero. It follows from this that

$$\nu(A) = \mu(\psi I_A) = \mu(\dot{\psi} I_A)$$

The measure ν is thus defined by the class of functions μ a.e. equal to the function ψ, which signifies that the function ψ is only μ a.e. defined.

We remark, in conclusion, that the measure μ is bounded if and only if the function ψ is integrable with respect to the measure μ.

4.1.5.2. Definitions

Let μ and v be two positive measures defined over a measurable space $(X, \mathscr{A})$. The measure v is **absolutely continuous** with respect to the measure μ when there exists a real measurable positive function ψ, defined a.e. over X, such that

$$\forall A \in \mathscr{A} \quad v(A) = \mu(\psi I_A) \tag{6.57}$$

The function ψ is the **density** of the measure v with respect to the measure μ and we write $v = \psi\mu$. It is also called the **Radon–Nikodym derivative** of the measure v with respect to the measure μ and is denoted by $\psi = dv/d\mu$. We shall therefore write, according to circumstances,

$$v = \psi\mu \quad \text{or} \quad \psi = \frac{dv}{d\mu} \tag{6.58}$$

4.1.5.3. Radon–Nikodym theorem

Let μ and v be two positive measures defined over a measurable space $(X, \mathscr{A})$. The measure v is absolutely continuous with respect to the measure μ if and only if every μ-negligible set is v negligible.

We shall accept this very important theorem (Guichardet, 1969, p. 217).

4.1.5.4. Integration with respect to an absolutely continuous measure

Let γ be the real measurable positive step function defined by Eqn (6.38). It follows on applying Eqns (6.40) and (6.57) that

$$v(\gamma) = \sum_{i=1}^{n} a_i \mu(\psi I_{A_i}) = \mu(\gamma\psi)$$

From this it follows, by reasoning as in the preceding sections, that the integral of a complex function g, measurable with respect to the measure v defined by Eqn (6.57), may be written

$$v(g) = \mu(g\psi) = \int_X g(x)\psi(x)\,d\mu(x) \tag{6.59}$$

When $X = \mathbb{R}$ and the measure μ is the Lebesgue measure μ_0, the integral of the function g may be written in the classical manner

$$v(g) = \int_\mathbb{R} g(x)\psi(x)\,dx \tag{6.60}$$

When the density ψ is the indicator function I_A of a measurable set A, we get

$$v(g) = \mu(gI_A) = \int_A g(x)\,d\mu(x) \tag{6.61}$$

The integral over X with respect to the measure v is then equal to the integral over A with respect to the measure μ.

Remark: To write Eqn (6.57) or (6.61) we must suppose that a function of the form gI_A is zero when I_A is zero, whatever the value taken by g.

Example 4.7 Relative density of two absolutely continuous measures

Let $(X, \mathcal{A}, \mu)$ be a measured space and let v_1 and v_2 be two bounded measures, absolutely continuous with respect to the measure μ.

Let ψ_1 and ψ_2 be the probability densities of the measures v_1 and v_2. These are μ-integrable functions and consequently (cf. Eqn (6.86))

$$\mu\{\psi_2 = +\infty\} = \mu\{\psi_1 = +\infty\} = 0$$

Let us consider the set $\beta_1 = \{\psi_1 = 0\}$. It is v_1 negligible since

$$v_1(\beta_1) = \mu(I_{\beta_1}\psi_1) = 0$$

It is also μ negligible, according to the Radon–Nikodym theorem. The same is true of the set $\{\psi_2 = 0\}$.

Let N be a v_1-negligible set. By definition there exists a measurable set A such that

$$N \subset A \qquad v_1(A) = \mu(\psi_1 I_A) = 0$$

The positive measurable function $\psi_1 I_A$ is therefore μ negligible. Let us write

$$\{\psi_2 I_A \neq 0\} = [\{\psi_2 I_A \neq 0\} \cap \{\psi_1 = 0\}] \cup [\{\psi_2 I_A \neq 0\} \cap \{\psi_1 \neq 0\}]$$
$$\subset \{\psi_1 = 0\} \cup \{\psi_2 = +\infty\} \cup \{\psi_1 I_A \neq 0\}$$

It follows from the above that the positive measurable function $\psi_2 I_A$ is μ negligible and that the set N is v_2 negligible. The measure v_2 is thus absolutely continuous with respect to the measure v_1 by virtue of the Radon–Nikodym theorem.

Let ψ be the density of the measure v_2 with respect to the measure v_1. It follows on applying Eqn (6.57) that

$$\forall A \in \mathcal{A} \qquad \mu\{(\psi_1\psi - \psi_2)I_A\} = 0$$

The function $\psi_1\psi - \psi_2$ is thus μ negligible (cf. Exercise 4.2). The function ψ can then be defined, since the set $\{\psi_1 = 0\}$ is μ negligible, by

$$\begin{aligned}\psi(x) &= \psi_2(x)/\psi_1(x) & x \in \{\psi_1 \neq 0\} \\ \psi(x) &= 0 & x \in \{\psi_1 = 0\}\end{aligned} \qquad (6.62)$$

Note that the function ψ is v_1 a.e. defined since the set $\{\psi_1 = 0\}$ is v_1 negligible. This allows us to write

$$v_2 = \frac{\psi_2}{\psi_1} v_1 \quad \text{or} \quad \frac{dv_2}{dv_1} = \frac{\psi_2}{\psi_1} \qquad (6.63)$$

4.1.6. The space $L^1(X, \mathscr{A}, \mu)$

4.1.6.1. Definitions

Complex measurable integrable functions defined over a measured space $(X, \mathscr{A}, \mu)$ constitute a vector space denoted by $\mathscr{L}^1(X, \mathscr{A}, \mu)$. The quotient vector space consisting of the classes of a.e. equal integrable functions is denoted by $L^1(X, \mathscr{A}, \mu)$. We generally identify a function with the class to which it belongs.

The vector space L^1 has the norm $\|.\|_1$ defined by

$$\|g\|_1 = \mu(|g|) \tag{6.64}$$

It defines the topology of convergence in the mean. With this norm, L^1 is a Banach space (Dieudonné, 1969, p. 91). This means that the necessary and sufficient condition that a sequence of functions $g_n \in \mathscr{L}^1$ converges in the mean to a function $g \in \mathscr{L}^1$ is that it be a Cauchy sequence, which may be written

$$\lim_{n \to \infty} \|g_n - g\|_1 = 0 \Leftrightarrow \lim_{m,n \to \infty} \|g_m - g_n\|_1 \tag{6.65}$$

4.1.6.2. Property

When the measure μ is bounded, the constant functions belong to the space $\mathscr{L}^1$.

In fact let g be the constant function $g: x \mapsto c$. It can be written $g = cI_X$. We then get from Eqns (6.39) and (6.49)

$$\mu(g) = c\mu(I_X) = c\mu(X) < +\infty \tag{6.66}$$

4.1.7. The space $L^2(X, \mathscr{A}, \mu)$

4.1.7.1. Definitions

A complex function g defined over a measured space $(X, \mathscr{A}, \mu)$ is square integrable when it is measurable and when the function $|g|^2$ is integrable.

Square integrable functions constitute a vector space denoted by $\mathscr{L}^2(X, \mathscr{A}, \mu)$. The quotient vector space consisting of the classes of a.e. equal square integrable functions is denoted by $L^2(X, \mathscr{A}, \mu)$. We generally identify the functions in practice with the classes to which they belong.

The vector space L^2 has the norm $\|.\|_2$ defined by

$$\|g\|_2 = \{\mu(|g|^2)\}^{1/2} \tag{6.67}$$

It defines the topology of convergence in mean square. With this norm, L^2 is a Hilbert topological vector space (Dieudonné, 1969, p. 117). This means that the necessary and sufficient condition that a sequence of functions $g_n \in \mathscr{L}^2$ converge in mean square to a function $g \in \mathscr{L}^2$ is that it be a Cauchy sequence, which may be written

$$\lim_{n\to\infty} \|g_n - g\|_2 = 0 \Leftrightarrow \lim_{m,n\to\infty} \|g_m - g_n\|_2 \qquad (6.68)$$

4.1.7.2. Schwarz's inequality

Let g and h be two square integrable functions. We can write

$$|g - h^*|^2 \geq 0 \Rightarrow |gh^*| \leq \tfrac{1}{2}(|g|^2 + |h|^2)$$

It follows from this, by application of Eqn (6.51), that the function gh^* is integrable. Its integral defines the **scalar product** $\langle g \mid h \rangle$ of the functions g and h:

$$\langle g \mid h \rangle = \mu(gh^*) \qquad (6.69)$$

The scalar product satisfies Schwarz's inequality:

$$|\langle g \mid h \rangle| \leq \|g\|_2 \|h\|_2 \qquad (6.70)$$

The equality holds if and only if there exists a constant $c \in \mathbb{C}$ such that the functions g and h are a.e. equal. Schwarz's inequality is an elementary but very useful property of Hilbert spaces. We shall accept it (Dieudonné, 1969, p. 120).

4.1.7.3. Property

When the measure μ is bounded, a square μ-integrable function is μ integrable.

In fact it follows from Eqns (6.70) and (6.69) that

$$\mu(|g|) = |\langle g \mid 1 \rangle| \leq \|g\|_2 \|1\|_2 = \|g\|_2 \mu(X) < +\infty$$

4.1.7.4. The orthogonal projection theorem

Let g be a complex function belonging to the space $L^2(X, \mathscr{A}, \mu)$ and let F be a closed subspace of L^2. There exists a unique function $g_F \in F$ such that

$$\forall h \in F \quad \langle g - g_F \mid h \rangle = 0 \quad \|g - g_F\|_2 \leq \|g - h\|_2 \qquad (6.71)$$

The function g_F is the orthogonal projection of the function g onto the subspace F and the norm $\|g - g_F\|_2$ is the shortest distance from the function g to the subspace F.

We shall accept this theorem, which plays a very important part in the theory of Hilbert spaces (Dieudonné, 1969, p. 121).

4.1.8. Projection of a function onto a function space

Let $(X, \mathscr{A}, \mu)$ be a measured space, μ being a positive measure, and let us consider the following:

- a sub-Borel field $\mathscr{B} \subset \mathscr{A}$ and the measured space $(X, \mathscr{B}, \mu_{\mathscr{B}})$, $\mu_{\mathscr{B}}$ being

the restriction of the measure μ to the sub-Borel field $\mathscr{B}$ (this means that every set $B \in \mathscr{B}$ has measure $\mu_\mathscr{B}(B) = \mu(B)$);
- a real function g, measurable with respect to the Borel field $\mathscr{A}$ and integrable (an $\mathscr{A}$-measurable and μ-integrable function).

The function g is not necessarily $\mathscr{A}$ measurable. In fact, the inverse image $g^{-1}(\beta)$ of a Borel set $\beta \in \mathscr{R}$ belongs to the Borel field $\mathscr{A}$ but can be outside the Borel field $\mathscr{B}$, which is smaller.

The object of this section is to define a $\mathscr{B}$-measurable function $g_\mathscr{B}$ which is a.e. equal, under certain conditions, to the $\mathscr{A}$-measurable function g. When the function g is also square integrable, the function $g_\mathscr{B}$ is the projection of the function g onto the Hilbert space generated by the square integrable $\mathscr{B}$-measurable functions. These properties are indispensable for the foundation of the theory of conditioning which is basic in the theory of detection and of probabilistic estimation.

(a) Suppose first that the function g is positive and let us consider the positive measure $g\mu$. We can show that the positive measure $(g\mu)_\mathscr{B}$ is absolutely continuous with respect to the positive measure $\mu_\mathscr{B}$. Let N be a $\mu_\mathscr{B}$-negligible set. There exists a set $B \in \mathscr{B}$ such that $N \subset B$ and $\mu_\mathscr{B}(B) = \mu(B) = 0$. We can write, moreover, since the function g is integrable (cf. Exercise 4.1)

$$\forall \varepsilon > 0 \quad \exists n \in \mathbb{N} \quad \int_{A_n} |g(x)| \, d\mu(x) < \varepsilon \quad A_n = \{x \mid g(x) \geq n\}$$

It follows from this, by application of Eqn (6.51), that

$$(g\mu)_\mathscr{B}(B) = \int_{A_n \cap B} g(x) \, d\mu(x) + \int_{\bar{A}_n \cap B} g(x) \, d\mu(x)$$

$$\leq \varepsilon + n\mu(B) = 0$$

The set N is therefore $(g\mu)_\mathscr{B}$ negligible, which implies the result stated, by virtue of the Radon–Nikodym theorem. Thus there exists a real positive $\mathscr{B}$-measurable function $g_\mathscr{B}$ such that

$$(g\mu)_\mathscr{B} = g_\mathscr{B} \mu$$

This is the density of the measure $(g\mu)_\mathscr{B}$ with respect to the measure $\mu_\mathscr{B}$ (cf. Eqn (6.58)).

When the function g is not positive, the same line of reasoning can be applied separately to the positive functions g^+ and $-g^-$ (cf. Eqn (6.46)). This enables us to conclude that there exists, for every integrable $\mathscr{A}$-measurable function g, a $\mathscr{B}$-measurable function $g_\mathscr{B}$ such that $(g\mu)_\mathscr{B} = g_\mathscr{B} \mu_\mathscr{B}$, which can be written in the equivalent form (cf. Eqn (6.57))

$$\forall B \in \mathscr{B} \quad \mu(g I_B) = \mu(g_\mathscr{B} I_B) \tag{6.72}$$

This relation is satisfied, in particular, for $B = X$. It follows from this that the function $g_\mathcal{B}$ is integrable and has the same integral as the function g:

$$g_\mathcal{B} \in \mathcal{L}^1(X, \mathcal{B}, \mu_\mathcal{B}) \qquad \mu(g_\mathcal{B}) = \mu(g) \tag{6.73}$$

Suppose that the $\mathcal{A}$-measurable function g is also $\mathcal{B}$ measurable. It follows from Eqn (6.72) that the $\mathcal{B}$-measurable function $g - g_\mathcal{B}$ is $\mu_\mathcal{B}$ negligible (cf. Exercise 4.2). The functions g and $g_\mathcal{B}$ are then a.e. equal:

$$g \in \mathcal{L}^1(X, \mathcal{B}, \mu_\mathcal{B}) \Rightarrow g_\mathcal{B} \underset{\text{a.e.}}{=} g \tag{6.74}$$

(b) Now let $g_\mathcal{A}$ be an $\mathcal{A}$-measurable function a.e. equal to the function $g_\mathcal{B}$. The $\mathcal{A}$-measurable function $g_\mathcal{A} - g_\mathcal{B}$ is therefore μ negligible and consequently (cf. Exercise 4.2)

$$\forall B \in \mathcal{B} \subset \mathcal{A} \qquad \mu(g_\mathcal{A} I_B) = \mu(g_\mathcal{B} I_B) = \mu(g I_B)$$

This implies that the relation (6.72) defines in fact the equivalence class in the space $L^1(X, \mathcal{A}, \mu)$ of the function $g_\mathcal{B}$ which itself belongs to the space $\mathcal{L}^1(X, \mathcal{B}, \mu_\mathcal{B}) \subset \mathcal{L}^1(X, \mathcal{A}, \mu)$.

(c) Suppose finally that the $\mathcal{A}$-measurable function $g_\mathcal{A}$ belongs to the space $\mathcal{L}^2(X, \mathcal{A}, \mu)$, which implies, the measure μ being bounded, that it belongs also to the space $\mathcal{L}^1(X, \mathcal{A}, \mu)$. The above results then apply, but it can be shown, in addition, that the function $g_\mathcal{B}$ belongs to the space $\mathcal{L}^2(X, \mathcal{B}, \mu_\mathcal{B})$. In fact let $L^2_\mathcal{B}(X, \mathcal{A}, \mu)$ be the set of equivalence classes of square integrable $\mathcal{A}$-measurable functions containing a $\mathcal{B}$-measurable function. This is the image in the space $L^2(X, \mathcal{A}, \mu)$ of the space $L^2(X, \mathcal{B}, \mu_\mathcal{B})$ through the mapping $h_\mathcal{B} \to h_\mathcal{A} \sim h_\mathcal{B}$. The mapping thus defined is linear and it obviously conserves the norm $\|.\|_2$. It follows from this that the space $L^2_\mathcal{B}(X, \mathcal{A}, \mu)$ is a complete, and therefore closed, subspace of the space $L^2(X, \mathcal{A}, \mu)$ since the space $L^2(X, \mathcal{B}, \mu_\mathcal{B})$ is complete. Let us apply the orthogonal projection theorem, denoting by $\hat{g}$ the projection of the function $g \in L^2(X, \mathcal{A}, \mu)$ onto the space $L^2_\mathcal{B}(X, \mathcal{A}, \mu)$ (cf. Section 4.1.7.2). Then

$$\forall h_\mathcal{B} \in L^2_\mathcal{B}(X, \mathcal{A}, \mu) \qquad \mu(gh_\mathcal{B}) = \mu(\hat{g}h_\mathcal{B})$$

Now, for every set $B \in \mathcal{B}$, the function I_B belongs to the space $L^2_\mathcal{B}(X, \mathcal{A}, \mu)$. Furthermore, there exists by construction a $\mathcal{B}$-measurable function $\hat{g}_\mathcal{B}$ a.e. equal to the function $\hat{g}$. It follows then from the above relation and from Eqn (6.72) that

$$\forall B \in \mathcal{B} \qquad \mu(gI_B) = \mu(\hat{g}_\mathcal{B} I_B)$$

The $\mathcal{B}$-measurable functions $g_\mathcal{B}$ and $\hat{g}_\mathcal{B}$ are therefore a.e. equal (cf. Exercise 4.2). We can therefore say, identifying classes with the functions which represent them, that the function $g_\mathcal{B}$ is the projection of the function g onto the subspace $L^2_\mathcal{B}(X, \mathcal{A}, \mu)$. The situation is illustrated in Fig. 6.4. From this the following inequalities result:

$$\|g\|_2^2 = \|g_\mathcal{B}\|_2^2 + \|g - g_\mathcal{B}\|_2^2 \geqslant \|g_\mathcal{B}\|_2^2 \tag{6.75}$$

$$\forall h_\mathcal{A} \in L^2_\mathcal{B}(X, \mathcal{A}, \mu) \qquad \|g - h_\mathcal{A}\|_2 \geqslant \|g - g_\mathcal{B}\|_2$$

314 *Elements of measure theory*

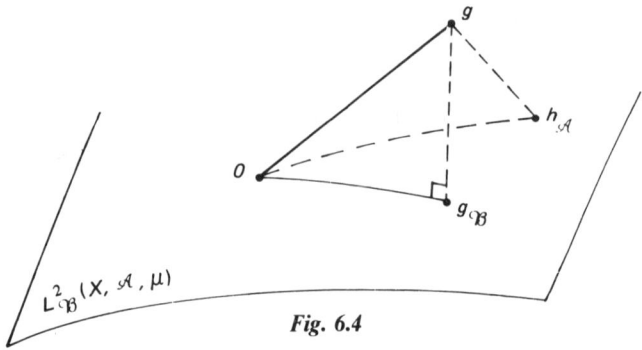

Fig. 6.4

4.2. INTEGRATION OVER A PRODUCT SPACE

4.2.1. *Generalities*

Let $(X_1 \times X_2, \mathscr{A}_1 \otimes \mathscr{A}_2)$ be a measurable product space and let μ be a positive measure over this space. The integral of a complex measurable function g defined over the product space is the double integral

$$\mu(g) = \int_{X_1} \int_{X_2} g(x_1, x_2) \, d\mu(x_1, x_2) \tag{6.76}$$

It has all the properties stated in preceding sections.

Let μ_1 and μ_2 be two positive measures defined over the measurable spaces $(X_1, \mathscr{A}_1)$ and $(X_2, \mathscr{A}_2)$ respectively and let $\mu_1 \otimes \mu_2$ be the product measure. The integral of a function g with respect to the product measure can be obtained, under certain conditions, through two simple integrals. These conditions are the subject of the following theorem.

4.2.2. *Fubini's theorem*

For a complex measurable function g, defined over the measurable space $(X_1 \times X_2, \mathscr{A}_1 \otimes \mathscr{A}_2)$, to be integrable with respect to the product measure $\mu_1 \otimes \mu_2$, it is necessary and sufficient that the function

$$x_1 \mapsto \int_{X_2} |g(x_1, x_2)| \, d\mu(x_2)$$

be μ_1 integrable over X_1 or that the function

$$x_2 \mapsto \int_{X_1} |g(x_1, x_2)| \, d\mu(x_1)$$

be μ_2 integrable over X_2. The integral is then given by the double equality

$$\mu_1 \otimes \mu_2(g) = \int_{X_1} \left\{ \int_{X_2} g(x_1, x_2) \, d\mu_2(x_2) \right\} d\mu(x_1)$$

$$= \int_{X_2} \left\{ \int_{X_1} g(x_1, x_2) \, d\mu_1(x_1) \right\} d\mu(x_2) \tag{6.77}$$

Integration 315

The functions within the braces are defined respectively a.e. over X_1 and X_2 and are integrable with respect to the measures μ_1 and μ_2 respectively.

We shall accept this theorem (Guichardet, 1969, p. 20). It enables us to change the order, under the conditions stated, of the two successive integrations.

Example 4.8 Convolution with a Dirac measure

Let μ be a positive bounded measure over the space $(\mathbb{R}, \mathscr{R})$ and let a be a point $a \in \mathbb{R}$. Let $\mu * \delta_a$ be the convolution of the measure μ and the Dirac measure δ_a. Let us put

$$\forall \beta \in \mathscr{R} \qquad A_\beta = \{(x_1, x_2) \,|\, x_1 + x_2 \in \beta\}$$

It follows from Eqns (6.35) and (6.39) that

$$\mu * \delta_a(\beta) = \mu \otimes \delta_a(I_{A_\beta})$$

Fubini's theorem can be applied to the product measure $\mu \otimes \delta_a$ and it follows that

$$\mu * \delta_a(\beta) = \int_\mathbb{R} \left\{ \int_\mathbb{R} I_{A_\beta}(x_1, x_2) \delta_a(x_2) \right\} d\mu(x_1)$$

$$= \int_\mathbb{R} I_{\beta - a}(x_1) \, d\mu(x_1)$$

$$= \mu(\beta - a)$$

where we put $\beta - a = \{x_1 \,|\, x_1 + a \in \mathscr{B}\}$. It follows from this that

$$\forall \beta \in \mathscr{R} \qquad \mu * \delta_a(\beta) = \mu(\beta - a) \qquad (6.78)$$

The convolution $\mu * \delta_a$ is therefore the measure μ translated by a (cf. Example 3.4). It thus follows, in the particular case of Lebesgue measure μ_0, that $\mu_0 * \delta_a = \mu_0$.

Example 4.9 Convolution with an absolutely continuous measure

Let μ be a positive bounded measure over the space $(\mathbb{R}, \mathscr{R})$, ψ be a positive borelian function, integrable with respect to Lebesgue measure μ_0, and v be the positive, bounded, absolutely continuous measure $v = \psi \mu$. The convolution $\mu * v$ is a positive bounded measure over the space $(\mathbb{R}, \mathscr{R})$. Let us put

$$\forall \beta \in \mathscr{R} \qquad A_\beta = \{(x_1, x_2) \,|\, x_1 + x_2 \in \beta\}$$

It follows applying Eqns (6.35) and (6.39) that

$$\mu * v(\beta) = \mu \otimes v(I_{A_\beta})$$

The function I_{A_β} is bounded by unity. It is therefore integrable with respect to the bounded measure $\mu \otimes v$ and Fubini's theorem applies. We get (cf. Eqn

(6.60))
$$\mu * v(\beta) = \int_{X_2} d\mu(x_2) \left\{ \int_{X_1} I_{A_\beta}(x_1, x_2) \psi(x_1) \, dx_1 \right\}$$

Next, noting that when x_2 is fixed we can write $I_{A_\beta}(x_1, x_2) = I_\beta(x_1 + x_2)$ and taking into account the invariance of Lebesgue measure under translation, we obtain

$$\mu * v(\beta) = \int_{X_2} d\mu(x_2) \left\{ \int_{X_1} I_\beta(x_1 + x_2) \psi(x_1) \, dx_1 \right\}$$
$$= \int_{X_2} d\mu(x_2) \left\{ \int_{X_1} I_\beta(x) \psi(x - x_2) \, dx \right\}$$

Fubini's theorem also enables us to write

$$\mu * v(\beta) = \int_{X_1} I_\beta(x) \, dx \left\{ \int_{X_2} \psi(x - x_2) \, d\mu(x_2) \right\}$$

It follows from this that the measure $\mu * v$ is absolutely continuous with respect to Lebesgue measure and has as density the positive integrable function h defined by the integral

$$h(x) = \int_{X_2} \psi(x - x_2) \, d\mu(x_2) \tag{6.79}$$

The function h can be considered as the convolution of the measure μ and the function ψ, where we write

$$\mu * (\psi \mu_0) = (\mu * \psi) \mu_0$$

In particular, the convolution of the measure μ with the Heaviside function Y, defined by Eqn (6.22), is the function

$$[\mu * Y](x) = \mu(]-\infty, x[) = \int_{-\infty}^{x} d\mu(x_2) \tag{6.80}$$

4.3. FOURIER TRANSFORMATION OF A BOUNDED MEASURE

4.3.1. Definitions

Let μ be a positive bounded measure defined over the space $(\mathbb{R}^n, \mathscr{R}^n)$. Put

$$x = (x_1, \ldots, x_n) \qquad f = (f_1, \ldots, f_n) \qquad \langle x, f \rangle = x_1 f_1 + \ldots + x_n f_n$$

The (direct) Fourier transform and the inverse Fourier transform of the measure μ are respectively the complex functions $\mathscr{F}\mu$ and $\bar{\mathscr{F}}\mu$ defined over $\mathbb{R}^n$ by the integrals

$$\mathscr{F}\mu(f) = \mu\{\exp(-i2\pi\langle x, f \rangle)\}$$
$$\bar{\mathscr{F}}\mu(f) = \mu\{\exp(i2\pi\langle x, f \rangle)\} \tag{6.81}$$

4.3.2. Properties

The functions $\mathscr{F}\mu$ and $\bar{\mathscr{F}}\mu$ are bounded and uniformly continuous.

To prove this property note that the moduli of the functions $\exp(\pm i2\pi\langle x, f\rangle)$ are bounded by unity. These functions are therefore integrable since the measure μ is bounded and it follows from Eqn (6.51) that

$$|\mathscr{F}\mu(f)| \leq \mu(\mathbb{R}^n) \qquad |\bar{\mathscr{F}}\mu(f)| \leq \mu(\mathbb{R}^n)$$

Next we write

$$|\mathscr{F}\mu(f) - \mathscr{F}\mu(f_0)| \leq 2 \int_{\mathbb{R}^n} |\sin(2\pi\langle x, f - f_0\rangle)| \, d\mu(x)$$

The function $\sin(2\pi\langle x, f - f_0\rangle)$ is bounded by the integrable function unity and tends to zero for every $x \in \mathbb{R}^n$ when f tends to f_0. The theorem of dominated convergence applies (cf. Section 4.1.3.5) and it follows from it that the first member of the above inequality tends to zero when f tends to f_0. The function $\mathscr{F}\mu$ is therefore indeed uniformly continuous and it follows immediately that the same is true of the function $\bar{\mathscr{F}}\mu$.

Example 4.10 Fourier transforms of a Dirac measure

The Fourier transforms of the Dirac measure δ_a, $a \in \mathbb{R}$, are obtained immediately from Eqns (6.81) and (6.53):

$$\mathscr{F}\delta_a(f) = \exp(-i2\pi af) \qquad \bar{\mathscr{F}}\delta_a(f) = \exp(i2\pi af) \qquad (6.82)$$

4.3.3. Property

Let μ_1 and μ_2 be two positive bounded measures defined over the space $(\mathbb{R}, \mathscr{R})$. The Fourier transforms of their convolution are the products of the transforms of the two measures:

$$\mathscr{F}(\mu_1 * \mu_2) = (\mathscr{F}\mu_1)(\mathscr{F}\mu_2) \qquad \bar{\mathscr{F}}(\mu_1 * \mu_2) = (\bar{\mathscr{F}}\mu_1)(\bar{\mathscr{F}}\mu_2) \qquad (6.83)$$

To prove this important property it is sufficient to write, applying Eqn (6.54),

$$\mathscr{F}(\mu_1 * \mu_2)(f) = \mu_1 \otimes \mu_2[\exp\{-i2\pi f(x_1 + x_2)\}]$$

Fubini's theorem applies to the product measure $\mu_1 \otimes \mu_2$ and Eqn (6.83) follows immediately from it.

4.3.4. Bochner's theorem

For a complex continuous function g, defined over $\mathbb{R}^n$, to be the Fourier transform of a positive bounded measure, defined over the space $(\mathbb{R}^n, \mathscr{R}^n)$, it is necessary and sufficient that it be positive definite, i.e. that it satisfy

$$\forall (c_i, c_j) \in \mathbb{C}^2 \qquad \forall (x_i, x_j) \in \mathbb{R}^2$$

$$\sum_{i \in I} \sum_{j \in J} c_i c_j^* g(x_i - x_j) \geq 0 \qquad (6.84)$$

for all finite subsets I and J of $\mathbb{N}$.

We will accept this theorem (Neveu, 1970).

Exercises

Exercise 4.1 Real integrable function

Let g be a real measurable integrable function, defined over a measured space $(X, \mathscr{A}, \mu)$, μ being a positive measure. Consider the sets of X defined by

$$A_n = \{x \,|\, |g(x)| \geq n\} \qquad A_\infty = \{x \,|\, |g(x)| = +\infty\}$$

(1) Prove that

$$\forall n \in \mathbb{N} \qquad \mu(A_n) \leq \frac{\mu(|g|)}{n} \qquad (6.85)$$

(Write, $\forall n \in \mathbb{N}$, $0 \leq n I_{A_n} \leq |g|$ and apply Eqn (6.50).)

(2) Deduce from this that

$$\mu(A_\infty) = 0 \qquad (6.86)$$

The set of points for which a μ-integrable function takes an infinite value is μ negligible. This property of integrable functions is very important. (Write

$$A_\infty \subset \bigcap_{n \in \mathbb{N}} A_n$$

and apply Eqn (6.27).)

(3) Prove that

$$\forall \varepsilon > 0 \qquad \exists N \in \mathbb{N} \qquad \forall n \geq N \qquad \int_{A_n} |g(x)| \, d\mu(x) < \varepsilon \qquad (6.87)$$

(Apply the theorem of dominated convergence to the sequence of functions $h_n = g I_{\bar{A}_n}$ showing that it converges a.e. to the function g, and apply Eqn (6.86).)

Exercise 4.2 Negligible function

Let g be a real measurable function defined over the measured space $(X, \mathscr{A}, \mu)$, μ being a positive measure. Prove that the function g is μ negligible if and only if

$$\forall A \in \mathscr{A} \qquad \mu(g I_A) = 0 \qquad (6.88)$$

(Argue in terms of the functions g^+ and g^- defined by Eqn (6.46) and apply proposition 4.1.4.2.)

Exercise 4.3 Absolutely continuous image measure

Let $(X, \mathcal{A}, \mu)$ be a measured space, let ν be a measure absolutely continuous with respect to the measure μ, and let k be a measurable mapping $k:(X, \mathcal{A}, \mu) \to (Y, \mathcal{B})$.

(1) Verify that the image measure ν_k is absolutely continuous with respect to the image measure μ_k. (Apply Eqn (6.28).)

(2) Let ψ and ψ_k be the densities of the measures ν and ν_k. Verify that they are connected by the relation

$$\forall B \in \mathcal{B} \quad \mu\{\psi(I_B \circ k)\} = \mu_k(\psi_k I_B) \qquad (6.89)$$

(Apply Eqn (6.57).)

Exercise 4.4 Limit of a scalar product

Let (g_m) and (h_n) be two sequences of functions converging in $L^2(X, \mathcal{A}, \mu)$ to the functions g and h respectively. Prove directly, using Schwarz's inequality, that

$$\lim_{m,n \to \infty} \langle g_m | h_n \rangle = \langle g | h \rangle \qquad (6.90)$$

Exercise 4.5 Limit of a scalar product

Let $g_n \in L^2(X, \mathcal{A}, \mu)$ be a sequence of functions. Show that, for the sequence to converge in L^2 to a function g, it is necessary and sufficient that

$$\lim_{m,n \to \infty} \langle g_m | g_n \rangle = \langle g | g \rangle$$

Exercise 4.6 Passage to the limit under the integration sign

Let μ be a bounded measure and (g_n) be a sequence of functions converging in $L^2(X, \mathcal{A}, \mu)$ to a function g. Prove that the sequence $\mu(g_n)$ converges in $\mathbb{C}$ to $\mu(g)$:

$$\lim_{n \to \infty} \mu(g_n) = \mu\left(\lim_{n \to \infty} g_n\right) \qquad (6.91)$$

(Apply Schwarz's inequality (6.70).)

Exercise 4.7 Integration

Let $(X, \mathcal{A}, \mu)$ be a space and $\mathcal{B} \subset \mathcal{A}$ be a sub-Borel field. Let g be a real $\mathcal{A}$-measurable function and let h be a $\mathcal{B}$-measurable function such that the function gh is integrable. Prove that

$$\mu(gh) = \mu(g_{\mathcal{B}} h) \qquad (6.92)$$

the function $g_{\mathcal{B}}$ being defined by Eqn (6.72).

Exercise 4.8 Dominated convergence

(1) Let g be the positive function defined over $\mathbb{R}$ by

$$g(x) = \left\{\frac{\sin(\pi a x)}{\pi a x}\right\}^2 \quad a \in \mathbb{R}$$

and let μ be a positive bounded measure, defined over $(\mathbb{R}, \mathscr{R})$. Prove that

$$\lim_{|a| \to \infty} \mu(g) = \mu(\{0\})$$

(2) Let μ be a positive bounded measure over $(\mathbb{R}, \mathscr{R})$ such that the positive function $1/x^2$ defined over $\mathbb{R}$ is μ integrable. Prove that

$$\lim_{|a| \to \infty} \int_{\mathbb{R}} \frac{1}{2a} \left[\int_{-a}^{+a} \left\{ \frac{\sin(\pi t x)}{\pi x} \right\}^2 dt \right] d\mu(x) = \int_{\mathbb{R}} \frac{1}{2\pi^2 x^2} d\mu(x)$$

(Note that $\mu(\{0\}) = 0$ and apply the theorem of dominated convergence.)

Exercise 4.9 Absolutely continuous measure

Let $(X, \mathscr{A})$ be a measurable space and let μ and ν be two positive measures. The positive measure $\lambda = \mu + \nu$ is defined by

$$\forall A \in \mathscr{A} \qquad \lambda(A) = \mu(A) + \nu(A)$$

(1) Verify that the measure μ is absolutely continuous with respect to the measure λ.

(2) Let ψ be the density of the measure μ with respect to the measure λ. Let A be the set $A = \{\psi \geq 1\}$. Prove that $\nu(A) = 0$. (Apply Eqn (6.57) to the set A and use the inequality $I_A \psi \geq I_A$.)

(3) Let B be the set $B = \{\psi > 1\}$. Prove that $\mu(B) = 0$ and then that $\lambda(B) = 0$. (Apply Eqn (6.57) to the set A and apply the increasing property of the integral to the inequality $I_A \psi \geq I_A$.)

Exercise 4.10 Absolutely continuous measure

Let $(X, \mathscr{A}, \mu)$ be a measured space and let ν be a measure absolutely continuous with respect to the measure μ, having as density the function ψ. Let $\mathscr{B} \subset \mathscr{A}$ be a sub-Borel field of the set X and let $\mu_{\mathscr{B}}$ and $\nu_{\mathscr{B}}$ be the restrictions of the measures μ and ν to the sub-Borel field $\mathscr{B}$.

(1) Verify that the measure $\nu_{\mathscr{B}}$ is absolutely continuous with respect to the measure $\mu_{\mathscr{B}}$.

(2) Verify that the density of the measure $\nu_{\mathscr{B}}$ with respect to the measure $\mu_{\mathscr{B}}$ is the function $\psi_{\mathscr{B}}$ associated with the function ψ by

$$\forall B \in \mathscr{B} \qquad \mu(\psi I_B) = \mu(\psi_{\mathscr{B}} I_B)$$

(Reason as in Section 4.1.8 and apply Eqn (6.72).)

Exercise 4.11 Absolutely continuous measures

With the assumptions and notation used in Section 4.1.8, suppose that the sub-Borel field $\mathscr{B}$ is generated by a measurable mapping $k: (X, \mathscr{A}, \mu) \to (Z, \mathscr{C})$. Let μ_k be the image measure of the measure μ through the mapping k. Prove that there exists a real measurable function $g_{\mathscr{C}}$, defined over the space $(Z, \mathscr{C})$, such that the real $\mathscr{B}$-measurable functions $g_{\mathscr{B}}$ and $g_{\mathscr{C}} \circ k$ are a.e. equal and that

$$\forall C \in \mathscr{C} \qquad \mu_k(g_\mathscr{C} I_C) = \mu\{g(I_C \circ k)\} \qquad (6.93)$$

We can say, identifying the equivalence classes with the functions which belong to them, that $g_\mathscr{A}$ is the composite function $g_\mathscr{C} \circ k$, the function $g_\mathscr{C}$ being defined by Eqn (6.93). (Argue as in Section 4.1.8, beginning by supposing that the function g is positive and showing that the image measure $(g\mu)_k$ is absolutely continuous with respect to the image measure μ_k. The corresponding density is then the function $g_\mathscr{C}$. Use $I_{k^{-1}(C)} = I_C \circ k$.)

References

BERTRANDIAS P. (1969): *Analyse Fonctionelle*, Colin, Paris.
DIEUDONNÉ J. (1969): *Eléments d'Analyse*, Vol. I, Gauthier-Villars, Paris.
GUICHARDET A. (1969): *Calcul Intégral*, Colin, Paris.
MARLE C.-M. (1974): *Mesures et Probabilités*, Hermann, Paris.
NEVEU J. (1964): *Bases Mathématiques du Calcul des Probabilités*, p. 25, Masson, Paris.
NEVEU J. (1970): *Cours de l'Ecole Polytechnique*, Ecole Polytechnique, Paris.

Appendix 1
Table of Fourier series

Table 1 gives the Fourier coefficients c_n, $n \in \mathbb{Z}$, of a certain number of periodic distributions of period T. They are denoted by U in the general case and by s when they are defined by a function. The distribution U_T (or function s_T) is obtained by truncation of the distribution U (or of the function s) in the interval $[-T/2, T/2[$, which enables us to write

$$U(t) = U_T(t) * \sum_{n \in \mathbb{Z}} \delta(t - nT)$$

The expansions in a Fourier series of the distribution U and its coefficients c_n are defined by

$$U(t) = \sum_{n \in \mathbb{Z}} c_n \exp\left(in\frac{2\pi t}{T}\right)$$

and

$$c_n = \frac{1}{T} \int_{\mathbb{R}} U_T(t) \exp\left(-in\frac{2\pi t}{T}\right) dt$$

or

$$c_n = \frac{1}{T} \int_a^{a+T} s(t) \exp\left(-in\frac{2\pi t}{T}\right) dt \qquad a \in \mathbb{R}$$

Table 2 gives, as a function of the coefficient c_n of the distribution U, the coefficients, of the same order, of a certain number of distributions defined in terms of the distribution U.

The coefficients a_n and b_n of the sine and cosine series are given in terms of the coefficients c_n by

$$a_0 = c_0 \qquad a_n = c_n + c_{-n} \qquad b_n = i(c_n - c_{-n})$$

Table 1

Expression or graph of U_T or s_T in the interval $[-T/2, T/2[$, $\vartheta \in]0, T]$		Coefficient c_n
①	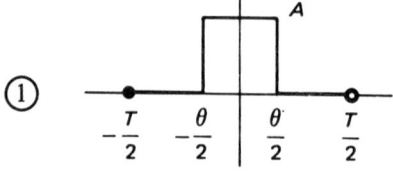	$c_n = \dfrac{A\vartheta}{T} \dfrac{\sin(n\pi\vartheta/T)}{n\pi\vartheta/T}$
②	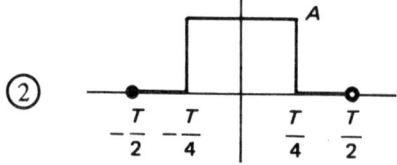	$c_0 = \dfrac{A}{2} \quad c_{2n} = 0$ $c_{2n+1} = (-1)^n \dfrac{A}{(2n+1)\pi}$
③	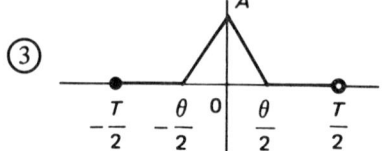	$c_n = \dfrac{A\vartheta}{2T} \dfrac{\sin^2(n\pi\vartheta/2T)}{(n\pi\vartheta/2T)^2}$
④	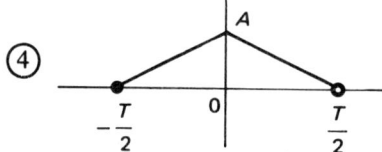	$c_0 = \dfrac{A}{2} \quad c_{2n} = 0$ $c_{2n+1} = \dfrac{2A}{\pi^2(2n+1)^2}$
⑤	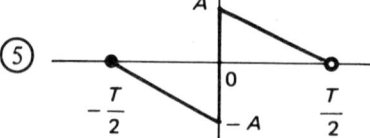	$c_0 = A$ $c_n = \dfrac{A}{in\pi}$
⑥	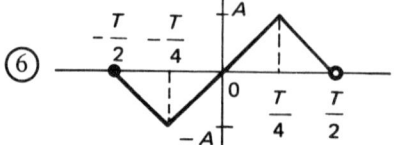	$c_0 = 0 \quad c_{2n} = 0$ $c_{2n+1} = (-1)^{n+1} \dfrac{i4A}{\pi^2(2n+1)^2}$

Table of Fourier series

Table 1—continued

	Expression or graph of U_T or s_T in the interval $[-T/2, T/2[$, $\vartheta \in \,]0, T]$	Coefficient c_n
⑦	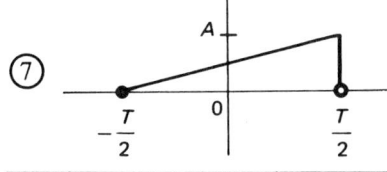	$c_0 = \dfrac{A}{2}$ $c_n = (-1)^{n+1} \dfrac{A}{in2\pi}$
⑧	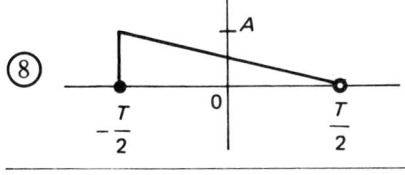	$c_0 = \dfrac{A}{2}$ $c_n = (-1)^n \dfrac{A}{in2\pi}$
⑨	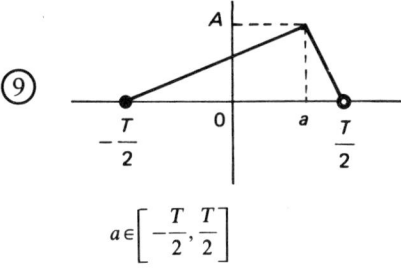 $a \in \left[-\dfrac{T}{2}, \dfrac{T}{2}\right]$	$c_0 = \dfrac{A}{2}$ $c_n = \dfrac{A}{n^2\pi^2} \dfrac{1}{1-(2a/T)^2}$ $\times \left\{\exp\left(-in\pi \dfrac{2a}{T}\right) - (-1)^n\right\}$
⑩	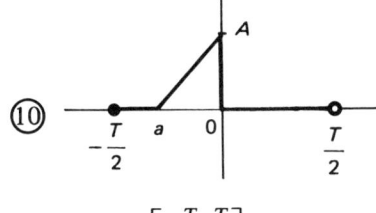 $a \in \left[-\dfrac{T}{2}, \dfrac{T}{2}\right]$	$c_0 = \dfrac{A\vartheta}{2T}$ $c_n = \dfrac{A}{in2\pi} \left\{\exp\left(in\pi \dfrac{a}{T}\right) \dfrac{\sin(n\pi a/T)}{n\pi a/T} - 1\right\}$
⑪	$s_T(t) = \begin{cases} A\cos(\pi t/\vartheta) & t \in [-\vartheta/2, \vartheta/2] \\ 0 & \text{otherwise} \end{cases}$ 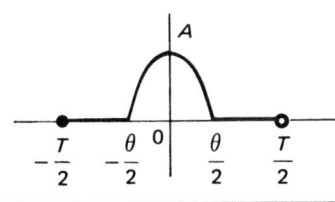	$c_n = -\dfrac{2AT}{\pi\vartheta} \dfrac{\cos(n\pi\vartheta/T)}{4n^2 - T^2/\vartheta^2}$

326 Appendix 1

Table 1—continued

Expression or graph of U_T or s_T in the interval $[-T/2, T/2[$, $\vartheta \in]0, T]$	Coefficient c_n

$s_T(t) = A \cos\left(\pi \dfrac{t}{T}\right)$

(12)

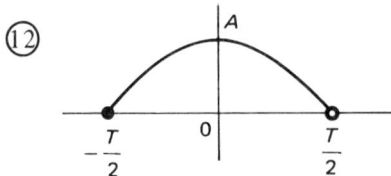

$$c_n = (-1)^{n+1} \dfrac{2A}{\pi} \dfrac{1}{4n^2 - 1}$$

$s_T(t) = \begin{cases} \dfrac{A}{1 - \cos \alpha}\left\{\cos\left(2\pi \dfrac{t}{T}\right) - \cos \alpha\right\} \\ \qquad\qquad t \in \left[-\dfrac{\vartheta}{2}, \dfrac{\vartheta}{2}\right] \\ \qquad\qquad \text{otherwise} \\ 0 \end{cases}$

(13) with $\alpha = \pi \dfrac{\vartheta}{T}$

$$c_n = \dfrac{A}{\pi}\left\{\dfrac{\sin(n\alpha)\cos \alpha - n\cos(n\alpha)\sin \alpha}{n(n^2 - 1)(1 - \cos \alpha)}\right\}$$

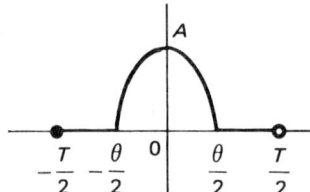

$s_T(t) = \begin{cases} \dfrac{A}{2}\left\{1 + \cos\left(2\pi \dfrac{t}{\vartheta}\right)\right\} & t \in \left[-\dfrac{\vartheta}{2}, \dfrac{\vartheta}{2}\right] \\ 0 & \text{otherwise} \end{cases}$

(14)

$$c_0 = \dfrac{A\vartheta}{2T}$$

$$c_n = \dfrac{A}{n 2\pi} \dfrac{(n\vartheta/T)^2}{1 - (n\vartheta/T)^2} \sin\left(n\pi \dfrac{\vartheta}{T}\right)$$

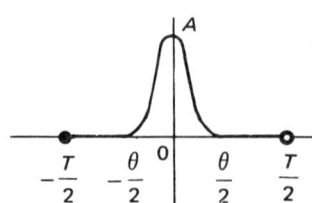

Table 1—continued

Expression or graph of U_T or s_T in the interval $[-T/2, T/2[$, $\vartheta \in]0, T]$	Coefficient c_n
$s_T(t) = A\left(\dfrac{2t}{T}\right)^2$ 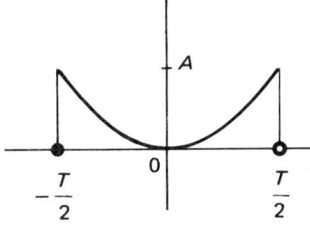	$c_0 = \dfrac{A}{3}$ $c_n = (-1)^n \dfrac{2A}{n^2\pi^2}$
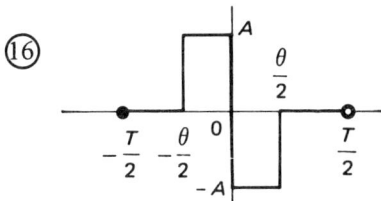	$c_n = i\dfrac{2A}{n\pi}\sin^2\left(\dfrac{n\pi}{2}\dfrac{\vartheta}{T}\right)$
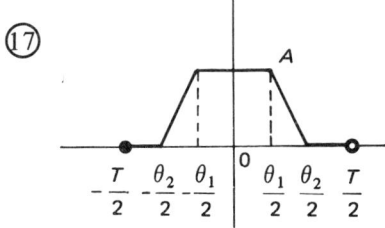	$c_n = \dfrac{A}{n^2\pi^2}\dfrac{2T}{\vartheta_2 - \vartheta_1}$ $\times \sin\left(n\pi\dfrac{\vartheta_2 - \vartheta_1}{2T}\right)\sin\left(n\pi\dfrac{\vartheta_1 + \vartheta_2}{2T}\right)$
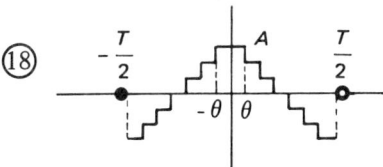 $T = 2(2N+1)\vartheta \qquad N \geqslant 1$	$c_{2n} = 0$ $c_{2n+1} = \dfrac{A}{(2n+1)\pi N}$ $\times \left[\tan\left\{(2n+1)\dfrac{\pi}{2N}\right\}\right]^{-1}$

⑮

⑯

⑰

⑱

328 Appendix 1

Table 1—continued

Expression or graph of U_T or s_T in the interval $[-T/2, T/2[,\quad \vartheta \in]0, T]$	Coefficient c_n

$s_T(t) = \exp(t)$

⑲

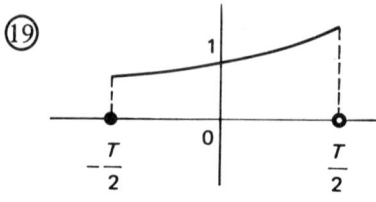

$$c_n = \frac{2}{T} \sinh\left(\frac{T}{2}\right) \frac{(-1)^n}{1 - in2\pi/T}$$

$s_T(t) = \exp(-t)$

⑳

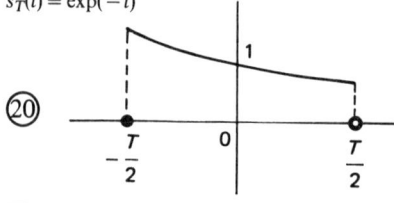

$$c_n = \frac{2}{T} \sinh\left(\frac{T}{2}\right) \frac{(-1)^n}{1 + in2\pi/T}$$

㉑

$$U_T = vp\,\frac{1}{\sin(2\pi t/T)}$$

$c_{2n} = 0$
$c_{2n+1} = -i \quad n > 0$
$c_{-(2n+1)} = i \quad n > 0$

㉒

$$U_T = vp\,\cot(\pi t/T)$$

$c_0 = 0$
$c_n = -i \quad n > 0$
$c_n = i \quad n < 0$

$U_T = \delta(t - a) \qquad a \in \left[-\dfrac{T}{2}, \dfrac{T}{2}\right[$

㉓

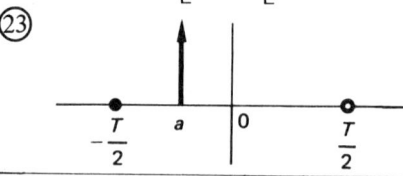

$$c_n = \frac{1}{T} \exp\left(-in2\pi\,\frac{a}{T}\right)$$

㉔

$s_T(t) = \exp\{ia \sin(2\pi t/T)\}$

$c_n = J_n(a) \qquad a \in \mathbb{R}$

㉕

$s_T(t) = \exp\{ia \cos(2\pi t/T)\}$

$c_n = i^n J_n(a) \qquad a \in \mathbb{R}$

㉖

$s_T(t) = \cos\{a \sin(2\pi t/T)\}$

$c_{2n} = J_{2n}(a) \qquad c_{cn+1} = 0 \qquad a \in \mathbb{R}$

㉗

$s_T(t) = \sin\{a \sin(2\pi t/T)\}$

$c_{2n} = 0 \qquad c_{2n+1} = J_{2n+1}(a) \qquad a \in \mathbb{R}$

Table 2

Distribution		nth coefficient	
$\check{U}(t) = U(-t)$		c_{-n}	
$U^*(t)$		$c_{-n}{}^*$	
$U(t - \vartheta)$	$\vartheta \in \mathbb{R}$	$c_n \exp\left(-in\dfrac{2\pi\vartheta}{T}\right)$	
$U(t) + k$	$k \in \mathbb{C}$	c_n	$n \in \mathbb{Z}^*$
		$c_0 + k$	$n = 0$
$U^{(m)}$	$m \in \mathbb{R}$	$\left(in\dfrac{2\pi}{T}\right)^m c_n$	

Appendix 2
Table of Fourier transforms

Table 1, read from left to right, gives the Fourier transforms $\mathscr{F}U(f)$ of a certain number of common tempered distributions $U(t)$. The inverse Fourier transforms $\bar{\mathscr{F}}U(f)$ can be deduced from them according to

$$\bar{\mathscr{F}}U(f) = \mathscr{F}U(-f)$$

Table 1, read from right to left, gives the inverse Fourier transforms $\bar{\mathscr{F}}V(t)$ of a certain number of tempered distributions $V(f)$. The Fourier transforms $\mathscr{F}V(t)$ can be deduced from them by using

$$\mathscr{F}V(t) = \bar{\mathscr{F}}V(-t)$$

The transformations $\mathscr{F}$ and $\bar{\mathscr{F}}$ are inverses:

$$U = \mathscr{F}\bar{\mathscr{F}}U = \bar{\mathscr{F}}\mathscr{F}U$$

The Fourier transform of the distribution $\check{U}(t) = U(-t)$ can be obtained from

$$\mathscr{F}\check{U}(f) = \mathscr{F}U(-f)$$

The Fourier transform of the conjugate distribution $U^*(t)$ can be obtained from

$$\mathscr{F}U^*(f) = \{\mathscr{F}U(-f)\}^*$$

The Fourier transform of the distribution

$$V(t) = U(at+b) \qquad a \in \mathbb{R}^*, b \in \mathbb{R}$$

can be obtained from

$$\mathscr{F}V(f) = \frac{1}{|a|} \exp\left(\frac{i2\pi bf}{a}\right) \mathscr{F}U\left(\frac{f}{a}\right)$$

Fourier transformation interchanges convolution and direct product (when each of these has a meaning):

$$U(t) * V(t) \underset{\mathscr{F}}{\overset{\mathscr{F}}{\rightleftarrows}} \mathscr{F}U(f)\mathscr{F}V(f)$$

Table of Fourier transforms

Notation

The notation $\mathring{I}_a, a \in \mathbb{R}_+^*$, denotes the indicator function of the interval $[-a/2, a/2]$. It is the gate function of width a, centred at the origin.

The Heaviside function Y is the indicator function of the interval $]0, +\infty[$.

The function Θ is the error function. It is also denoted by erf:

$$\Theta(t) = \text{erf}(t) = \frac{2}{\pi^{1/2}} \int_0^t \exp(-x^2)\, dx$$

The notations p.v. and p.f. signify respectively Cauchy principal value and pseudofunction.

γ is Euler's constant:

$$\gamma = -\int_0^\infty \ln u \exp(-u)\, du = 0.57721\ldots$$

$$\psi(1) = \gamma \qquad \forall n \in \mathbb{N}^* \qquad \psi(n+1) = -\gamma + 1 + \ldots + \frac{1}{n}$$

$$\cos \vartheta = f/a \qquad f \in [-a, +a] \qquad \vartheta \in [0, \pi] \qquad a \in \mathbb{R}_+^*$$

$$\sin \vartheta = \left(1 - \frac{f^2}{a^2}\right)^{1/2}$$

Table 1

$U(t)$		$\xrightarrow{\mathscr{F}}$	$\mathscr{F}U(f)$
$\bar{\mathscr{F}}V(t)$		$\xleftarrow{\mathscr{F}}$	$V(f)$
$\delta(t)$			1
$\delta(t-a)$	$a \in \mathbb{R}$		$\exp(-i2\pi a f)$
$\delta^{(n)}(t)$	$n \in \mathbb{N}^*$		$(i2\pi f)^n$
$\sum_{n \in \mathbb{Z}} \delta(t - nT)$	$T \in \mathbb{R}_+^*$		$\frac{1}{T} \sum_{n \in \mathbb{Z}} \delta\left(f - \frac{n}{T}\right) = \sum_{n \in \mathbb{Z}} \exp(-i2\pi n T f)$
$Y(t)$			$\frac{1}{2}\delta(f) + \frac{1}{i2\pi f} \quad \left(\text{p.v.}\left(\frac{1}{f}\right)\right)$
sgn t		(p.v)	$\dfrac{1}{i\pi f}$
t^n	$n \in \mathbb{N}^*$		$\dfrac{(-1)^n}{(i2\pi)^n} \delta^{(n)}(f)$

Table 1—continued

	$U(t)$			$\mathscr{F}U(f)$
		$\xrightarrow{\mathscr{F}}$		
	$\mathscr{\bar F}V(t)$	$\xleftarrow{\mathscr{F}}$		$V(f)$
	$Y(t)t^n$	$n \in \mathbb{N}^*$	(p.f.)	$\dfrac{1}{2}\left(\dfrac{i}{2\pi}\right)^n \delta^{(n)}(f) + \dfrac{n!}{(i2\pi f)^n}$
	$(\operatorname{sgn} t)t^n$	$n \in \mathbb{N}^*$	(p.f.)	$\dfrac{2(n)!}{(i2\pi f)^{n+1}}$
	$\|t\|^{2n+1}$	$n \in \mathbb{N}$	(p.f.)	$(-1)^{n+1}\dfrac{2(2n+1)!}{(2\pi f)^{2n+2}}$
(p.f.)	$Y(t)t^v$	$\operatorname{Re} v \notin \mathbb{Z}$	(p.f.)	$\dfrac{\Gamma(v+1)}{(i2\pi f)^{v+1}}$
(p.f.)	$\dfrac{1}{t^{n+1}}$	$n \in \mathbb{N}$		$(-1)^{n+1}i\pi(\operatorname{sgn} f)\dfrac{(i2\pi f)^n}{n!}$
(p.f.)	$\dfrac{Y(t)}{t^{n+1}}$	$n \in \mathbb{N}$		$(-1)^{n+1}\dfrac{(i2\pi f)^n}{n!} \times$ $\times \left\{\ln(2\pi\|f\|) + i\dfrac{\pi}{2}(\operatorname{sgn} f) - \psi(n+1)\right\}$
(p.f.)	$\dfrac{1}{\|t\|^{2n+1}}$	$n \in \mathbb{N}$		$(-1)^{n+1}2\dfrac{(2\pi f)^{2n}}{(2n)!} \times \{\ln(2\pi\|f\|) - \psi(2n+1)\}$
	$\exp(-a\|t\|)$	$a \in \mathbb{R}_+^*$		$\dfrac{2a}{a^2 + 4\pi^2 f^2}$
	$\exp(-a^2 t^2)$	$a \in \mathbb{R}$		$\dfrac{\pi^{1/2}}{a}\exp\left(-\dfrac{\pi^2 f^2}{a^2}\right)$
	$\exp(i2\pi a t^2)$	$a \in \mathbb{R}$		$\dfrac{1+i}{2a^{1/2}}\exp\left(-\dfrac{i\pi}{2a}f^2\right)$
	$\cos(2\pi f_0 t + \varphi_0)$	$f_0, \varphi_0 \in \mathbb{R}$		$\dfrac{1}{2}\{\exp(i\varphi_0)\,\delta(f - f_0) + \exp(-i\varphi_0)\,\delta(f + f_0)\}$
	$\sin(2\pi f_0 t + \varphi_0)$	$f_0, \varphi_0 \in \mathbb{R}$		$\dfrac{1}{2i}\{\exp(i\varphi_0)\,\delta(f - f_0) - \exp(-i\varphi_0)\,\delta(f + f_0)\}$
(p.v.)	$\dfrac{1}{\sin(2\pi f_0 t)}$	$f_0 \in \mathbb{R}$		$\displaystyle\sum_{n \in \mathbb{N}}[i\delta\{f + (2n+1)f_0\} - i\delta\{f - (2n+1)f_0\}]$
(p.v.)	$\dfrac{1}{\cos(2\pi f_0 t)}$	$f_0 \in \mathbb{R}$		$\displaystyle\sum_{n \in \mathbb{N}}[(-1)^n\delta\{f + (2n+1)f_0\} + (-1)^n\delta\{f - (2n+1)f_0\}]$

Table of Fourier transforms

Table 1—continued

$U(t)$ $\quad\overset{\mathscr{F}}{\to}\quad$ $\mathscr{F}U(f)$		
$\bar{\mathscr{F}}V(t)$ $\quad\overset{\bar{\mathscr{F}}}{\leftarrow}\quad$ $V(f)$		
$Y(t)\exp(-ct)\dfrac{t^{n-1}}{(n-1)!}$	$n\in\mathbb{N}$ $\operatorname{Re}c\in\mathbb{R}_+^*$	$\dfrac{1}{(c+i2\pi f)^n}$
$Y(t)\exp(-ct)\cos(2\pi f_0 t)$	$f_0\in\mathbb{R}$ $\operatorname{Re}c\in\mathbb{R}_+^*$	$\dfrac{c+i2\pi f}{(c+i2\pi f)^2+4\pi^2 f_0^2}$
$Y(t)\exp(-ct)\sin(2\pi f_0 t)$	$f_0\in\mathbb{R}$ $\operatorname{Re}c\in\mathbb{R}_+^*$	$\dfrac{2\pi f_0}{(c+i2\pi f)^2+4\pi^2 f_0^2}$
$\mathring{I}_{2T}(t)$	$T\in\mathbb{R}_+^*$	$\dfrac{\sin(2\pi Tf)}{\pi f}$
$(\operatorname{sgn}t)\mathring{I}_{2T}(t)$	$T\in\mathbb{R}_+^*$	$\dfrac{2\sin^2(\pi Tf)}{i\pi f}$
$\sin\left(\dfrac{\pi t}{T}\right)\mathring{I}_{2T}(t)$	$T\in\mathbb{R}_+^*$	$\dfrac{2T\sin^2(\pi Tf)}{i\pi(1-4f^2T^2)}$
$\cos\left(\dfrac{\pi t}{2T}\right)\mathring{I}_{2T}(t)$	$T\in\mathbb{R}_+^*$	$\dfrac{4T\cos(2\pi Tf)}{\pi(1-16f^2T^2)}$
$\dfrac{1}{2}\left\{1+\cos\left(\dfrac{\pi t}{T}\right)\right\}\mathring{I}_{2T}(t)$	$T\in\mathbb{R}_+^*$	$\dfrac{\sin(2\pi Tf)}{2\pi f}\dfrac{1}{1-4f^2T^2}$
$\left(1-\dfrac{|t|}{T}\right)\mathring{I}_{2T}(t)$	$T\in\mathbb{R}_+^*$	$T\left\{\dfrac{\sin(\pi Tf)}{\pi Tf}\right\}^2$
$\left(1-\dfrac{|t|}{T}\right)(\operatorname{sgn}t)\mathring{I}_{2T}(t)$	$T\in\mathbb{R}_+^*$	$\dfrac{1}{i\pi f}\left\{1-\dfrac{\sin(2\pi Tf)}{2\pi f}\right\}$
$\ln|t|$		$-(\ln 2\pi+\gamma)\delta(f)-\dfrac{1}{2|f|}\left(\text{p.f.}\dfrac{1}{|f|}\right)$
$\Theta(at)$	$a\in\mathbb{R}_+^*$ (p.v.)	$\dfrac{1}{i\pi f}\exp\left(-\dfrac{\pi^2 f^2}{a^2}\right)$
$J_{2n}(2\pi at)$	$n\in\mathbb{N}\quad a\in\mathbb{R}_+^*$	$\dfrac{(-1)^n}{\pi a}\dfrac{\cos(2n\vartheta)}{\sin\vartheta}\mathring{I}_{2a}(f)$
$J_{2n+1}(2\pi at)$	$n\in\mathbb{N}\quad a\in\mathbb{R}_+^*$	$\dfrac{(-1)^{n+1}i}{\pi a}\dfrac{\cos\{(2n+1)\vartheta\}}{\sin\vartheta}\mathring{I}_{2a}(f)$

Appendix 3
Table of Laplace transforms

Table 1 gives the Laplace transforms $\mathscr{L}U(p)$, where

$$p = \sigma + i2\pi f$$

of a certain number of common tempered distributions $U(t)$. They are two-sided transforms. They are reduced to one-sided transforms when the initial distribution is multiplied by the Heaviside function $Y(t)$.

The function $\mathscr{L}U(p)$ is holomorphic within a band specified by a condition to be satisfied by $\sigma = \operatorname{Re} p$.

Laplace transformation is connected with Fourier transformation by the relations

$$\mathscr{L}U(p) = \mathscr{F}\{\exp(-\sigma t)\, U(t)\}(f) \qquad p = \sigma + i2\pi f$$

The Laplace transform of the distribution $\check{U}(t) = U(-t)$ can be obtained from

$$\mathscr{L}\check{U}(p) = \mathscr{L}U(-p)$$

The Laplace transform of the conjugate distribution $U^*(t)$ can be obtained from

$$\mathscr{L}U^*(p) = \{\mathscr{L}U(p^*)\}^*$$

The Laplace transform transforms the convolution into a direct product (when each of these has a meaning, which is always the case for distributions whose support is contained in $\mathbb{R}_+$ (distributions belonging to $\mathscr{D}'_+$)):

$$U(t) * V(t) \underset{\mathscr{L}^{-1}}{\overset{\mathscr{L}}{\rightleftarrows}} \mathscr{L}U(p)\mathscr{L}V(p)$$

Notation
The notation is that of Appendix 2.

Table 1

$U(t)$		$\mathcal{L}U(p)$	
$\delta(t)$		1	$\forall \sigma$
$\delta(t-a)$	$a \in \mathbb{R}$	$\exp(-ap)$	$\forall \sigma$
$\delta^{(n)}(t)$	$n \in \mathbb{N}^*$	p^n	$\forall \sigma$
$\sum_{n \in \mathbb{N}} \delta(t-nT)$	$T \in \mathbb{R}_+^*$	$\dfrac{1}{1-\exp(-pT)}$	$\sigma > 0$
$Y(t)$		$\dfrac{1}{p}$	$\sigma > 0$
$\sum_{n \in \mathbb{N}} Y(t-nT)$	$T \in \mathbb{R}_+^*$	$\dfrac{1}{p\{1-\exp(-pT)\}}$	$\sigma > 0$
$\sum_{n \in \mathbb{N}} Y(t-nT) \dfrac{(t-nT)^n}{n!}$	$T \in \mathbb{R}_+^*$	$\dfrac{1}{p-\exp(-pT)}$	$\sigma > 0$
$Y(t) \dfrac{t^{n-1}}{(n-1)!} \exp(ct)$	$n \in \mathbb{N}^* \quad c \in \mathbb{C}$	$\dfrac{1}{(p-c)^n}$	$\sigma > \operatorname{Re} c$
$Y(t) t^\nu \exp(ct)$	$\operatorname{Re} \nu > -1 \quad c \in \mathbb{C}$	$\dfrac{\Gamma(\nu+1)}{(p-c)^{\nu+1}}$	$\sigma > \operatorname{Re} c$
$Y(t) \exp(i\omega_0 t)$	$\omega_0 \in \mathbb{R}$	$\dfrac{1}{p-i\omega_0}$	$\sigma > 0$
$Y(t) \exp(ct) \cos(\omega_0 t + \varphi_0)$	$c \in \mathbb{C}$	$\dfrac{(p-c)\cos\varphi_0 - \omega_0 \sin\varphi_0}{(p-c)^2 + \omega_0^2}$	$\sigma > \operatorname{Re} c$
$Y(t) \exp(ct) \sin(\omega_0 t + \varphi_0)$	$c \in \mathbb{C}$	$\dfrac{(p-c)\sin\varphi_0 + \omega_0 \cos\varphi_0}{(p-c)^2 + \omega_0^2}$	$\sigma > \operatorname{Re} c$
$Y(t) \exp(ct) \dfrac{1}{(\pi t)^{1/2}}$		$\dfrac{1}{(p-c)^{1/2}}$	$\sigma > \operatorname{Re} c$
$\dfrac{Y(t)}{t}$	(p.f.)	$-\ln p - \gamma$	$\sigma > 0$
$\dfrac{Y(t) \sin t}{t}$	(p.f.)	$\dfrac{\pi}{2} - \arctan p$	$\sigma > 0$
$\dfrac{Y(t) \cos t}{t}$		$-\gamma - \ln\{(1+p^2)^{1/2}\}$	$\sigma > 0$
$\dfrac{1}{a\pi^{1/2}} Y(t) \exp\left(-\dfrac{t^2}{4a^2}\right)$	$a \in \mathbb{R}_+^*$	$\exp(a^2 p^2)\{1 - \Theta(ap)\}$	$\sigma > 0$

Table 1—continued

$U(t)$		$\mathscr{L}U(p)$	
$Y(t)\Theta(t)$		$\dfrac{1}{p}\exp\left(\dfrac{p^2}{4}\right)\left\{1-\Theta\left(\dfrac{p}{2}\right)\right\}$	$\sigma>0$
$Y(t)\ln t$		$-\dfrac{\gamma+\ln p}{p}$	$\sigma>0$
$Y(t)J_n(at)$	$n\in\mathbb{N}$, $a\in\mathbb{R}_+^*$	$\dfrac{\{(p^2+a^2)^{1/2}-p\}^n}{a^n(p^2+a^2)^{1/2}}$	$\sigma>0$
$Y(t)I_n(at)$	$n\in\mathbb{N}$, $a\in\mathbb{R}_+^*$	$\dfrac{\{(p^2-a^2)^{1/2}-p\}^n}{a^n(p^2-a^2)^{1/2}}$	$\sigma>0$
$Y(t)\Theta\{(at)^{1/2}\}$	$a\in\mathbb{R}_+^*$	$\dfrac{a^{1/2}}{p(p+a)^{1/2}}$	$\sigma>0$
$\dfrac{Y(t)}{t^{n+1}}$	$n\in\mathbb{N}^*$ (p.f.)	$(-1)^{n+1}\dfrac{p^n}{n!}\{\ln p-\psi(n+1)\}$	
$\exp(-a\lvert t\rvert)$	$a\in\mathbb{R}_+^*$	$\dfrac{2a}{a^2-p^2}$	$\sigma\in\,]-a,+a[$
$\mathring{I}_T(t)$	$T\in\mathbb{R}_+^*$	$\dfrac{\sinh(pT/2)}{p/2}$	$\forall\sigma$

# Appendix 4
Table of numerical values of the function $\Phi(x)$

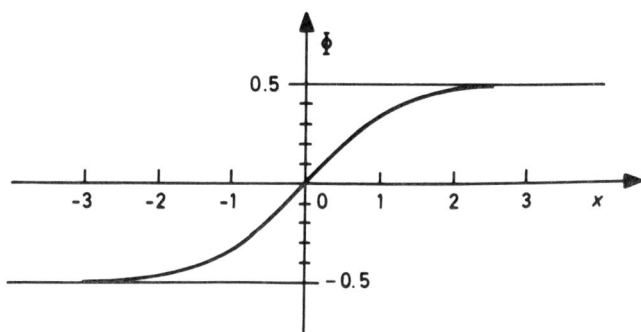

$$\Phi(x) = \frac{1}{(2\pi)^{1/2}} \int_0^x \exp\left(-\frac{u^2}{2}\right) du$$

$$= \frac{1}{2} \operatorname{erf}\left(\frac{x}{2^{1/2}}\right)$$

The series expansion is

$$\Phi(x) = \frac{1}{(2\pi)^{1/2}} \sum_{n=1}^{\infty} (-1)^{n+1} \frac{x^{2n-1}}{2^{n-1}(2n-1)(n-1)!}$$

The asymptotic expansion is

$$\Phi(x) \approx \frac{1}{2} - \frac{1}{x(2\pi)^{1/2}} \left\{ 1 + \sum_{n=1}^{\infty} (-1)^n \frac{(2n)!}{2^n n!} \frac{1}{x^{2n+1}} \right\} \exp\left(-\frac{x^2}{2}\right)$$

338 Appendix 4

x	0	1	2	3	4	5	6	7	8	9
0.0	0.00000	00399	00798	01197	01595	01994	02392	02790	03188	03586
0.1	08983	04380	04776	05172	05567	05962	06356	06749	07142	07535
0.2	07926	08317	08706	09095	09483	09871	10257	10642	11026	11409
0.3	11791	12172	12552	12930	13307	13683	14058	14431	14803	15173
0.4	15542	15910	16276	16640	17003	17364	17724	18082	18439	18793
0.5	19146	19497	19847	20194	20540	20884	21226	21566	21904	22240
0.6	22575	22907	23237	23565	23891	24215	24537	24857	25175	25490
0.7	25804	26115	26424	26730	27035	27337	27637	27935	28230	28524
0.8	28814	29103	29389	29673	29955	30234	30511	30785	31057	31327
0.9	31594	31859	32121	32381	32639	32894	33147	33398	33646	33891
1.0	34134	34375	34614	34850	35083	35314	35543	35769	35993	36214
1.1	36433	36650	36864	37076	37286	37493	37698	37900	38100	38298
1.2	38493	38686	38877	39065	39251	39435	39617	39796	39973	40147
1.3	40320	40490	40658	40824	40988	41149	41309	41466	41621	41774
1.4	41924	42073	42220	42364	42507	42647	42786	42922	43056	43189
1.5	43319	43448	43574	43699	43822	43943	44062	44179	44295	44408
1.6	44520	44630	44738	44845	44950	45053	45154	45254	45352	45449
1.7	45543	45637	45728	45818	45907	45994	46080	46164	46246	46327
1.8	46407	46485	46562	46638	46712	46784	46856	46926	46995	47062
1.9	47128	47193	47257	47320	47381	47441	47500	47558	47615	47670
2.0	47725	47778	47831	47882	47932	47982	48030	48077	48124	48169
2.1	48214	48257	48300	48341	48382	48422	48461	48500	48537	48574
2.2	48610	48645	48679	48713	48745	48778	48809	48840	48870	48899
2.3	48928	48956	48983	49010	49036	49061	49086	49111	49134	49158
2.4	49180	49202	49224	49245	49266	49286	49305	49324	49343	49361
2.5	49379	49396	49413	49430	49446	49461	49477	49492	49506	49520
2.6	49534	49547	49560	49573	49585	49598	49609	49621	49632	49643
2.7	49653	49664	49674	49683	49693	49702	49711	49720	49728	49736
2.8	49744	49752	49760	49767	49774	49781	49788	49795	49801	49807
2.9	49813	49819	49825	49831	49836	49841	49846	49851	49856	49861
3.0	0.49865		3.1	49903	3.2	49931	3.3	49952	3.4	49966
3.5	49977		3.6	49984	3.7	49989	3.8	49993	3.9	49995
4.0	499968		4.75	499999						
4.5	499997									
5.0	4999997									

Index

A
additivity, 5, 7, 9, 36
algebra, 4, 271, 276, 278, 288, 294
  boolean, 270, 271
almost everywhere true property, 292
almost surely true property, 4
alphabet, 43
approximation in quadratic mean, 40, 52, 268
atom, 2, 274, 279–280, 286
average
  moving, 156–157, 158
  time, 231–232
axiom, 270, 271, 272, 273

B
Bernoulli experiment, 17, 53
binary word, 18, 28
Boltzmann's constant, 229
Borel–Cantelli lemma, 16–17, 106
Borel field, 2, 4, 270
  complete, 4
  generated, 165, 273, 278, 283, 285
    by a family of subsets, 272–274, 278, 279
    by a mapping, 271–272
    by a set, 273
  induced by a mapping, 272
  product, 275
  separable, 273
  of subsets, 2, 7, 271, 273, 279
  trace, 278
Borel set, 23, 27, 29, 67, 69, 274, 287
brownian motion, 203

C
causality relations, 114, 162–163
change of frequency, 142–143
channel
  Nyquist, 159–162
  partial response, 161–162

composition
  of a function, 301–302
  of a mapping, 172, 281
compression of a sequence of random variables, 268–269
conditional probability, 9–10, 183
conditioning
  on a discrete random variable, 50, 85
  on an event, 9–13
  on a sub-Borel field, 47–54, 178, 312
continuity, 290–291, 294
convergence
  almost sure, 96–97, 98–99, 106
  in law, 97, 100, 106
  in mean square, 97, 98–99, 100, 106–107
  in probability, 97, 99, 100
convolution, 92, 113, 139, 148, 168, 189, 220, 234, 259, 297, 300, 304, 315, 317, 330, 333
correlation, 74–78, 268
covariance, 74–78, 193, 194, 195
criterion of choice, 13, 32, 45

D
demodulator, linear envelope, 135
density of a measure, 308, 309, 319
deterministic signal, 108–163
  analytic, 113–114, 126–127
  band-limited, 112, 132, 139–141, 142
  causal, 109, 114, 117
  of finite energy, 117–118
  of finite mean power, 120–121, 122
  ideal sampled, 140, 146, 148, 159
  periodic, 123–124
diagonalization, 78
discrete information, 1, 42–45
dispersion, 39, 41, 52, 269
distortion, 132–133, 144, 240–241
distribution
  Dirac, 109, 212, 214, 216
  periodic, 123, 146, 256, 323–329

distribution (*continued*)
  tempered, 36–37, 46, 108, 109, 215, 255, 330–333
drawing (sample), 6–7

E

entropy, 42–43
  conditional, 43
envelope
  complex, 137–138
  real, 135
equation
  Chapman–Kolmogoroff, 180
  convolution, 129, 134, 138
  Wiener–Hopf, 47
ergodicity, 232–233
erlang, 188
error
  prediction, 268
  quantification, 62
  sampling, 263–264
estimation, 52–53, 94, 238, 267–268
event(s), 1, 2, 5, 6, 23, 24, 171, 176, 177
  almost surely certain, 4, 7, 14, 19
  almost surely impossible, 4, 7, 14, 19
  certain, 2, 7
  conditionally independent, 13, 186
  contrary, 2, 7
  elementary, 2, 7
  equiprobable, 5
  generated by a random signal, 178–180, 234
  impossible, 2, 7
  incompatible, 2, 11, 14, 28
  independent, 13–19, 21
  point, 167, 183, 184, 190, 244
expectation
  conditional, 48, 50–54, 58, 62, 70, 71, 72, 178, 268
  mathematical, 38

F

fading, 32
filter
  adapted, 137–138
  complementary, 236–237
  cosine-squared, 161
  ideal band-pass, 136, 251–252
  ideal low-pass, 133, 242
  numerical recursive, 150, 152–155, 157–159, 161, 266–267
  numerical transversal, 150, 156–157, 161
  RC low-pass, 133–134, 153, 156
  realizable, 116–117, 128, 131, 138, 237–238
  RLC band-pass, 138–139, 250
  stable analogue, 128, 130, 131–132
  stable numerical, 150, 151, 152, 153, 156
filtering
  analogue linear, 127–139, 234–241, 260
  distortion free, 132–133, 240–241
  numerical linear, 148–163, 258–260
formula
  Bayes', 10–11
  Poincaré, 20–21, 22
  Taylor–Young, 46
  of total probability, 7–8
Fourier coefficient, 123, 197, 256, 323–329
Fourier series, 123, 125, 146, 162, 196, 256
frequency
  cut-off, 112, 119, 137, 140, 141, 160, 161, 215, 228, 268
  relative, 4–5, 103, 188
  resonance, 139, 158
  sampling, 140, 141, 142, 144
function
  almost everywhere equal, 307–308
  autocorrelation
    of an analogue filter, 117, 134
    of a discrete signal, 116, 118, 121, 122, 124, 125, 138
    of a numerical filter, 155–156, 259
    of a random signal, 181, 182, 183, 193, 195, 197, 198, 200, 201, 206, 210–211, 212, 218, 221, 222, 223, 235, 254–255, 259–260
  Bessel, 90
  Borel, 58, 60, 218, 283–284, 292, 301, 315
  characteristic
    of a random variable, 45–46, 56, 84, 92, 103, 105, 190
    of a random vector, 78–80
  counting, 167, 183
  cross-correlation
    of two discrete signals, 116, 118, 235–236
    of two random signals, 182, 195, 196, 206, 211, 215
  distribution, 35–38, 58, 90–91, 97
  error (erf), 37, 333
  Heaviside, 36, 92, 114, 122–123, 228, 252, 287, 333, 334
  holomorphic, 46, 47, 147, 334
  indicator, 24, 50, 93, 118, 138, 169, 268, 284, 333
  integrable, 24, 49, 231, 254, 302–305, 318
  measurable, 13, 34, 280–286
  moment generating, of a random variable, 46–47, 84
  negligible, 29, 305–307, 318
  Φ, 37, 38, 333
  sign, 113, 136
  square integrable, 117, 212, 238, 310, 311
  transfer
    of an analogue filter, 116, 128, 130, 139, 161, 162, 237, 240, 242, 264
    of a numerical filter, 151, 153, 161

G
group propagation time, 117, 132, 138

H
Hilbert conjugate, 113–114, 127, 239–240, 252
hypothesis test, 1, 11–13, 32–33

I
image
   inverse, 23, 67, 272, 273, 281
   of a measure, 26, 34, 174, 291–292
increasing property
   of an integral, 41, 60, 299, 303
   of a measure, 289
independence
   conditional, of a family of random variables, 84–85
   of a family of events, 14–15, 19
   of a family of variables, 81–84
   of a family of random variables, 81–84
   of two events, 13–14, 21
   of two random signals, 174, 224
inequalities, Chernoff, 61–62
inequality
   Markov, 41, 100
   Tchebycheff, 41–42
   Schwarz, 60, 75, 107, 118, 121, 198, 207, 211, 241, 311, 319
integral, 38, 40, 299–300, 302, 303, 304
integration, 298–321
interference, 160, 161
interpolation, 143, 262, 264
inversion of a filter, 152, 154

J
jacobian, 69

L
Laurent series, 146
law
   Bernoulli, 28–29, 89
   binomial, 28–29, 53, 104–105, 184
   χ^2, 94–95
   exponential, 30–31, 35, 56, 57, 59, 91, 92
   gaussian, 29–30, 103, 104–105, 204
   marginal probability, 68
   multinomial, 82–83
   normal, 29–30
   Poisson, 28, 59–60, 89, 105
   Rayleigh, 31–32, 55
   standard normal, 103
   strong, of large numbers, 102, 105–106
   weak, of large numbers, 100–101
length of life, 31
likelihood ratio, 1, 11–13, 32

linearity of the expectation, 39, 40, 50, 51, 74, 75, 258
linearity of the integral, 299, 303
linear transformation, 80, 87
line of regression, 75–76

M
mapping
   borelian, 69, 283
   denumerably additive, 287, 288
   inverse, 69
   measurable, 23, 165, 170, 173, 280, 286, 291
martingale, 186–187
matrix
   correlation, 76–78
   covariance, 76–78, 80, 173
   cross-correlation, 175
   random, 68, 76, 78, 79, 80
mean square value, 38, 39
measurable block, 171, 276, 282, 294
measure(s)
   absolutely continuous, 12, 29, 212, 307–309, 315–316, 319
   complete, 293, 297
   counting, 5, 287
   Dirac, 5, 28, 287, 299–300, 304, 315, 317
   image, 26, 67, 170, 174, 176, 291, 304–305, 319
   Lebesgue, 29, 69, 115, 288. 292, 295, 315
   positive, 287–298
   positive bounded, 121, 256, 287, 297, 300, 310, 311, 316–318
   positive normed, 2, 4, 7, 9, 26, 172, 173, 174, 287, 294
   probability, 2, 5
   product, 17, 18, 294–297, 314, 315
   projective family of, 172, 173, 174, 176
   translated, 292, 315
   uniform, 5, 7, 15, 21
model, mathematical, 1, 2, 11, 17, 24, 187, 244
modulation
   amplitude, 137
   angular, 137

N
noise
   gaussian, 33
   measuring, 30, 267
   shot, 249–250
   thermal, 173, 226–228
   white
      in continuous time, 226
      in discrete time, 257–258, 266–267
      gaussian, 224, 226–227, 228, 237–238, 239, 250, 251, 267

P

partition, 8–9, 25, 28, 33, 50, 278, 279, 298
photodetector, 28
Planck's constant, 229
Poisson process, 183–186, 189, 191, 249–250
power
   finite mean, 120, 124, 212, 213
   instantaneous mean, 166, 211, 227, 231
   time average, 212, 213, 214, 231–232
power factor, 139
power series, 149, 258
prediction, 238–239
principal value, 113, 333
probabilistic judgment, 22
probability density, 29, 32, 35, 37, 47, 68–69, 173
   conditional, 58–59, 70–74, 88–89, 183, 202
   marginal, 69, 70, 89, 179
probability law
   of a pair of random signals, 174, 180
   of a random signal, 170–173, 176
   of a random variable, 26–35, 36, 40, 45, 46, 59, 84, 103
   of a random vector, 67–71, 171, 172, 174, 179, 192, 193
product
   scalar, 198, 200, 311, 319
   tensor, 174, 295
projection, 66, 68, 165, 275, 281, 282, 295, 311
propagation, 32

Q

quadratic detection, 241–243
quantification, 62–63
quantity of information, 44
queueing theory, 30, 169

R

radar, 11
Radon–Nikodym derivative, 12–13, 308
random signal(s)
   almost surely zero, 192
   autocorrelation coefficient of, 181, 218, 219–220, 229, 268
   band-limited, 214–215, 223, 248, 265–266, 268
   binary, 187–188, 245–246
   causal, 165
   centred, 180, 183, 253
   complex, 165, 210, 212
   continuous in mean square, 198–199, 203, 206, 208, 211, 218
   in continuous time, 165, 198, 199, 210, 212, 215, 234, 265
   cross-correlation coefficient of two, 182, 252
   differentiable in mean square, 199–201, 203, 206, 207, 224, 225, 228

random signal(s) (*continued*)
   in discrete time, 96, 165, 186, 254–269
   equivalence of two, 176–177
   ergodic, 232, 253
   of finite mean power, 167, 197–208
   gaussian, 173, 188, 194, 202, 207, 217–220, 229, 234, 238–239, 241–243
   independent, 174, 176, 183, 224, 237, 246–247, 250, 251
   with independent increments, 185, 202
   integrable in mean square, 204–206, 230–232, 247–248, 253–254
   markovian, 178–180, 186, 202, 217–218
   mean value of, 180, 193, 195, 196, 197, 200–201, 205, 206, 218, 225, 231, 235, 242, 258–259
   moments associated with a pair of, 180–182
   mutually gaussian, 174–176, 224
   mutually weakly stationary, 195, 196, 225, 236
   narrow-band, 224
   of order 2, 167
   passage through zero of, 228–229
   periodic, 196–197
   Poisson, 167–169, 208, 244
   real, 164–165, 173, 215, 216
   realization of, 165, 166, 170, 176, 192, 194, 196, 208, 212, 213, 232, 233, 240, 241, 258
   representation of, 176, 177, 221–224, 243, 251
   sampled, 261–265
   single side-band, 252–253
   sinusoidal, 167, 195–196, 256–257
   smoothing of, 248
   strongly equivalent, 176–177, 240, 241, 265
   uncorrelated, 182
   weakly equivalent, 176, 177, 207
   weakly stationary, 193, 194, 195, 232, 234, 244
   Wiener–Levy, 201–204, 207, 227
random variable(s)
   absolute moment of, 41, 45, 46
   almost surely equal, 26, 27, 39, 176
   binary, 24, 98, 203, 244
   centred, 38, 39
   complex, 23–24, 25, 165
   constant, 25, 51, 98, 99, 101
   continuous, 37, 41
   correlation coefficient of two, 74, 76
   discrete, 25, 27, 28, 42, 50
   equivalent, 26
   gaussian, 30, 37–38, 46, 47, 55, 57, 72, 84, 92–93, 107, 177, 267
   indicator, 24–25, 53, 58, 101, 168
   mean value of, 38, 39, 42, 48, 50, 51
   moment of a, 1, 38–42, 45, 46
   of order 2, 39

random variable(s) (*continued*)
   Poisson, 28, 56, 169
   product of two, 93
   positive, 30, 47, 51, 62, 185, 189
   Rayleigh, 56
   real, 23–24
   square integrable, 39, 41, 52, 75, 97, 166
   uncorrelated, 74, 83
   uniformly distributed, 35, 53, 55, 95, 195, 243, 261
random vector(s), 65–107, 168, 169
   almost surely equal, 67
   gaussian, 65, 68, 79–80, 87, 88–90, 96
   uncorrelated, 78
random walk, 203
rational fraction, 130, 151, 153, 154, 157–158, 238
receiver, 33, 43, 44
renewal process, 168, 189–191, 249
response
   impulse, 116–117, 127–128, 130, 135, 138, 234, 260
   of a numerical filter, 149, 155, 158, 258, 260
   permanent, 131, 151–152
   step, 135, 156
   transient, 129

S

sampling, 110, 139–163, 261–266
   analogue, 145
   with hold, 143–144
sequence
   binary random, 17–19, 21, 22–23, 65, 101–103
   Cauchy, 310
   causal numerical, 148, 149, 162
   of events
      decreasing, 8, 10, 15, 16, 19, 277, 298
      increasing, 8, 277, 290, 298
   independent, of random variables, 81–82, 83, 84, 85, 90, 98, 102, 168, 187, 208, 267
   limit of, 277, 284
      lower, 96, 277, 297–298
      upper, 97, 277, 297–298
   multisymbol, 22
   numerical, 96, 147–148, 255, 298
set
   measurable, 271, 272, 284
   negligible, 292–294, 309
signal
   clock, 126–127, 145
   cosine-squared, 120
   deterministic, see Deterministic signal
   gate, 118–119, 125, 127
   ideal sampling, 124–125, 139, 143, 197, 261–262
   random, see Random signal
   sampled, 146, 166, 261–265

signal (*continued*)
   telegraph, 244–245
   unit step, 121–122, 135, 156
singleton. 2, 233, 254, 271, 274, 286
space
   Banach, 310
   Hilbert, 39, 167, 198, 311
   measurable, 170, 172, 270–280
   measurable product, 164, 173, 274–276, 279, 280, 283, 314
   measured, 1, 287, 291
   measured product, 295
   probability, 1–23, 65, 101, 164, 170, 172, 173, 176
   topological, 274
spectral band, 112, 119, 120, 132, 133, 161, 215, 220, 247
spectral density, 118, 121, 212, 214, 226, 227, 237, 257
spectral line, 123, 216, 257
spectrum
   complex, 111–114, 115, 118, 119, 123–124, 125, 136, 140, 142, 144, 146, 147, 162, 163, 212
   cross-correlation, of two random signals, 215, 236
   energy, 118
   line, 111, 124, 216
   power, 121–122, 124
   of a random signal in continuous time, 212, 215–220, 223, 225, 235, 256
   of a random signal in discrete time, 255–256, 257
standard deviation, 30, 38, 39
stationarity
   of order 2, 193
   strong, 192
   weak, 193–194
subadditivity, 8
sub-Borel field(s), 23–24, 25, 48, 178, 186, 187, 238, 272, 311, 312
   independence of a family of, 19
support, 256
   frequency, 112, 137, 140, 220, 221, 256
   time, 112, 120
symmetric difference, 4, 20, 27, 58

T

telephone exchange, 28
telephonic communication, 17, 31, 59
theorem
   Bochner's, 212, 256, 317–318
   central limit, 30, 103–104, 204, 227
   for a change of variables, 34
   of differentiation under the summation sign, 46, 47, 60
   of dominated convergence, 231, 233, 248, 304, 318, 320
   extension, 288, 296

theorem (*continued*)
  final value, 188
  Fubini's, 70, 82, 230, 314–315, 317
  Kolmogoroff's, 172, 176, 296–297
  of monotonic convergence, 302, 305
  orthogonal projection, 75, 311, 313
  Parseval's, 117, 213
  Radon–Nikodym, 308, 312
  of residues, 147
  sampling, 144, 264–265
  Wiener–Kintchine, 212
theory
  of modulations, 251
  of reliability, 30
time average value, 231
topology, usual, 274
traffic, 188
transform
  Fourier, 45, 79, 109, 111, 123, 128, 129, 130, 212, 215, 316–318, 330–333
  Hilbert, 113
  Laplace, 46, 109, 111, 128, 129, 130, 134, 189, 190, 334–336

transform (*continued*)
  z, of a numerical filter, 152, 157–158, 258, 259
  z, of a numerical sequence, 146–147, 255
transition probability, 179
translation, 34, 114
transmission
  noisy, 1, 42–45
  numerical, 141
transmission medium, 44
trial, 1–2, 5, 6, 7, 11, 24, 165

U
uncertainty, 42, 44

V
variance, 29, 40, 77, 94, 180, 193
variance, conditional, 62

W
whitening, 237–238